Frege's Philosophy of Mathematics

Frege's Philosophy of Mathematics

Edited by William Demopoulos

Harvard University Press
Cambridge, Massachusetts
London, England

First Harvard University Press paperback edition, 1997

This book was typeset using AMS–TEX, the typesetting package of
the American Mathematical Society.

Library of Congress Cataloging-in-Publication Data

Frege's Philosophy of Mathematics / edited by William Demopoulos
      p.   cm.
  Includes bibliographical references and index.
  ISBN 0-674-31942-7 (cloth)
  ISBN 0-674-31943-5 (pbk.)
    1. Mathematics—Philosophy.   2. Mathematics—Foundations.
  3. Logic, Symbolic and mathematical.   I. Demopoulos, William.
  QA8.6.F74   1995
  510′.1–dc20   94–34381
                                CIP

Contents

II. The Mathematical Content
of *Begriffsschrift* and *Grundlagen*

III. *Grundgesetze der Arithmetik*

Preface

Widespread interest in Frege's general philosophical writings is, relatively speaking, a fairly recent phenomenon. Precisely when it became widespread is difficult to say, but its occurrence, at least in Britain and North America, is hard to imagine without J. L. Austin's elegant translation of *Grundlagen*,[1] or Max Black and Peter Geach's *Translations from the philosophical writings of Gottlob Frege*.[2] It is also evident that the publication, in 1973, of Michael Dummett's landmark study, *Frege: philosophy of language*,[3] contributed very substantially to the extensive and rapidly growing interest in Frege's metaphysics and philosophy of language. Nevertheless, I think it fair to say that until only quite recently, a collection of essays devoted to Frege's philosophy of *mathematics* would have seemed a rather unlikely event. Crispin Wright's book, *Frege's conception of numbers as objects*,[4] has been a decisive factor in initiating much of the attention now being devoted to this topic. In addition, the past several years have seen the publication of a number of excellent studies

---

[1] Second rev. ed., Oxford: Basil Blackwell, 1953.

[2] Ithaca: Cornell University Press, 1960.

[3] London, Duckworth. Second ed., Cambridge: Harvard University Press, 1991.

[4] Aberdeen: Aberdeen University Press, 1983.

which have deepened our understanding of the development of the analytic tradition. Given Frege's central position in this tradition, interest in the scientific background and philosophical origins of the analytic movement has naturally led to reassessments, both of his logicism and of our picture of its intellectual context.

There are three main developments in recent work on Frege's philosophy of mathematics which this collection of essays addresses:

- The emerging interest in the intellectual background to Frege's logicism
- The rediscovery of Frege's theorem
- The reevaluation of the mathematical content of *Grundgesetze*

The Introduction presents an overview of some of this work. I hope it suffices to orient the reader who has not followed these developments, while maintaining the interest of one who has.

With the exception of Charles Parsons's prescient "Frege's theory of number," first published in 1964, all of the essays collected here date from 1981 or later. Each paper attempts a sympathetic, if not uncritical, reconstruction, evaluation, or extension of one or another facet of Frege's thought, a consideration that has served as a major principle governing the process of selection. This consideration, together with a more practical one deriving from limitation of size, has meant that many worthwhile papers on Frege's philosophy of mathematics have not found their way into this volume. I can only hope that satisfaction with what is included here will outweigh any disappointment occasioned by my omissions.

*   *   *

I wish to record my deep appreciation to John Bell, Michael Friedman, Anil Gupta, Michael Hallett, and Mark Wilson, for their patient advice—both practical and intellectual. I wish to thank George Boolos and Richard Heck, both of whom were very generous with their attention to this project, especially during its early stages. Thanks also are very much owing to my several contributors, and, in the case of Alberto Coffa, to Linda Wessels, literary executor of his estate, for so cheerfully agreeing to the reprinting of their papers. My research associated with this project was supported by the Social Sciences and Humanities Research Council of Canada. I am grateful to the Council for its financial support; I am grateful as well to James Good, Dean of the Faculty of Arts of the University of Western Ontario, for so expeditiously securing financial assistance for the preparation of the manuscript. Dr. David DeVidi is responsible for the thoroughly excellent job of setting the manuscript in $\mathcal{A}_{\mathcal{M}}\mathcal{S}$-TEX.

While I assume full responsibility for the final form of this volume, I would like it to be generally known that the idea for such a collection originated with Michael Dummett. I owe him a special debt of gratitude for suggesting this project to me and for entrusting me with its execution. He was, throughout the duration of its preparation, a ready and willing source of sage advice and generous encouragement, and it is a pleasure to acknowledge this very considerable personal debt. This is not my only debt to Dummett: I am sure I speak for most philosophers of my generation when I say that his writings on Frege have set for us an intellectual standard to which we all aspire.

August 1994   W. D.

WILLIAM DEMOPOULOS

Introduction

My purpose in this Introduction is to give an overview of the rediscovery of Frege's theorem, and of certain of the issues that this
rediscovery has raised. The series of developments I will be reviewing by no means exhausts the considerations which have led to
the current, renewed, interest in Frege's philosophy of mathematics.
Nonetheless, the topic of Frege's theorem has been a major factor
underlying this event, and its centrality is such that by reviewing it,
we will be led to touch upon virtually all the topics covered in this
collection. If, in coming to appreciate Frege's achievement, we are
not tempted to revive his program of accounting for our knowledge
of infinity on the basis of our mastery of principles having the status
of Euclid's Common Notions, we will at least have come to a better
understanding of his goals and of how close he came to achieving
them.

I. FREGE'S THEOREM

The contextual definition of the cardinality operator, suggested in
§ 63 of *Grundlagen*—what, after George Boolos, has come to be

known as Hume's principle[1]—asserts

The number of Fs = the number of Gs if, and only if, $F \approx G$,

where $F \approx G$ (the Fs and the Gs are in one-to-one correspondence) has its usual, second-order, explicit definition. The importance of this principle for the derivation of Peano's second postulate ('Every natural number has a successor') was emphasized by Crispin Wright who presented an extended argument showing that, in the context of the system of second-order logic of Frege's *Begriffsschrift*, Peano's second postulate is derivable from Hume's principle.[2] And in a review of Wright's book, John Burgess[3] proved Wright's conjecture that Hume's principle is consistent.

The significance of Wright's proof, as he himself observed, is that the argument proceeds without any appeal to a theory of sets; in particular it does not rely on the inconsistent theory of extensions of concepts implicit in *Grundlagen*.[4] Boolos[5] later showed in detail how, in *Grundlagen* §§ 68–83, Frege had already established Wright's result, and proposed that we call Frege's discovery that, in the context of second-order logic, Hume's principle implies the infinity of the natural numbers, *Frege's theorem*.

In *Grundgesetze* concepts are treated as a special type of function (after the proposal of Frege's lecture, *Function and concept*), and what in *Grundlagen* is called the extension of a concept is, in *Grundgesetze*, represented by the notion of the value range of the

---

[1] So called because Frege introduces it with a quote from the *Treatise*, I, III, 1, para. 5: "When two numbers are so combined as that one has always an unite answering to every unite of the other, we pronounce them equal ... ." See George Boolos, "The consistency of Frege's *Foundations of arithmetic*," and "The standard of equality of numbers," reprinted here as Chapters 8 and 9 respectively.

[2] See *Frege's conception of numbers as objects*, Aberdeen: Aberdeen University Press, 1983, pp. 154–69, where Hume's principle is called '$N^=$.'

[3] Cf. *The philosophical review*, **93** (1984) 638–40. Burgess's observation on the consistency of Hume's principle is elaborated by Boolos in Chapters 8 and 9, below.

[4] This observation and the appreciation of its significance can already be found in § VI of Charles Parsons's paper, "Frege's theory of number," reprinted here as Chapter 7.

[5] In the papers cited in Note 1.

function corresponding to the concept.[6]   Basic Law V, explicitly formulated in *Grundgesetze* I, § 20, implies that the extension of the concept F = the extension of the concept G if, and only if, everything falling under F falls under G, and conversely, everything falling under G falls under F.[7] Richard Heck has recently investigated how, in *Grundgesetze*, Hume's principle may be deployed in the proof of Frege's theorem without the use of Basic Law V.[8]

In the Appendix to this Introduction, it is shown how Frege's theorem can be viewed as a contribution to set theory. Frege in effect discovered an 'axiom of infinity' which does not explicitly postulate the existence of an infinite set, but, within a consistent fragment of his theory of concepts and their extensions, allows both for its derivation and for the proof of the existence of an infinite cardinal number. This undeservedly neglected achievement is comparable to Zermelo's proof of the Well-ordering theorem: just as Zermelo's formulation of Choice makes no explicit mention of a well-ordering, so Frege's formulation of Infinity—viz., Hume's principle—makes no explicit mention of an infinite set or an infinite cardinal. From a contemporary point of view, Zermelo's proof of Well-ordering and Frege's proof of the existence of a Dedekind-infinite set are very closely related: both can be shown to follow from the same fundamental lemma of Bourbaki's analysis of the Well-ordering theorem.

There can be no question that this series of developments, which might be referred to as 'the rediscovery of Frege's theorem'—the theorem that the second-order theory consisting of Hume's principle implies the infinity of the natural numbers—is of the greatest

---

[6]For a comparison with the development of the function concept in analysis, see John Burgess, "Frege on arbitrary functions," Chapter 4.

[7]Basic Law V is generally regarded as the source of the contradiction exposed by Russell's paradox. The situation is, however, rather subtle, as the papers by Terence Parsons, "The consistency of the first-order portion of Frege's logical system," and John Bell, "Fregean extensions of first-order theories," clearly show; these two papers appear here as Chapters 16 and 17, respectively.

[8]See his paper, "The development of arithmetic in Frege's *Grundgesetze der Arithmetik*," Chapter 10 of the present volume. In Chapter 11, "Definition by induction in Frege's *Grundgesetze der Arithmetik*," Heck continues this investigation, showing how the notion of the extension of a concept may be avoided elsewhere in *Grundgesetze*, especially in connection with Frege's presentation of an analogue of Dedekind's categoricity theorem.

importance for our appreciation of Frege's mathematical achievement. It nevertheless remains unclear whether Frege's theorem can
be marshalled in support of a possibly revised formulation of his logicism. The difficulties with such a development of Frege's view have
clustered around three questions: (i) To what extent does Hume's
principle yield an *analysis* of number? (ii) How securely does the
principle fix the reference of numerical singular terms? And finally,
(iii) with what justification can we say that our knowledge of the
truth of Hume's principle is 'independent of experience or intuition'?

II. THE ANALYSIS OF STATEMENTS OF NUMBER

Frege's belief that Hume's principle expresses the preanalytic meaning of assertions of numerical identity is based on his conviction
that the equivalence relation $\approx$ is 'conceptually prior' to any notion of number. His argument to this effect is indirect, being first
presented[9] in connection with the notion of the direction of a line.
Frege claims that we attain the concept of direction only when we
have grasped the relation of parallelism: although the direction of
l = the direction of m if, and only if, l is parallel to m, our understanding of direction depends on our grasp of parallelism, rather
than the other way round, and this is what accounts for the fact
that direction can be given an analysis in terms of parallelism. It is
clear that Frege intends to extend this consideration to the analysis
of 'the number of Fs = the number of Gs' in terms of the existence
of a one-to-one correspondence, and that he regards the two cases
as analogous in all relevant respects.[10] Indeed, since in the projective geometric tradition of the 19th century, a tradition with which
Frege was certainly familiar, 'the direction of l' meant *the point at
infinity associated with l*, the analogy with numbers appears more
perfect than we might otherwise have imagined.[11] Frege, however,
chooses not to pursue this connection, and asks instead whether 'we
distinguish in our intuition between [a] straight line and something

---

[9]In *Grundlagen* §§ 64–67.

[10]See, for example, the first footnote to § 65 of *Grundlagen*.

[11]See Mark Wilson, "Frege: The royal road from geometry," reprinted here
as Chapter 5. These issues are currently being further investigated by Jamie
Tappenden in what promises to be an important series of papers.

else, its direction?';[12] although we may perhaps take his remark that while contextual definition ' . . . seems to be a very odd kind of definition, to which logicians have not paid enough attention, . . . it is not altogether unheard of . . . ' (§ 63) as an allusion to the geometrical use of this notion. In any case, if one grants the correctness of Frege's analysis of 'the number of Fs = the number of Gs,' then Hume's principle acquires the status of a condition of adequacy, a characterization of the principle that comports well with Frege's practice: Once the explicit definition of a cardinal number as a class of equinumerous concepts has been shown to imply Hume's principle, the explicit definition is no longer appealed to, and the entire account of the arithmetic of natural numbers, given in *Grundlagen*, is developed from this principle.

Frege's proof of Peano's second postulate from Hume's principle proceeds by establishing the existence of an appropriate family of 'representative' concepts. The proof exploits what Michael Dummett[13] has called the 'indefinite extensibility' of the natural number concept by showing how, from any initial segment of the natural number sequence, it is possible to characterize the next member of the sequence. With the exception of the first, each representative concept is characterized as holding of precisely the cardinal numbers less than its number. The sequence of extensions of such concepts therefore anticipates the construction of the finite von Neumann ordinals. Judged from the standpoint of their centrality to the key mathematical argument of *Grundlagen*, it could be argued that this family of extensions (rather than the family of equivalence classes of equinumerous concepts) form the true Frege finite cardinals. Following another suggestion of Dummett,[14] we may say that Frege perceived that the domain of the theory of natural numbers is 'in-

---

[12] *Grundlagen*, § 64. Quotations from *Grundlagen* (1884) are in the translation by J. L. Austin, *The foundations of arithmetic: a logico-mathematical enquiry into the concept of number*, second revised edition, Evanston: Northwestern University Press, 1980.

[13] *Frege: philosophy of mathematics*, Cambridge: Harvard University Press, 1991, p. 317. Dummett traces the notion to Russell's paper, "On some difficulties in the theory of transfinite numbers and order types," *Proceedings of the London Mathematical Society* (Series 2), **4** (1906) 29–53, reprinted in D. Lackey, ed., *Bertrand Russell: essays in analysis*, London: Allen and Unwin, 1973.

[14] *Frege: philosophy of mathematics*, p. 317.

trinsically infinite' in the sense that once we are given a part of the domain, we can always characterize a new element in terms of those already given. Similarly, Cantor's proof of the non-denumerability of the domain of the theory of the real numbers exploits the fact that this domain is intrinsically infinite, and the concept of a real number, indefinitely extensible: Given any denumerable set of real numbers, Cantor's diagonal construction provides a means of characterizing a new real number.

III. CONTEXTUAL DEFINITION AND THE CONTEXT PRINCIPLE

Hume's principle, the contextual definition of number, or more exactly, the contextual definition of the cardinality operator, 'specifies' a second level concept which establishes a many-one correspondence between first level concepts and certain objects, namely, the cardinal numbers. It follows that if the contextual definition restricts the reference of the cardinality operator to a particular mapping from first level concepts to objects, then, assuming that the reference of the relevant concept-expression has been fixed, the reference of a wide class of numerical singular terms will also have been settled. This is why the equivocation between 'contextual definition of number' and 'contextual definition of the cardinality operator' is not so important as it might at first appear: To the extent that the contextual definition successfully picks out the cardinality operator, it must also determine the cardinal numbers.

The context principle was first formulated in the Introduction to *Grundlagen* as one of three 'fundamental principles' guiding the enquiry: ' ... never ... ask for the meaning of a word in isolation, but only in the context of a proposition.' The relation between this principle and contextual definition is by no means a simple matter, but when the context principle is interpreted as a principle governing reference, it suggests that reference to abstract objects can be achieved once we have established the truth of the propositions into which they enter. The relevant contrast is with a naive Platonism which would suppose that we can refer to numbers independently of our knowledge of such truth conditions. On such a view, knowledge of reference precedes knowledge of truth. In § 62 of *Grundlagen*, Frege makes it clear that he sees his use of the context principle as an inversion of this traditional ordering of truth and reference.

So understood, the context principle is still very much a part of *Grundgesetze*; if in *Grundlagen* the point of the principle is that truth takes precedence over reference, then in the later work, its point is that *reference to truth* (to the True) takes precedence over reference to other objects.[15]

For Frege the propositions that are accorded a special role in settling questions of reference are what he calls 'recognition statements'; in the case of number words (and other singular terms) recognition statements involve the relation of identity, so that for this case, recognition statements are said to provide a 'criterion of identity,' or 'standard of equality,' for numbers. Hume's principle is an example of such a recognition statement. The 19th century added to the idea of a criterion of identity given by an equivalence relation the practice of passing to the quotient structure of equivalence classes determined by the relation—what has come to be known as definition by (logical) abstraction.[16] Frege combines his use of the context principle in *Grundlagen* with definition by abstraction. Having determined what are the recognition statements appropriate to a particular class of numerical expressions, he proceeds to infer from the truth of such statements to the referential character of the expressions of this class. This inference is then supplemented by an appeal to the theory of definition by abstraction, with the equivalence classes providing the referents of the numerical singular terms. By contrast with Frege, one might proceed *without* resorting to definition by abstraction.[17] But it should be clear that on either

---

[15] The proposal that the context principle is concerned with the relative priority of the semantic categories of truth and reference should be contrasted with Wright's suggestion that the context principle asserts 'the priority of syntactic over ontological categories'; this is evidently a different (though not necessarily incompatible) claim. See *Frege's conception of numbers as objects*, p. 51.

[16] The terminology can be confusing. Thus, in *The principles of mathematics*, § 210, Russell attributes to Peano a notion of 'definition by abstraction' that corresponds to what we are calling contextual definition. For an account of the role of abstraction in the 19th century and its relation to the Greek mathematical tradition, see Howard Stein, "Eudoxus and Dedekind: on the ancient Greek theory of ratios and its relation to modern mathematics," partially reprinted here as Chapter 12.

[17] Such an approach to the arithmetic of the natural numbers is considered by Charles Parsons in Chapter 7, below. This approach has been elaborated by Wright in his book, *Frege's conception of numbers as objects*; see also his

approach the context principle is vindicated as a heuristic or guiding principle in the philosophy of arithmetic if the contextual definition succeeds in 'specifying' the cardinal numbers. It is in this connection that 'the Julius Caesar problem' has come to be regarded as an issue of central importance for the contextual definition of number.

IV. THE JULIUS CAESAR PROBLEM AND HUME'S PRINCIPLE

In § 56 Frege considers an inductive definition of the numbers based on the conclusion reached in § 46—what, in *Grundgesetze* (p. ix), Frege describes as 'the most fundamental' of *Grundlagen*'s results— that a statement of number makes an assertion about a concept. The definition begins by defining 0 as the number of the concept under which *no* object falls, and has the inductive clause:

> $n + 1$ is the number of the concept F if there is an object a falling under F and n is the number of the concept expressed by 'x is F and $x \neq a$.'

The problem Frege raises in § 56—what we are calling the Julius Caesar problem—is, to a first approximation, that the definition does not determine what objects the numbers are:

> ... we can never—to take a crude example—decide by means of our definitions whether any concept has the number Julius Caesar belonging to it, or whether that conqueror of Gaul is a number or is not.

The context in which Frege subsequently places his discussion of criteria of identity (§ 62) suggests that by supplementing the inductive definition with Hume's principle it might be possible to resolve this difficulty. It is therefore natural to ask whether the Julius Caesar problem can be 'iterated,' to ask, in other words, whether the problem it raises for the inductive clause isn't also a problem for Hume's principle. And indeed, in *Grundlagen* § 66 Frege raises a similar difficulty for his proposed criterion of identity for *the direction of l*, a criterion of identity which, as we noted earlier, he takes to be exactly analogous to Hume's principle.

paper, "Why numbers can believably be," *Revue internationale de philosophie*, **42** (1988) 425–73, and "Field and Fregean Platonism," in Andrew Irvine, ed., *Physicalism in mathematics*, Dordrecht and Boston: Kluwer, 1989.

It is evidently a presupposition of Frege's objection (a presupposition of the Julius Caesar problem) that numbers are objects. There are two considerations that constitute Frege's main support for this assumption. First, there is the *observation* that the statements of number theory—as opposed to most everyday statements of number—employ singular terms for numerical expressions; if taken at face value, such terms require objects as their referents. That Frege would rely so unapologetically on this form of argument strongly suggests that he never seriously doubted that the analysis of statements of number must accord such evidence considerable weight.

Secondly, there is the *theorem*, which occupies the mathematical core of *Grundlagen*, that Hume's principle, when understood to incorporate the assumption that numbers are objects, implies the infinity of the natural numbers. Granted, this does not *prove* the thesis; it nevertheless lends considerable methodological support to Frege's view. *Grundlagen*, taken as a whole, both justifies this thesis, and clarifies its sense, by showing that its key implication is that numbers are possible arguments to first level concepts. This is the central difference between Frege's conception of numbers as objects and the conception of the cardinal numbers that emerges from Russell and Whitehead. If, working within the framework of the simple theory of types, we suppose that a cardinal number is the extension of a concept of second level, the type restrictions preclude a number from falling under a concept of first level; and the extensional character of all concepts (including those that are numbers) means that to guarantee the infinity of the numbers, we will require an axiom of infinity governing the number of *non*–logical objects.[18] That Frege is able to avoid such an axiom is one respect in which the rediscovery of Frege's theorem can be cited in support of his logicism.

Returning to our discussion of the Julius Caesar problem, it seems reasonable to demand of an account of Frege's presentation of the problem, in connection with Hume's principle, that it explain how

---

[18] Although, as Boolos has observed, Russell and Whitehead do not postulate a *Dedekind*–infinity of non-logical objects; rather, the Dedekind infinity of the finite cardinal numbers is shown to follow—without the Axiom of Choice—from the assumption that the set of individuals is 'non-inductive.' See his paper, "The advantages of honest toil over theft," in Alexander George, ed., *Mathematics and mind*, Oxford: Oxford University Press, 1994.

the introduction of extensions overcomes the difficulty which the Julius Caesar problem posed for numbers while not itself succumbing to a similar objection. In short, the Julius Caesar problem must not iterate to extensions. This requirement may not seem especially pressing for an account of *Grundlagen*, where extensions are simply assumed; but it certainly needs to be addressed in connection with *Grundgesetze*, where extensions are introduced (I, §§ 3, 9 and 20) in a manner that is formally exactly analogous to the contextual definition of number. Any interpretation which fails to satisfy this demand implies not only that Frege's mature theory of number is ungrounded, but that Frege must have known this; clearly, an interpretation which avoids such an implication is to be preferred.

The difficulty is that once we take the Julius Caesar problem to show that the contextual definition fails to specify the referents of numerical expressions, we seem committed to interpreting *Grundgesetze* I, § 10 as raising precisely the same problem for extensions of concepts. In this section, which is entitled, 'The value range of a function more exactly specified,' Frege complains

> ... We have only a means of always recognizing a value range if it is designated by a name like '$\grave{\varepsilon}\Phi(\varepsilon)$,' by which it is already recognizable as a value range. But we can [not] decide, so far, whether an object is a value range that is not given to us as such ... .[19]

But when Frege comes to resolve this issue, he does not (as this passage might lead one to expect) address the question of how extensions and other 'logical objects,' such as the truth values, are to be singled out from non-logical objects. Instead, he begins his discussion with the observation that, as we might put it, the reference of extension-expressions is fixed only up to arbitrary permutations of the domain of any 'model' of *Grundgesetze*. He then shows how an elaboration of this argument (an elaboration that has come to be known as the permutation argument) establishes that the truth-values (the only 'primitive' logical objects which are not 'given to

---

[19] *Grundgesetze* I, § 10. I have followed the translation of Furth, with the exception that I have replaced 'course of values' with 'value range.' Recall that in *Grundgesetze* the extension of a concept is represented by the value range of the function which corresponds to the concept.

us' as extensions), can be consistently identified with their own unit classes. This eliminates any problem of distinguishing, from among the logical objects, those which are not extensions, since it makes extensions of *all* the primitive logical objects. It would, however, be highly artificial to try to extend this solution to the case of non-logical objects. Moreover, the long footnote of § 10 shows that the key idea of the section—the possibility of replacing the truth values with their unit classes—cannot be extended to the case of every extension which is not given to us *as* an extension (as the reference of an extension-expression) without conflicting with Frege's earlier stipulations regarding value-range expressions, i.e., essentially, without conflicting with Basic Law V. Clearly, neither the observation that the truth values may be taken to be extensions, nor the observation that any 'model' of the system of *Grundgesetze* is determined only 'up to permutations,' settles the question whether an arbitrary extension is, or is not, identical with Julius Caesar.

The analogy between *Grundlagen*'s discussion of the Julius Caesar problem and the discussion in *Grundgesetze* I, § 10 would seem to run as follows. Both in the case of the numbers and in the case of extensions, Frege finds that the truth value of certain identity statements has been left indefinite by a statement having the form of a contextual definition, Hume's principle or Basic Law V, as the case may be. And in both cases he proposes to remedy the deficiency by a stipulation: the explicit definition of the cardinal numbers as classes, in the first case, and the identification of the truth values with their unit classes, in the second. He then has the problem of showing that his stipulations are compatible with what has gone before. In *Grundlagen* this is established by the proof that Hume's principle is a logical consequence of the explicit definition of a cardinal number as a class of concepts, a procedure that would be unexceptionable as a proof of consistency were it not for the fact that the explicit definition rests on Frege's inconsistent theory of concepts and extensions. In the case of *Grundgesetze*, the proof of consistency is the point of the permutation argument, and modulo the inconsistency of *Grundgesetze*, the permutation argument *does* show[20] that it is

---

[20] See Terence Parsons's paper (Chapter 16) for a clarification of the valid core of Frege's argument.

consistent to stipulate that the True and the False are value ranges, thus enabling Frege to 'resolve' the question, whether either of these objects is a value range, by the adoption of a convention. The interpretive difficulty arises from the fact that the question whether an extension is identical with Julius Caesar is one that could be posed for any expansion of the language of *Grundgesetze* which included names for non-logical objects and the fact that Frege's discussion, of what appears to be an obvious analogue of the Julius Caesar problem for extensions, fails even to address this issue.

V. THE ROLE OF EXTENSIONS OF CONCEPTS

Of Frege's derivation of Peano's second postulate from Hume's principle, we can say that it requires that numbers be objects. But if the derivation of Peano's second postulate from Hume's principle is to support Frege's logicism, numbers must not only be *objects*, they must be *logical* objects; i.e., they must not only be possible arguments to first level functions, but our knowledge of them must be capable of being shown to be independent of the principles of any 'special science.' In particular, our knowledge of the numbers cannot depend on geometry or kinematics without renouncing the claim that it is free of intuition. This suggests an interpretation of Frege's use of the Julius Caesar problem which allows for a certain asymmetry between Basic Law V and Hume's principle. In order to see this, it is necessary to distinguish two accounts of the role of extensions: one holding them to be necessary to complete the argument that our knowledge of the numbers has the requisite autonomy and independence, and that Hume's principle is a logical principle; the other holding that even if the notion of an extension is not necessary to insure the logical character of the numbers or of Hume's principle, it, or something like it, *is* needed to fix the reference of numerical singular terms. As we shall see, the first account has much to recommend it as an interpretation of Frege's final view of the role of extensions in his theory of number. However, the second account, which is perhaps in greater conformity with the position of *Grundlagen*, is more likely to find applications to contemporary problems.

To begin with, notice that, as interpretations of Frege's methodology, both accounts need to explain why Frege believed himself justified in holding that our knowledge of extensions is independent of experience or intuition, the former, in order that his appeal to extensions might *justify* his logicism, the latter, in order that his use of extensions should be *compatible* with his logicism. But on the latter view Frege required more than a characterization of the numbers that would show our knowledge of them to be independent of the principles of any special science; he also required a characterization that would enable us to settle in a unitary way—i.e., by a single stipulation, general enough to cover all cases—the truth of every identity involving numerical singular terms. To this extent, the introduction of extensions of concepts formed part of a general explanation of our ability to refer to the natural numbers. A central difficulty, according to this view, is that Frege's identification of numbers with extensions is simply insufficient to settle all such questions of identity. For example, even if we know that the number of all objects (the number of the concept *is self-identical*) is the extension of a (second level) concept, it is still not determined by Frege's principles whether this number is identical with or distinct from the first infinite cardinal.[21]

It is, however, not clear why the failure of a *general* stipulation, one intended to cover, once and for all, all questions of identity, should have a decisive effect on Frege's program. Once we allow that the truth of the identities left indeterminate by Hume's principle may be resolved by *stipulating* their truth value, why should it matter whether just *one* stipulation suffices for all cases? It is also unclear why the introduction of extensions is required to insure that certain identities are false: Even if Hume's principle fails to settle whether (say) Julius Caesar is identical with the number of the planets, the ordinary notion of number is surely as unequivocal as any notion of extension in deciding this identity. Although it has often been taken for granted that Frege's reconstruction must considerably exceed the precision of our preanalytic notion, this assumption needs to be reconsidered in light of recent suggestions that Frege was not particularly committed to the rejection of informal reasoning.[22]

[21]See Dummett, *Frege, philosophy of mathematics*, p. 227, and Boolos, Chapter 8, below.

[22]In this connection, see the papers by Coffa, Benacerraf, Boolos, and myself, reprinted here as Chapters 1, 2, 6, and 3, respectively.

Assuming that it achieves a level of precision sufficient to show its independence from intuition, it is an open question to what extent Frege's reconstruction of the numbers must exhibit more precision than the preanalytic notion.

But if neither the number-theoretic nor the ordinary notion of number seems capable of resolving whether or not our knowledge of the numbers—in particular our knowledge of the fact that they form an infinite series—depends on experience or intuition, perhaps Frege's point, in introducing extensions, was to vindicate the *logical* character of the numbers and of Hume's principle. On the surface, this interpretation seems to be excluded by Frege's formulation of the question which opens § 62: 'How, then, are numbers to be given to us, if we cannot have any ideas or intuitions of them?' The formulation suggests that Frege takes himself to have established—before Hume's principle has even been stated—that our knowledge of the numbers is not based on intuition. If this were right, it would have been quite unnecessary for Frege to have introduced extensions for the purpose of showing that arithmetic, in general, and Hume's principle, in particular, owe nothing to intuition.

Certainly, in the sections preceding § 62, Frege has succeeded in refuting a variety of accounts of number which depend on some notion of intuition. By and large, the views targeted for criticism in these early, polemical, sections of *Grundlagen*—even the position attributed to Kant in § 5—have an obviously psychologistic ring to them. At this point in the argument of the book, all Frege may justifiably claim to have established is that arithmetic exhibits the generality we associate with logic and that our knowledge of the numbers is not intuitive in the sense of one or another of the psychologistic theories he has reviewed. But this review of accounts of number based on intuition is hardly exhaustive—presumably the notion of intuition Frege thinks relevant to geometry is not tainted with psychologism, and so is not among those canvassed. In the Introduction to *Grundgesetze* (I, p. vii) Frege makes essentially this point:

> Of course the pronouncement is often made that arithmetic is merely a more highly developed logic: yet that remains disputable so long as transitions are made in proofs that are not

made according to acknowledged laws of logic, but seem rather to be based upon something known by intuition.

It therefore remains to be shown that our knowledge of arithmetic is not, in *any* sense, based on intuition, and that arithmetic is, in fact, an autonomous science. That this is what the introduction of extensions was supposed to achieve is suggested by Frege's own final evaluation of his project. If in *Grundlagen* Frege expressed a certain indifference toward extensions of concepts,[23] by the time of *Grundgesetze* he had come to regard their use as essential to the goal of showing the independence of arithmetic from intuition:

> ... I myself was long reluctant to recognize ranges of values and hence classes; but I saw no other possibility of placing arithmetic on a logical foundation. But the question is, How do we apprehend logical objects? And I found no other answer to it than this, We apprehend them as extensions of concepts .... I have always been aware that there are difficulties connected with them: but what other way is there?[24]

Clearly, if extensions are self-evidently logical objects, the contextual definition's inability to decide the logical character of the numbers does not 'iterate' to the case of extensions and Basic Law V, and our interpretive problem is solved. We are, however, left with the task of showing that there is an explanation of the notion of the extension of a concept which serves the purposes of Frege's logicism, while not depending on the principles of any special science.[25] It is the apparent intractability of this problem that has recommended the development of Frege's theory of the natural numbers without recourse to the notion of the extension of a concept. And it is in this connection, rather than for its value as an interpretation, that the

[23] In the footnote to § 69 and in § 107's summary of the argument of the book.

[24] Letter to Russell of 28.vii.1902, in Brian McGuinness, ed., Gottlob Frege: *Philosophical and mathematical correspondence*, Chicago: University of Chicago Press, 1980, Hans Kaal, tr., pp. 140f.

[25] For a 'mathematically natural' development of Frege's notion of the extension of a concept, one which connects Frege's extensions with approaches to the paradoxes based on the principle of limitation of size, see George Boolos's "Saving Frege from contradiction," Chapter 18, below.

view of the Julius Caesar problem as primarily a problem of reference acquires its main support and its chief relevance to current work.

To put our discussion into sharper focus let us distinguish between two closely related, but distinct goals which the elimination of appeals to intuition might establish. I have suggested that Frege's first objective was to show that arithmetic—both the theory of the natural numbers and the theory of the reals—is autonomous from geometry and kinematics; this objective is what we normally think of as the refutation of Kant. Another objective, one emphasized by Dummett,[26] and one that is perhaps of greater interest to us, was to solve the problem of reference to abstract objects without resorting to a naive 'Platonist' picture, according to which abstract objects are 'ostended in intuition,' where intuition may (but need not) be tied to the Kantian tradition and the issue of the autonomy of arithmetic. The difficulty which the paradoxes presented for Frege's notion of the extension of a concept was an obstacle in the way of his refutation of Kant because it showed his justification of the *logical* character of Hume's principle to be unsound; and the chief effect of this failure was to throw into doubt the logical character of his demonstration of the natural number concept. But what of the second objective, that of avoiding naive Platonism?

For Frege, the basis for introducing abstract terms is the determination of truth conditions for the identity statements into which they enter; Frege believed that he could fix the truth conditions of such identity statements while (as we might put it) simultaneously specifying the domain of quantification of the first-order quantifiers. A mathematical theory like arithmetic or set theory is distinguished from one like Euclidian geometry by the fact that, in the latter case, we are content to speak only of truth in the appropriate class of structures; by contrast, arithmetic, at least, is generally thought to be true *simpliciter*. Frege seems to have been exclusively concerned with problems of reference for mathematical theories which, like arithmetic, are, in this sense, 'absolutely' true.[27] He sought an account of the notions of truth and reference radically different

---

[26]See especially Chapter 18 of *Frege: philosophy of mathematics*.

[27]Of course Frege held that Euclidian geometry is *also* true *simpliciter*. Cf. his correspondence with Hilbert, collected in Gottlob Frege: *Philosophical and mathematical correspondence*.

from one modelled on our notion of truth-in-a-structure, different, that is, from an account which begins by giving a characterization of the domain, a characterization that is external to the theoretical assumptions of the theory, and then proceeds to specify reference and truth with respect to this domain. If we were to follow this model for our account of the reference of the terms of arithmetic, then, since the truth of any proposition of the theory—including any recognition statement—would depend upon the reference of its constituent terms having first been given, we could hardly turn round and propose to account for reference by an appeal to the context principle and the truth of the appropriate class of recognition statements. Indeed, it would be possible to employ this model for our account of the truth of recognition statements even if we had not progressed beyond a naive Platonist explanation of how the objects of arithmetic are, in Frege's phrase, 'given to us.' This is why the truth of Hume's principle is insufficient for Frege's purposes so long as this is understood to depend on our having *first* specified the referents of its constituent expressions: this would defeat the whole point of an account of the reference of numerical expressions by a subsequent appeal to the context principle.[28] Had Frege succeeded in establishing the truth of Hume's principle independently of such a procedure, he would have given a complete answer to our concerns regarding reference to the numbers. Had he achieved this goal by the derivation of Hume's principle from logic, he would have succeeded in answering both Kantians and Platonists.

VI. DEFINITION BY ABSTRACTION

Frege's use of extensions had another purpose which is perhaps best brought out by comparison with Dedekind, with whom the practice of 'abstracting' to a set-theoretical object in order to characterize the domain of a particular mathematical theory seems to have originated.[29] Dedekind's first important application of this method

---

[28]Cp. Michael Dummett, "The context principle: centre of Frege's philosophy," in Werner Stelzner, ed., *Logik und Mathematik: Frege Kolloquium, Jena 1993*, Berlin: de Gruyter, 1995.

[29]Although, as Wilson argues in Chapter 5, there are importantly similar precedents in the earlier geometric tradition of von Staudt.

occurred in his initial formulation of Kummer's theory of ideal factorization.[30] Kummer had relied on the concept of an ideal prime
factor of 'cyclotomic integers,'[31] and although he had adequately
characterized a 'divisibility test' which would allow for the determination of when α is m-times divisible by P, where α is a cyclotomic
integer and P is an ideal prime factor,[32] he left open the question of
what an ideal prime factor *is*; a fortiori, Kummer's ideal numbers,
insofar as they were understood in terms of ideal prime factors, were
also left unspecified. To quote Edwards:

> ... This state of affairs was extremely unsatisfactory to Dede-
> kind who thought of *identifying ideal numbers* (and, in partic-
> ular ideal prime factors) *with the set* (system) *of integers they
> divide*. He then found ... that such subsets are characterized
> by the fact that they are closed under addition and multipli-
> cation by ring elements. ... "This situation led me to found
> the entire theory of numbers of the ring ... [of integers in a
> number field] on this simple definition [of 'ideal'] *delivered of
> all obscurity and the admission of ideal numbers*."[33]

There is an evident similarity between Dedekind's use of abstraction—his 'ascent' to set-theoretical objects—in the case of ideals and
his use of 'cuts' in connection with his characterization of the real
numbers: just as an 'ideal number in a ring is "concretely" repre-

---

[30]See Harold M. Edwards, "Dedekind's invention of ideals," *Bulletin of the
London Mathematical Society*, **15** (1983) 8–17, and "The genesis of ideal theory," *Archive for history of exact sciences*, **23** (1980) 321–78, whose account I
follow.

[31]Cyclotomic integers initially arose in the theory of complex integers where
they are also known as 'n-th roots of unity'; α is a *cyclotomic integer* or an *n-th
root of unity* if it is a root of the equation: $\alpha^{n-1} + \alpha^{n-2} + \cdots + \alpha + 1 = 0$.
Gauss made the use of these numbers prominent in his treatment of geometric
construction.

[32]Ideal prime factors were introduced by Kummer in order to recover a weakened form of the prime decomposition theorem for cyclotomic integers. See the
previously cited papers by Edwards for a full discussion.

[33]"Dedekind's invention of ideals," p. 9. The quotation from Dedekind is
from Vol. 1 of his *Gesammelte mathematische Werke*, Braunschweig: Vieweg,
1932; reprinted by Chelsea Publ. Co., New York, 1969. The translation (from
the French) is by Edwards; the emphases are all either Edwards's or have been
added by Edwards.

sented by the set of all *actual* numbers of the ring it divides, [so also, a] real number is "concretely" represented by the set of all *rational* numbers that it is greater than (or by those it is less than).'[34] Moreover, both applications bear an evident similarity to Frege's use of this method in connection with the cardinal numbers. Nevertheless, I think we would do well to distinguish these three uses, at least with respect to the goals they subserve. Dedekind's use of abstraction in the case of ideals was principally motivated by the desire to obtain a general mathematical theory of factorization. Dedekind sought a theory that, unlike Kummer's (which applied only to cyclotomic integers), would be applicable to *any* number field.[35] But Dedekind's introduction of cuts was intended to show that the the theory of the real numbers is autonomous and owes nothing to geometric ideas, something that seems to have formed no part of the motivation for his theory of ideals:

> For our immediate purpose, however, another property of the system $\mathbb{R}$ [of real numbers] is still more important; it may be expressed by saying that the system $\mathbb{R}$ forms a well-arranged domain of one dimension extending to infinity on two opposite sides. What is meant by this is sufficiently indicated by my use of expressions borrowed from geometric ideas; but just for this reason it will be necessary to bring out clearly the corresponding purely arithmetic properties *in order to avoid even the appearance [that] arithmetic [is] in need of ideas foreign to it*.[36]

Similarly, for Frege, the definition of the cardinal numbers in terms of extensions was, as we have seen, intended to show that our knowledge of their properties is independent of experience or intuition. With Frege, however, abstraction was also exploited in order

[34] "Dedekind's invention of ideals," p. 13.

[35] Dedekind also objected to the fact that Kummer's formulation 'depended on a particular explicit representation of the factor ... rather than being based on *intrinsic properties* of the factors.' (Edwards, "The genesis of ideal theory," p. 343.) On the importance of representation-free formulations for Dedekind, see also § X of Wilson's paper (Chapter 5).

[36] Dedekind, "Continuity and irrational numbers," in R. Dedekind, *Essays on the theory of numbers*, New York: Dover, 1963, W. W. Beman tr., p. 5, emphasis added.

to provide representations of the natural and the real numbers capable of encompassing all our applications of them. This is a goal that goes beyond the proof that arithmetic is an autonomous science. In this respect, Frege's plan for a theory of the real numbers was continuous with his theory of the natural numbers. In both cases he sought to make the applications of the numbers—of the natural numbers to counting, and of the real numbers to measuring—essential to their characterization.[37] And in contrast to Russell and Whitehead, Frege hoped to make the fulfilment of this plan the central point of difference between his theory of the reals and the theories of Dedekind and Cantor. Frege envisaged defining the real numbers directly as equivalence classes of ratios of *quantities*, in contradistinction to Dedekind and Cantor for whom the reals are defined in terms of (classes of) rational *numbers*, and in this respect, his approach was closer to the traditional Eudoxan theory of proportion than even Dedekind's.[38] Although Frege would have had to resort to the example of a 'quantitative domain' built up from the rational numbers in order to prove the existence of the reals, the rational numbers would not have entered essentially into his characterization of them.[39] However, it is not clear how the fulfilment of the goal of making the applications of the numbers essential to their characterization could proceed without classes; nor is it clear how Frege could have avoided their use in the technical development of his theory of the real numbers.

[37] This is a central thesis of Peter Simons's paper, "Frege's theory of real numbers," reprinted as Chapter 13 below. The thesis is also endorsed by Dummett in his discussion of Frege's theory of the real numbers in Chapter 14, below.

[38] Although as Dummett remarks, 'Frege nowhere calls explicit attention in *Grundgesetze* to the fact that, unlike Cantor and Dedekind, he is proposing to define the real numbers without taking the rationals as already known' (*Frege: philosophy of mathematics*, p. 270). Dummett notes that Frege does explicitly call attention to the difference in his letter to Russell of 21.v.1903.

[39] Cf. Simons, Chapter 13, §§ IV and VI. The paper of Peter M. Neumann, S. A. Adeleke and Michael Dummett, "On a question of Frege's about right-ordered groups," (Chapter 15) explains how, in the course of his characterization of the notion of a quantitative domain, Frege was led to a number of results of a group-theoretic nature. A further exposition of some of the topics dealt with there will be found in Dummett, Chapter 14, below.

APPENDIX: FREGE'S THEOREM
AND THE ZERMELO-BOURBAKI LEMMA[40]
JOHN L. BELL

This Appendix establishes the existence of an infinite well-ordering as a (hitherto unremarked) consequence of a general version of Zermelo's Well-ordering theorem. We also indicate how this fact can be derived along 'Fregean' lines within a certain system $\mathbb{F}$ of many-sorted first-order logic whose sorts correspond to Frege's domains of objects, relations, and first and second level concepts. We show that the system of axioms we formulate within $\mathbb{F}$ constitutes a consistent fragment of Frege's original (inconsistent) system sufficient for the development of arithmetic.

We begin by specifying the basic constituents of the system $\mathbb{F}$.

Sorts (or domains)

$\mathcal{O}$ — objects

$\mathcal{B}$ — basic (first level) concepts

$\mathcal{R}$ — relations

$\mathcal{S}_b$ — second level concepts

$\mathcal{S}_r$ — second level relational concepts

Variables and Constants

Sort	Variable	Constant
$\mathcal{O}$	$x, y, z, \ldots$	$a, b, c, \ldots$
$\mathcal{B}$	$X, Y, Z, \ldots$	$A, B, C, \ldots$
$\mathcal{R}$	$\underline{X}, \underline{Y}, \underline{Z}, \ldots$	$\underline{A}, \underline{B}, \underline{C}, \ldots$
$\mathcal{S}_b$	$\underset{\sim}{X}, \underset{\sim}{Y}, \underset{\sim}{Z}, \ldots$	$\underset{\sim}{A}, \underset{\sim}{B}, \underset{\sim}{C}, \ldots$
$\mathcal{S}_r$	$\underline{\underline{X}}, \underline{\underline{Y}}, \underline{\underline{Z}}, \ldots$	$\underline{\underline{A}}, \underline{\underline{B}}, \underline{\underline{C}}, \ldots$

A *term* is a variable or a constant or one of the concept or relation or extension terms to be introduced shortly. A variable of sort $\mathcal{B}$ or $\mathcal{S}_b$ will be called a *concept variable* for brevity.

We assume the presence of an *identity* sign = yielding atomic statements of the form $s = t$ where s and t are terms of the *same*

[40]Abstracted from the Appendix to William Demopoulos and John L. Bell, "Frege's theory of concepts and objects and the interpretation of second-order logic," *Philosophia mathematica* (Series III), **1** (1993) 139–56. Reprinted by kind permission of Robert Thomas, editor of that journal, and John L. Bell.

sort. On all domains except $\mathcal{O}$, $=$ is to be thought of as *intensional* equality.

We also assume the presence of a *predication* sign η yielding atomic statements of the form $s\,\eta\,t$, $(s't')\,\eta\,u$ where s is of sort $\mathcal{O}$, $\mathcal{B}$, $\mathcal{R}$ and t is of sort $\mathcal{B}$, $\mathcal{S}_b$, $\mathcal{S}_r$ respectively; and s', t' are both of sort $\mathcal{O}$ and u is of sort $\mathcal{R}$. We read '$s\,\eta\,t$' as 's falls under t.'

We shall assume the following comprehension scheme for concepts:

Corresponding to any formula $\Phi(x)$, $\Phi(x,y)$, $\Phi(X)$ or $\Phi(\underline{X})$ we are given a term s of sort $\mathcal{B}$, $\mathcal{R}$, $\mathcal{S}_b$, $\mathcal{S}_r$, respectively, for which we adopt as an axiom the formula

$$\forall \left\{\begin{matrix} x \\ xy \\ X \\ \underline{X} \end{matrix}\right\} \left[\left\{\begin{matrix} x \\ (xy) \\ X \\ \underline{X} \end{matrix}\right\} \eta\, s \leftrightarrow \Phi \left\{\begin{matrix} x \\ xy \\ X \\ \underline{X} \end{matrix}\right\} \right].$$

We write $\widehat{x}\Phi$, $(xy)\widehat{}\Phi$, $\widehat{X}\Phi$, $\widehat{\underline{X}}\Phi$ for s, as the case may be. A term of the first, third and fourth types is called the *concept* (term) determined by Φ, and a term of the second type the *relation* (term) determined by Φ.

We define the relation $\equiv$ of *extensional equality* on the domains $\mathcal{B}$, $\mathcal{R}$, $\mathcal{S}_b$, $\mathcal{S}_r$ by

$$X \equiv Y \Longleftrightarrow_{df} \forall x(x\,\eta\,X \leftrightarrow x\,\eta\,Y)$$

$$\underline{X} \equiv \underline{Y} \Longleftrightarrow_{df} \forall x \forall y[(xy)\,\eta\,\underline{X} \leftrightarrow (xy)\,\eta\,\underline{Y}]$$

$$\underset{\sim}{X} \equiv \underset{\sim}{Y} \Longleftrightarrow_{df} \forall X[X\,\eta\,\underset{\sim}{X} \leftrightarrow X\,\eta\,\underset{\sim}{Y}]$$

$$\underline{X} \equiv \underline{Y} \Longleftrightarrow_{df} \forall \underline{X}[\underline{X}\,\eta\,\underline{X} \leftrightarrow \underline{X}\,\eta\,\underline{Y}].$$

Clearly concepts are determined uniquely by formulas up to extensional equality. We assume that $\mathbb{F}$ contains

- a term e such that $e(\mathfrak{X})$ is well-formed and of sort $\mathcal{O}$ for any concept variable $\mathfrak{X}$;
- a predicate symbol E such that $E(\mathfrak{X})$ is well-formed for any concept variable $\mathfrak{X}$.

We finally assume the axioms

(1) $\forall\mathfrak{X}\forall\mathfrak{Y}[E(\mathfrak{X}) \wedge E(\mathfrak{Y}) \rightarrow [e(\mathfrak{X}) = e(\mathfrak{Y}) \leftrightarrow \mathfrak{X} \equiv \mathfrak{Y}]]$
(2) $\forall\mathfrak{X}\forall\mathfrak{Y}[E(\mathfrak{X}) \wedge \mathfrak{X} \equiv \mathfrak{Y} \rightarrow E(\mathfrak{Y})]$

where in both (1) *and* (2) $\mathfrak{X}$ *and* $\mathfrak{Y}$ *are concept variables of the same sort.*

If we think of $e(\mathfrak{X})$ as an object *representing* $\mathfrak{X}$, Axiom 1 above expresses the idea that *extensional equality* of any concepts satisfying E is equivalent to *identity* of their representing objects. That is, for any concept $\mathfrak{X}$ satisfying E, $e(\mathfrak{X})$ may be regarded as the *extension* of $\mathfrak{X}$. And the predicate E itself represents the property of *possessing an extension*. For these reasons Axiom 1 will be called the *Axiom of Extensions*. As for Axiom 2, it states the reasonable requirement that any concept extensionally equivalent to a concept possessing an extension *itself* possesses one (that is, $\equiv$ is a *congruence* relation with respect to E).

A straightforward Russell type argument in $\mathbb{F}$ enables us to infer $\neg\forall\mathfrak{X}E(\mathfrak{X})$,[41] that is, *not every concept possesses an extension*. This being the case, what concepts do we need to (consistently) assume possess extensions in order to enable an infinite well-ordering to be constructed? It was Frege's remarkable *discovery* that for this it suffices to assume just that extensions be possessed by the members of a certain class of simple and natural *second-order* concepts—those that, following Boolos,[42] we shall term *numerical*.

Numerical concepts are defined as follows. First, we formulate the relation $\approx$ of equinumerosity or equipollence on $\mathcal{B}$ as usual:

$$X \approx Y \Longleftrightarrow_{df} \exists \underline{Z}[\forall x \forall y[(xy)\,\eta\,\underline{Z} \to x\,\eta\,X \wedge y\,\eta\,Y]$$
$$\wedge\, \forall x \forall y \forall z[(xy)\,\eta\,\underline{Z} \wedge (xz)\,\eta\,\underline{Z} \to y = z]$$
$$\wedge\, \forall x[x\,\eta\,X \to \exists y\,(xy)\,\eta\,\underline{Z}]$$
$$\wedge\, \forall y[y\,\eta\,Y \to \exists x\,(xy)\,\eta\,\underline{Z}]$$

With any basic concept X we associate the second level concept

$$\|X\| =_{df} \widehat{Y}\,[X \approx Y].$$

Concepts of the form $\|X\|$ are called *numerical*.

---

[41]To be explicit, define $A =_{df} \hat{x}[\forall X[e(X) = x \wedge E(X) \to \neg x\,\eta\,X]]$. Then $\neg E(A)$ is inferrable in $\mathbb{F}$.

[42]"The standard of equality of numbers," Chapter 8, below.

If we assume that every numerical concept possesses an extension (i.e., $\forall X E(\|X\|)$), then the extension

$$|X| =_{df} e(\|X\|)$$

is called the (cardinal) *number* of X. Objects of the form $|X|$ are called (cardinal) *numbers*. Under these assumptions it is easy to derive *Hume's principle*, viz.

$$\forall X \forall Y \, [X \approx Y \leftrightarrow |X| = |Y|\,].$$

We shall call a concept X (Dedekind) *infinite* if $\exists Y \, [Y \subsetneqq X \wedge X \approx Y]$, where $Y \subsetneqq X$ of course stands for $\forall x(x \, \eta \, Y \rightarrow x \, \eta \, X) \wedge Y \not\equiv X$. Objects of the form $|X|$ with X infinite are called *infinite numbers*.

We are going to show how, in $\mathbb{F}$, the existence of an infinite well-ordering (i.e. an infinite well-ordered concept) may be derived as a special case of a general set-theoretic result—formulable and provable in $\mathbb{F}$—which is normally used to derive Zermelo's Well-ordering theorem. In its original form this result is what we shall call the

Zermelo-Bourbaki lemma. [43] *Let E be a set, $\mathcal{F}$ a family of subsets of E and $p : \mathcal{F} \to E$ a map such that $p(X) \notin X$ for all $X \in \mathcal{F}$. Then there is a subset M of E and a well-ordering $\leqslant$ of M such that, writing S_x for $\{y : y < x\}$,*

(i) $\forall x \in M \, [S_x \in \mathcal{F} \wedge p(S_x) = x]$
(ii) $M \notin \mathcal{F}$.

Bourbaki employs this result to construct an elegant derivation of Zermelo's Well-ordering theorem from the Axiom of Choice. In the present context, however, it will be used to produce an equally elegant proof of what we shall call, following a suggestion of Boolos,

---

[43]Lemma 3, § 2, Ch. 3 of N. Bourbaki, *Théorie des ensembles*, 2nd ed. Paris: Hermann, 1963. Bourbaki's proof is a generalization of Zermelo's argument for his Well-ordering theorem in his "Proof that every set can be well-ordered" (1904) in Jean van Heijenoort, ed., *From Frege to Gödel: a sourcebook in mathematical logic, 1879–1931*, Cambridge: Harvard University Press, 1967, Stefan Bauer-Mengelberg, tr.

Frege's theorem. *Suppose given a set E and a map $n : PE \to E$ such that*

$$(*) \qquad \forall X \subseteq E \, \forall Y \subseteq E \, [n(X) = n(Y) \leftrightarrow X \approx Y].$$

Then E has an infinite well-ordered subset.

Proof. We apply the Zermelo-Bourbaki lemma with $\mathcal{F}$ the family of all subsets X of E for which $n(X) \notin X$ and p the map n. We obtain $M \subseteq E$ and a well-ordering $\leqslant$ of M such that *(i)* $n(S_x) = x$ for all $x \in M$, *(ii)* $n(M) \in M$. Writing m for $n(M)$ we have $m \in M$ by *(ii)*, whence $n(S_m) = m = n(M)$ by *(i)*. Condition $(*)$ now yields $S_m \approx M$. Since $m \notin S_m$, S_m is a proper subset of M and it follows that the latter is infinite. $\square$

Now both of these results can be translated into and proved within $\mathbb{F}$. Carrying this out for the Zermelo-Bourbaki lemma yields the

Zermelo-Bourbaki lemma in $\mathbb{F}$. *Let $\underset{\sim}{S}$ be any second level concept with respect to which $\equiv$ is a congruence relation and t a term such that $t(X)$ is an object for all basic concepts X and satisfies*

$$\forall X \forall Y \left[X \equiv Y \wedge X \, \eta \, \underset{\sim}{S} \to t(X) = t(Y) \right]$$

$$\forall X \left[X \, \eta \, \underset{\sim}{S} \to \neg t(X) \, \eta \, X \right].$$

Then there is a relation R such that R is a well-ordering and, writing M for its field, and R_x for $\widehat{y}\,[(yx) \, \eta \, R \wedge y \neq x]$,

    (i) $\forall x \left[x \, \eta \, M \to R_x \, \eta \, \underset{\sim}{S} \wedge t(R_x) = x \right]$
    (ii) $\neg M \, \eta \, \underset{\sim}{S}.$

In the case of Frege's theorem, the same process yields

Frege's theorem in $\mathbb{F}$. *Suppose that every numerical concept has an extension. Then there exists an infinite well-ordered concept and hence an infinite number.*

Since, as is well known, Frege's original system in the *Grundgesetze* was inconsistent, we should assure ourselves that the axioms of

$\mathbb{F}$, together with the hypothesis of Frege's theorem—that every numerical concept has an extension—are consistent. The easiest way to see this is by noting that the following set-theoretic interpretations yield a model of the axioms of $\mathbb{F}$ in which the hypothesis of Frege's theorem holds. To wit, interpret $\mathcal{O}$ as $\omega+1$, $\mathcal{B}$ as $P(\omega+1)$, $\mathcal{R}$ as $P((\omega+1) \times (\omega+1))$, $\mathcal{S}_b$ as $PP(\omega+1)$, $\mathcal{S}_r$ as $PP((\omega+1) \times (\omega+1))$, E as the subset of $PP(\omega+1)$ consisting of all elements of the form $\|X\| =_{df} \{Y \in P(\omega+1) : X \approx Y\}$ with $X \in P(\omega+1)$, and e as the map $PP(\omega+1) \to \omega+1$ which sends each $\|X\|$ with $X \in P(\omega+1)$ to its cardinality $|X|$ ($\in \omega+1$) and everything else to 0. Thus the axioms of $\mathbb{F}$ — together with the hypothesis of Frege's theorem — may be regarded as a consistent fragment of Frege's original system.

PART I

The Intellectual Background
to Frege's Logicism

1

Kant, Bolzano and the Emergence of Logicism

There are many ways to look at logicism. My aim today is to present one of those versions, embedding it in a broader movement which includes the rigorization of the calculus, Frege's and Russell's theories of arithmetic, and Poincaré's and Hilbert's geometric conventionalism. I shall try to convince you that these episodes belong to an as yet unnamed natural kind which, for lack of a better word, I will call 'conceptualism.'[1]

Conceptualism is defined by an enemy, a goal, and a strategy: the enemy was Kant, the goal was the elimination of pure intuition from scientific knowledge, and the strategy was the creation of semantics as an independent discipline. Although this third component is by far the most important, here I shall have time to discuss in some detail only the other two.

I. THE ENEMY

Semantics, properly so called, was born as a response to problems which were implicit in the (tacit) semantics of rationalism and em-

From *The journal of philosophy*, **74** (1982) 679–89. Reprinted by kind permission of the editors of *The journal of philosophy* and the estate of Alberto Coffa.

[1]'Conceptualism' is just as inadequate a name as the name of Frege's *Begriffsschrift*.

piricism and which first came to light in Kant's philosophy. In order to contrast these semantic doctrines, the first question we must raise is this: How does one find out what Kant and his predecessors thought about meanings?

Concepts, not essences, is what meanings were before they were wedded to the word. If we want to know what our ancestors thought about meanings we should consult not their books about essences but their logic books, where they talked about concepts and judgments, propositions and meanings. What Kant was saying in that famous remark on the permanent value of Aristotelian logic was not that syllogistics was fine—nothing could have been further from his mind[2]—but that "Aristotelian" (i.e., Wolffian, Lambertian, Crusian, etc.) semantics was good enough for him.

The basic notion of that crypto-semantics is that of representation. Knowledge is, of course, expressed in judgments, and for Kant even experience consists of judgments (Ak 23, 24). But judgments consist of the union or separation of representations or of other judgments (e.g., Ak 16, 631/2, Refl. 3046 & 3049; Ak 9, 101).

Since Christian Wolff the word 'representation' had become a technical term designed to play a role analogous to that of the earlier 'idea' in French and English philosophy. Representations or concepts were conceived in terms of what we might call the "chemical" picture, according to which representations are to be thought of on the model of chemical elements, for they are usually complexes (or, in Kant's words, manifolds) and, like them, are subject to analysis. According to this picture, representations have a number of constituents, themselves representations, which might, in turn, be complex. Since Descartes, clarity and distinctness had become the highest virtues of the ethics of representation, and in this tradition distinctness (*Deutlichkeit*) was defined as the outcome of the process through which we identify all the constituents of a representation. The process itself was called *analysis*, and the complete, distinct version of a given concept, its *definition* [e.g., Ak 24, 571 (25-31)]. The ultimate, simple constituents were the indefinables or unanalyzable simples.

---

[2]Cf. Ak 2, 47–61; Ak 29, 32. 'Ak x, y' will mean *Kant's Academy Edition*, volume x, page(s) y.

Whereas most of Kant's German predecessors had identified representation and concept, Kant departed from this doctrine in his *Dissertation* of 1770, where he drew a sharp distinction between two different faculties of representation: the sensible and the intelligible. Soon he came to think there could be no individual concepts. Concepts, he said, are the product of the understanding to which individual objects are never given. Objects are given to human beings only in sensibility, and the representations of sensibility therefore deserve a different name; Kant's choice was 'intuition.' Thus he concluded that representations are of two radically different sorts: concepts and intuitions.

In spite of these innovations Kant remained firmly committed to the traditional chemical picture of representations. Wolff had praised "the great use of magnifying glasses toward gaining distinct notions."[3] Following this lead, Kant explained in his logic classes that, when we look at the Milky Way with our bare eyes, we have a clear but indistinct representation of it, since we see clearly not each star but only a continuous band of light. When we look through a telescope, however, our representation becomes distinct. The same is true of concepts, which usually have constituents (*Bestandstuecke*, *constitutiva*; Ak 8, 229; Ak 24, 753; Ak 11, 34) of which we are not clearly aware. Conceptual analysis is the intellectual telescope with which we come to see, e.g., that the concept of Freedom is contained in the concept of Virtue (Ak 24, 511/2).

As we all know, Kant defined an analytic judgment as one whose predicate concept is "tacitly thought" in the subject concept. One might think that it is a very small step from the chemical picture of the concept to this idea and to its complement, that of a synthetic judgment. Kant did not think so. He was more than willing to grant that many philosophers had recognized the division between the a priori and the a posteriori, but he insisted that no one before had seen the significance of the analytic-synthetic distinction. When Eberhard challenged Kant's originality and said, in effect, that his distinction was old hat, Kant was furious. In an effort toward irony he replied that everything new in science to which nothing can be opposed is eventually "discovered to have been known by the An-

---

[3] *Logic* (*deutsche Logik*), ch. I, sec. XXII.

cients" (Ak 8, 243).

Why didn't Kant think that his distinction was an utterly trivial consequence of the familiar notion of conceptual analysis? The answer I should like to propose is this: when Kant conjoined his notion of synthetic judgment with his casual understanding of semantics—specifically, of concepts—the outcome was an extremely powerful and truly original philosophical thesis, the "principle of synthetic judgments." The most striking proof of the power of this principle lies in the fact that, when conjoined with reasonable assumptions, it implies the existence of a nonempirical kind of intuition. Let us see how.

In the first *Critique* Kant had explained that, just as all analytic judgments are grounded on a single principle, that of identity, or, as he also called it, "the principle of analytic judgments," there is also a single principle involved in the grounding of synthetic judgments; and he sensibly called it "the principle of synthetic judgments." When Eberhard complained that no reader of the *Critique* could tell what the principle in question was, Kant obligingly formulated it once again for him as follows: synthetic judgments "are only possible under the condition that an intuition underlies the concept of their subject" (Ak 8, 241).[4] In fact, he had already explained in the *Critique* that the synthesis of two disjoint concepts must always be mediated by a third representation,[5] an X (A9, B13) not directly present in the judgment (as a constituent). Since he also thought that "from mere concepts only analytic knowledge ... can be derived" (A47, B64; see also A155, B194), this third representation couldn't be a concept; but, since representations are either concepts or intuitions, the ground of the synthesis in a synthetic judgment must lie in intuition. This proves the principle of synthetic judgments. If we now add the assumption that there are synthetic a

---

[4]Kant's formulation of the principle in A158, B197, in that inimitable style of his, may have contributed to the widespread neglect of the import of this principle. Yet his remarks immediately preceding that impenetrable formulation and elsewhere in the *Critique* (e.g., A155, B194; B289) as well as statements in *Prolegomena* and in the reply to Eberhard (e.g., Ak 8, 239/40) should have sufficed to make his unimpressive message clear.

[5]Kant also calls it "das dritte der Anschauung"; see *Textemendationen* to A259 in Ak 23, 49.

priori judgments, we can infer that there *is* pure intuition. For, qua synthetic, such judgments must be grounded in intuition, and, qua a priori, they must be grounded in an intuition entirely unlike the empirical sort recognized by empiricists, since it must have the power to confer universality and necessity upon the corresponding judgments.

II. THE GOAL

I can now be more specific about my proposal. I am inviting you to look at a number of nineteenth-century developments as stages in a process that showed Kant wrong in thinking that his pure intuition plays any role in science. That it plays no role in empirical science was not so hard to see. It was much harder to show that it plays no role either in the calculus or in arithmetic, or—hardest of all—that it plays no role even in geometry. Logicism, as I see it, is one stage in that complex process through which it was finally established that conceptual representations suffice for the construction of every pure a priori science.

One is tempted to compare Kant's discovery of pure intuition as it emerged from the argument above with Leverrier's discovery of Vulcan; but the comparison is unfair to Kant in that there were mathematical "data" that could well be interpreted as justifying Kant's conclusion. During the eighteenth century mathematics and in particular its most productive branch, the calculus, appeared to involve an essential appeal to spatial and temporal intuitions. Newton had presented functions as concerning the motion of points in time, and arithmetic was routinely described as involving processes such as counting. Thus, anyone trying to show Kant wrong in his conclusion that mathematics requires an appeal to pure intuition had two rather considerable tasks to perform, one in philosophy and the other in mathematics. The first was to determine what was wrong with Kant's "deduction" of pure intuition, the second, to show that one can actually construct mathematics in a way that by-passes intuition. It isn't often noticed that a single person initiated the developments that led to the fulfillment of both tasks; his name was Bernard Bolzano.

Let me examine first Bolzano's analysis of Kant's "deduction" of pure intuition. As we saw, the argument depended on two premises:

the assumption that there is synthetic a priori knowledge, and the principle of synthetic judgments. Those determined to reject Kant's pure intuition had to deny one of those premises.

It is widely thought that one of the common features of the anti-Kantian analytic tradition is its rejection of synthetic a priori knowledge. On the contrary, virtually all the members of the semantic tradition we are considering here, from Bolzano, Frege, and Russell to the leading members of the Vienna Circle, recognized the existence and decisive significance of synthetic a priori judgments in Kant's definitional sense. It was not until Quine started The Long March back to J. S. Mill that the analytic tradition seriously questioned the possibility of necessary knowledge that is not analytic in Kant's sense. The challenge emerging from the conceptualist movement against Kantian dogma focused on the *other* premise, the principle of synthetic judgments. Wherever the members of this movement may have stood vis-à-vis the principle of synthetic judgments and whatever they may have said or implied about it, my point is this: that everything these people did concerning foundational issues which was of lasting value, tended to undermine Kant's principle.

Bolzano was the first to recognize the fallacy behind the principle of synthetic judgments. The crucial step in Kant's inference for the need to appeal to intuition in synthetic judgments was the premise that from concepts alone only analytic knowledge can be derived. Astonishingly, there isn't a single argument in the *Critique* for this claim; all Kant says about it is that "it is evident" (A47, B64).[6] What is evident, instead, is that Kant had confused *true in virtue of concepts* with *true in virtue of definitions*, or, in his own language, he had erroneously identified judgments whose predicate is not contained in their subject-concept with judgments that extend our knowledge (*Erweiterungsurteile*). Against this, Bolzano was the first to make a point that even Frege would miss: that Kant's analytic judgments, far from exhausting the grounding power of the conceptual resources of our language, mobilize only a very modest fraction of them, the logical concepts. Bolzano's characterization of analyticity is well known, and it has often been noted that it anticipates not Frege's proof-theoretic treatment but the more modern

---

[6]For a very modest effort toward an argument, see Ak 20, 400.

semantic approach by means of interpretations. What is less well known is the reasoning that led Bolzano to this proposal. After reviewing a number of attempts to explain the point of Kant's notion of analyticity, Bolzano comments that "none of these explanations singles out what makes these [analytic] propositions important. I believe that this consists in the fact that their truth or falsity does not depend upon their constituent representations but remains unaltered, whatever changes one may make in some of these representations .... This is the ground of my preceding definition."[7] Thus, the *reason* why Bolzano came to his celebrated insight on the semantic characterization of logical truth is that he saw that Kant's analytic judgments, far from being those grounded on the information implicit in the constituent concepts, were grounded on only a few of those concepts, thus concluding that a proper definition of analyticity should emphasize the extent to which all other concepts are to be *ignored*.

Having been led by a confused semantics to think that all conceptual information available in a judgment is to be used up in the grounding of analytic judgments, Kant was naturally led to appeal to intuition to ground the rest. Bolzano, on the other hand, having recognized that analytic judgments are judgments grounded on a handful of concepts, was prepared to explore the possibility that *all* of our pure a priori knowledge—including synthetic a priori knowledge—could be stated and grounded on concepts alone. And he proceeded to put his ideas to the test in the field of mathematics. The development known as the rigorization of the calculus is often presented as the first, longest leg of a reductionist program, and it is certainly that. But there is another side to that development which becomes apparent when we look at matters from our present perspective and which is less easy to accommodate within the now fashionable methodologies.

During the eighteenth century the calculus was a Kuhnian's paradise: lots of wonderful exemplars fuelled a staggeringly successful puzzle-solving tradition. The growing ranks of practitioners shared several basic symbolic generalizations, formulas expressing equations and inequalities, which were fruitfully applied to all sorts of prob-

---

[7] *Wissenshcaftslehre*, Hamburg: Meiner, 1929, Vol. II, § 148, p. 88.

lems. To make matters even better, no one really knew what exactly was going on. Even though there was wide agreement on which symbolic generalizations one should assent to when prompted by appropriate stimuli, most people had their own strange interpretation of what these formulas said, and most or all of these interpretations made only modest sense. Moreover, as Kuhn might have predicted, it was a philosopher, the good Bishop Berkeley, who decided to issue a public complaint (in *The analyst*) about the incoherence of the whole enterprise, and specifically about the fact that no meanings were attached to the symbols most crucial to the calculus. (As a "good" philosopher, he wanted definitions, he was worried about "essences," he was worried about meaning rather than theory, etc. etc. In short, he violated every rule in the book of post-positivistic methodology.) As Kuhn would also have predicted, Berkeley's complaints did not stop anyone from following the proto-Kuhnian advice attributed to d'Alembert: "Allez en avant et la foi vous viendra!"

But around 1800 Kuhnian predictions started to go awry. For, this time, mathematicians themselves started to talk about meaning, to try to figure out what exactly was the meaning of each of the basic expressions of the calculus, the meaning of continuity, differentiability, infinitesimal, function, and so on. The ultimate purpose of this emerging project was not to produce some new theorems or to solve some new puzzles but to figure out what exactly it is that the calculus *says*, what its content is, what information it conveys. The question was clearly a semantical question. It is therefore not entirely surprising that the person who took semantics out of the swamp in which it had been sinking since Descartes was also the one who took the first decisive step in the philosophico-mathematical project known as the rigorization of the calculus. That first step was taken in Bolzano's paper of 1817 on the intermediate-value theorem. It is worth pausing for a moment to consider what Bolzano did in that paper.

Bolzano's problem was to prove that a continuous real function that takes values above and below zero, must also take a zero value somewhere in between. A Kantian would probably regard this as trivial: if the point whose path we are considering is moving from the negative to the positive quadrant and the path is continuous it must surely intersect the x axis at some point.[8] Bolzano's problem looks like a problem only to someone who has already understood that intuition is not an indispensable aid to mathematical knowledge, but rather a cancer that has to be extirpated in order to make mathematical progress possible.

Bolzano began by defining continuity in the now standard manner, in terms of epsilons and deltas (not his notation); he then defined convergence for series and gave the criterion now wrongly attributed to Cauchy; then he stated and proved the Bolzano-Weierstrass theorem (modulo a theory of real numbers) and, on that basis, he finally proved the intermediate-value theorem. It was the first time anyone had introduced these concepts and proofs. If Kant had known about Bolzano's paper there can be little doubt that he would have regarded it as a philosophically incoherent effort to prove the obvious. The paper was, instead, one of the landmarks of nineteenth-century mathematics.

Bolzano made many more brilliant contributions to this project of conceptualizing analysis—most of them ignored; but other mathematicians, including Cauchy, Weierstrass, Dedekind, and Cantor, eventually carried the project to completion, establishing that there need be in mathematics no more intuition than there is in arithmetic.

But, how much intuition *is* there in arithmetic? This is the question Frege raised about 1880.

Like Bolzano's, Frege's project started with the decision to take meanings, content seriously. Like Bolzano's, the semantic project of *Begriffsschrift* was put to the test in a theory of mathematics, which is sketched in *Grundlagen*. The most fundamental result of *Grundlagen*, as Frege explained it years later,[9] was his treatment of what he had called "number-statements" (*Zahlangaben*), statements in which we say that there is a certain number of objects of a certain sort. Frege wondered what these statements were about: About counting? About putting things together in the synthetic unity of apperception? About multitudes? The answer was no, in each case. Frege's own verdict, as you recall, is that they are all about

---

[8] For a Kantian "proof" of a theorem in the calculus, see Ak 2, 400.

[9] In *Grundgesetze*, Hildesheim: Olms, 1962, p. ix.

concepts. For example, he argued, when I say that Jupiter has four moons, I am actually talking about the concept 'x is a moon of Jupiter' and saying of that concept that it has an instance, and then another, and so on. In the end, objects came to play a decisive role in Frege's picture of arithmetic, but pure intuition was consistently kept at arm's length. There are many ways of looking at Frege's marvelous book; I prefer to think of it as a gigantic fly-swatter, ominously surveying the whole field of arithmetic, ready to squash pure intuition as soon as it comes in. My version of logicism is therefore this: Bolzano and his followers maneuvered pure intuition out of analysis and into arithmetic where Frege's fly-swatter finally finished it off.[10]

Actually, not quite; for there was still geometry to contend with. Of this complex, final chapter of the saga of pure intuition I can give here only the barest outline.[11]

Unlike Bolzano, Frege accepted the principle of synthetic judgments (cf. *Grundlagen* § 12). That explains why he felt forced to go from the premise that geometry is not analytic (in either Kant's sense or his own) to the conclusion that it must be grounded in intuition. It also explains, partly, why it was so hard for him to see the point that Poincaré and Hilbert were struggling to formulate, with only partial success.

Their key insight is contained in Poincaré's aphorism: geometric axioms are definitions in disguise. Like Frege and Russell, Poincaré had understood that syntactic form is often a poor guide to content, a "disguise"; like Helmholtz, he had recognized that one role of geometric axioms is of the sort Kantians called "constitutive." Helmholtz had noted that "until you have them, you cannot test them," and Poincaré had added that once you have them, it is too late to test them. Having rid himself of most of the old semantic confusions, Poincaré understood that what geometric axioms constitute is not experience or its objects, as some neo-Kantians hoped, but only the concepts in terms of which we think about experience,

[10]My version of Russell's contribution to logicism is in "Russell and Kant," *Synthese*, **46** (1981) 247–63.

[11]I have given a detailed account in my "From geometry to tolerance," in Robert Colodny, ed., *Pittsburgh series in the philosophy of science*, Vol. 7, Pittsburgh: University of Pittsburgh Press, 1986.

our conceptual framework, if you will. By calling them "definitions," Poincaré expressed the fact that their constitutive role belonged not to ontology but to semantics. It was, of course, on this very point that Frege and Russell raised their objections to the new picture of geometry, unwittingly displaying the limitations of their approach to semantics.

To put it paradoxically, Poincaré's claim that geometric axioms are conventional was intended to give an intuition-free explanation of why they are necessary or a priori. The paradox dissolves once we recognize that, in matters of meaning, convention is not the opposite of necessity but merely its opposite side: what looks conventional from the outside, looks necessary from the inside. Is it necessary that bachelors are unmarried male adults? If you ask me before I've made a commitment on the meaning of the word 'bachelor,' of course, I am free then to answer yes or no; if you ask after I have made the appropriate commitment, the answer is yes. Is it necessary that physical objects are colored through and through and through? That bodies not subject to external forces move inertially? That there be exactly one parallel? After we have made certain meaning choices embodied in the endorsement of these principles the answer is necessarily affirmative. Poincaré's decisive contribution is to have noted that one way to "define" certain expressions[12] is by endorsing certain sentences. The ensuing necessity—and the corresponding theory of the synthetic a priori—is thoroughly anti-Kantian in that it allows for seemingly incompatible alternatives—as in the case of geometry, which Poincaré's theory was designed to explain. Decades later Carnap and Wittgenstein would extend the project to the whole of epistemology. First Wittgenstein echoed Poincaré in observing that "the axioms of Euclidean geometry are rules of syntax in disguise";[13] then Carnap's principle of the conventionality of language forms (or principle of tolerance) generalized to language frameworks what Poincaré had observed for alternative geometries. Carnap finally saw what Bolzano almost (but not quite) saw, and what Frege almost (but not quite) saw: that logical truth is truth in virtue of

[12]Not in the Kantian-Russellian sense of finding an analysis, but in the dictionary sense of explaining what the meaning is.

[13]*Philosophische Bermerkunge*, Frankfurt am Main: Suhrkamp, 1964, p. 216.

logical concepts, that is, in virtue of the meanings of logical words, and that, just as the meanings of logical words can ground claims, so can the meanings of all other words.

By the end of this process the death of pure intuition had left only concepts in charge of the job of grounding pure a priori knowledge, either through explicit definition, as Frege wanted, or by the more complex methods displayed by Poincaré and Hilbert. When concepts were finally wedded to the word, a priori knowledge turned from true in virtue of concepts to true in virtue of meanings—as Carnap put it—or true *ex vi terminorum*—as Wilfrid Sellars puts it. Semantics, finally freed from its psychologistic fetters, had produced a coherent, appealing theory that explained a priori and necessary knowledge in a way no longer subject to the idealist temptation.

And then came Quine.

2

Frege: The Last Logicist

When I was young I was taught a number of fundamental propositions: Frege was the father of logicism—he showed that arithmetic
was really only logic (ingeniously disguised), and consequently that
it was really analytic, which was really why it was a priori, all of
which showed where Kant had gone wrong about arithmetic, and
probably about the rest of the alleged synthetic a priori as well.

I was told too that Frege had invented the logic that arithmetic
was really only—or at the very least that he was the father of modern
logic. Had I stopped to think, it might have occurred to me to
question this all too happy coincidence of discovery and invention.
At the very least, a decent interval should have been allowed to

From *Midwest studies in philosophy VI*, Peter French et al., eds., Minneapolis: University of Minnesota Press, 1981, 17–35. Reprinted by kind permission
of the author.

The first version of this paper was prepared for and delivered to the Chapel
Hill Colloquium at the University of North Carolina in October 1976. It also
served as the basis for two seminars offered by the author at the University of
Minnesota at Morris in February 1980. I am particularly indebted to Steve Wagner, Fabrizio Mondadori, Glenn Kessler, Ian Hacking, David Kaplan, Jim Van
Aken, and Hide Ishiguro for their helpful comments on earlier drafts. The final
draft was completed while the author was a Fellow at the Center for Advanced
Study in the Behavioral Sciences. The support of the Center, the Sloan Foundation, the National Endowment for the Humanities, and Princeton University
are gratefully acknowledged.

elapse between the discovery (invention) of the laws of logic and the further (?) discovery that they were just what had been needed to show that the basic laws of arithmetic had really been basic laws of logic all along.

We have a tangle of problems here

- concerning what Frege thought he was doing
- concerning what we should take him to have done, both for logic and for arithmetic
- concerning the proper assessment of the broader philosophical import of these achievements.

This paper addresses itself to some of these issues. My most immediate concern will be to examine the role of the logicist doctrine in early twentieth-century empiricism in order to determine whether the view adopted (or adapted) from Frege by the positivists was a view that he himself had held. I would like my discussion to serve as a starting point for a deeper understanding of Frege's own views and ultimately for the discussion of the fascinating philosophical issues themselves. To broach these, I must return to the fundamental propositions I was taught as a youth. They concern the content and philosophical import of the logicist thesis.

I. LOGICISM

As *I* learned it, logicism was a philosophical view closely allied with empiricism: it was heralded by Carnap, Hempel, members of the Vienna Circle, Ayer, and others as the answer to Kant's doctrine that the propositions of arithmetic were synthetic a priori. Focusing on the sentences that express mathematical propositions, logicists conceded that *these* were a priori—that they could be known independently of experience (except, of course, for what experience may be required to formulate them). But, in reply to Kant, logicists claimed that these propositions are a priori because they are *analytic*—because they are true (or false) merely "in virtue of" the meanings of the terms in which they are cast. Thus, to know their meanings is to know all that is required for a knowledge of their truth. No empirical investigation is needed. The philosophical *point* of advancing the view was nakedly epistemological: logicism, if it could be established, would show that our knowledge of mathematics could be accounted for by whatever would account for our knowledge of language. And of course it was assumed that knowledge of language could *itself* be accounted for in ways consistent with empiricist principles, that language was itself entirely *learned*.[1] Thus, following Hume, all our knowledge could once more be seen as concerning either "relations of ideas" (analytic and a priori) or "matters of fact"" (synthetic and a posteriori). Kant's challenge to that dichotomy was turned back by showing that his most challenging counterexample—mathematics—though admittedly a priori, had mistakenly been classified by him as synthetic.

Logicism comes packaged in a number of different versions, each with its own wrinkles; but most have the following general structure:

(1) The truths of arithmetic are *translatable* into truths of logic.
(2) (1) is demonstrated by
    a. providing definitions for the "extra-logical" vocabulary (concepts) of arithmetic in "purely logical" terms; and
    b. noting that the translations induced by these definitions carry arithmetical truths into logical truths and arithmetical falsehoods into logical falsehoods.
(3) This arithmetical demonstration is then claimed to establish the *analyticity* of the mathematical propositions, because (a) since the definitions supposedly preserve meaning, the logical translations have the same meaning as the arithmetic originals and (b) the logical truths are themselves thought to be true in virtue of meaning, in this case, of the meanings of the logical particles occurring in them (and thus analytic).

Whether this is a viable view is something that has been much discussed. I engaged in that discussion myself some years ago[2] in a piece that can be read as arguing that either the definitions of the

---

[1] Recent controversies in the foundations of linguistic theory have indicated that, even if granted the linguistic nature of mathematical truth, empiricists are some distance from home. I think these arguments are problematic, but their very existence shows the issues are not crystal clear. I have in mind here the work of Noam Chomsky.

[2] P. Benacerraf, "What numbers could not be," *The philosophical review*, **74** (1965) 47–73.

mathematical terms do not preserve their meaning, or their meaning does not determine their reference, since different and equally adequate definitions assign different referents to the mathematical vocabulary. I will argue later, contrary to what I formerly thought Frege to hold, that he and I speak with one voice. Definitions adequate to *his* purposes need not preserve reference. But more about this later. Right now, I am primarily interested in outlining the logicist's position for the purpose of comparison with Frege's own.

Relatively little has been written on the issue of whether logic itself is analytic, so a word might be in order here simply to locate the question and some of its possible answers in the spectrum of positions under examination.

If, as W. V. Quine has done,[3] one defines analytic truth as transformability into a logical truth by meaning-preserving definitions, it becomes a trivial matter that the laws of logic are analytic; but such a definition, as applied to logic, bears little ostensible relation to the traditional account of analyticity as truth-in-virtue-of-meanings. Yet it was *this* latter explanation which bore the epistemological burden of persuading us that analytic propositions were also a priori. It should also be mentioned in favor of Quine's definition that it bears clear lines of ancestry to Frege and through him, back to Kant. For Frege's account of analyticity, to be found early in the *Grundlagen*, was:

> The problem whether a mathematical proposition is analytic or not becomes in fact that of finding the proof of the proposition, and of following it up right back to the primitive truths. If in carrying out this process we come only on general logical laws and on definitions, then the truth is an analytic one ... (*Grundlagen* § 3).[4]

Thus, a proposition is analytic if its proof makes use only of general logical laws and definitions.

We will examine this definition in detail below. The aspect to notice at this time is that "proof" from general logical laws and def-

---

[3]W. V. Quine, "Two dogmas of empiricism," *The philosophical review*, **60** (1951) 20–43.

[4]All quotations from *Grundlagen* are from the translation of J. L. Austin, *The foundations of arithmetic*, Evanston: Northwestern University Press, 1968.

initions is *sufficient* for analyticity. Of course, this does not make Frege's account equivalent with the one suggested (but not advocated) by Quine, since Quine and Frege differ on what the logic is to be and probably also on the role of definitions. For Quine, logic is first-order quantification theory plus identity, whereas for Frege it was considerably more extensive than that. Since Quine's narrower version of logic does not suffice for the "proofs" of the laws of arithmetic, mathematics is *not* analytic in Quine's sense, though it still might be in Frege's. I say "might be" because when Frege's logic failed to be consistent (and presumably, thereby also failed to be logic), his view was left with a certain indefiniteness; whether arithmetic turns out to be analytic in Frege's sense will have to depend at least on what logic one substitutes for Frege's ill-starred version.

The connection with Kant is made by Frege himself when he likens his own account to Kant's but chides Kant for the narrowness of his conception (*Grundlagen* § 88). He criticizes Kant for giving a definition that applies only to universal general propositions—those which can be construed as having a subject-predicate form—and for having too narrow a conception of definition, one presumably fashioned for use in showing that the concept of the subject term of a proposition contains the concept of the predicate.

So Frege offered an account of analyticity designed to improve Kant's in two respects: (a) it classified *all* propositions as analytic or synthetic, i.e., it was purportedly exhaustive, and (b) it broadened the concept of definition beyond Kant's notion (which Frege refers to as definition "by a simple list of characteristics" (*Grundlagen* § 88)) into one that encompassed "the really fruitful definitions in mathematics" (*Grundlagen* § 88).[5]

I wish to stress the epistemological motivation of the twentieth-century logicist. This is, of course, manifest in the third component

---

[5] I cannot resist noting in Kant's defense, that given Kant's notion of analyticity his concept of definition was just fine. It is only when you broaden the notion *à la* Frege that definition "by a simple list of characteristics" becomes too constricting—particularly if you are defining functions as well as predicates. So, Frege's argument at the end of § 88 to the effect that Kant would erroneously regard as synthetic certain conclusions drawn from his (Frege's) new kind of definition by purely logical means is a bit of a *petitio principii* on Frege's part; since it is only on Frege's enlarged notion of analyticity that they turn out to be analytic, regardless of the kind of definition employed in the proof.

of his view, in which he tries to reap a rich philosophical harvest from the seeds that Frege has sown. A striking example is the position advanced by C. G. Hempel in an article[6] that, although breaking no really new ground, presented a nuclear point of view as only Hempel can. According to Hempel, the Frege-Russell definitions of number, 0, successor, and related concepts have shown the propositions of arithmetic to be analytic because they follow by stipulative definitions from logical principles. What Hempel has in mind here is clearly that in a constructed formal system of logic (set theory or second-order logic plus an axiom of infinity), one may introduce by stipulative definition the expressions 'Number,' 'Zero,' 'Successor' in such a way that sentences of such a formal system using these introduced abbreviations and which are formally the same as (i.e., spelled the same way as) certain sentences of arithmetic—e.g., 'Zero is a Number'—appear as theorems of the system. He concludes from that undeniable fact that these definitions show *the theorems of arithmetic* to be mere notational extensions of theorems of logic, and thus analytic.

He is not entitled to that conclusion. Nor would he be *even* if the theorems of logic in their primitive notations were themselves analytic. For the only things that have been shown to follow from theorems of logic *by stipulation* are the abbreviated theorems of the logistic system. To parlay that into an argument about the *propositions of arithmetic*, one needs an argument that the sentences of arithmetic, in their preanalytic senses, *mean the same (or approximately the same) as their homonyms in the logistic system*. That requires a separate and longer argument. I bring this up here not to berate Hempel but to use his view as an illustration of the epistemological motivation that drives twentieth-century logicists. The *point* of logicism was to make it intelligible on empiricist principles how we might have a priori knowledge of mathematics. "By stipulation," says Hempel. If this much were correct, it would at least reduce the problem to the analyticity of logic, a problem I will not tackle here, although I will point out some obvious ways in which certain answers

[6]C. G. Hempel, "On the nature of mathematical truth," *American mathematical monthly*, **52** (1945) 543–56. Reprinted in Hilary Putnam and Paul Benacerraf, eds., *The philosophy of mathematics: selected readings*, second ed., Cambridge: Cambridge University Press, 377–93.

affect the proper assessment of the logicist's philosophical position.

Matters become particularly tangled when the analyticity of logic is in turn discussed in the context of defending such a logicist position; for such a logic must comprise enough set theory (or suitable equivalent) to yield enough mathematics. The only solution that seems to be offered in such a case is that the *axioms* constitute an implicit definition of the concepts. This is a form of conventionalism that construes the axioms as stipulations that are to govern the use of the terms they contain: *So use/understand this language that its sentences come out true!* This is difficult enough to understand when the interpretation of the "logical vocabulary" is kept fixed; as an instruction applicable to an entire language it makes no sense at all; as an explanation of how sentences of logic *in fact* get their truth values it is worthless, as Quine[7] and others have made abundantly clear. Logic is needed to apply such a rule to individual cases. Of course, it may *in fact* be the case that we use the language in such a way that the sentences in question do come out true. But what we seek is an *explanation* of that fact which will at the same time make the truth-values of these sentences knowable a priori. For this is the task that the twentieth-century logicist has set for himself.

So far I have concentrated on explaining the complex of philosophical views, taken in the philosophical setting of meeting Kant's challenge to empiricism, which has been the logicism of this century. It might seem curiously eccentric of me in a discussion of twentieth-century logicism to have made no mention of the most famous twentieth-century logicist: Russell Whitehead. I have omitted him for two reasons. (1) Because no brief account will encompass his shifting and differing stands; and (2) because, perhaps for this very reason, what I have taken the liberty to call "logicism" has fed more on his (and Frege's) technical achievements than on his fluctuating philosophical assessment of those achievements. (I might add, parenthetically, that to say even this much is already to make an untenable distinction between his "technical achievements" and his "philosophical views"—as anyone who has tried to puzzle out

---

[7]W. V. Quine, "Truth by convention," in O. H. Lee, ed., *Philosophical essays for A. N. Whitehead*, New York: Longmans, Green, 1936. Reprinted in Putnam and Benacerraf, 329–54.

the *Principia mathematica* concept of propositional function well knows.)

Such, then, is the received view: Kant's challenge has been met. Mathematics is really analytic, not synthetic. This was shown by Frege when he showed how mathematical propositions have the same meaning as logical propositions, which themselves are analytic (and therefore knowable a priori). Frege showed this by analyzing the "extra-logical" vocabulary of arithmetic and providing definitions that preserved meaning (and therefore, reference and truth). Frege was, therefore, the first logicist.

II. FREGE

If Frege was the first logicist, then he was also the last. If it is appropriate to call what Frege *actually* believed "logicism," and if he was the first to believe it, then, most probably he was also the last. To my knowledge, no one since Frege—and certainly no twentieth-century "logicist"—has held precisely the position that Frege advocated in the *Grundlagen* and that moved him to write that philosophical masterpiece. Although the views summarized in the previous paragraph are fairly widely held (I think most philosophers with a view on the subject are closet "logicists" in this sense), they were not Frege's. There are various points of contact which make it tempting to think that Frege held such a position, but I will argue that he did not—that his view was a much more intriguing one and in its spirit directly antithetical to the philosophical motivation of his twentieth-century "followers."[8]

---

[8]The view I have been calling "logicism" is evidently an amalgam of two views: a semantical thesis to the effect that *arithmetic is a definitional extension of logic* and an epistemological claim about *how this explains the a priori character of arithmetic*. Evidently, one can (and perhaps should) reserve the title for the semantical thesis alone, in which case Frege was certainly as much of a logicist as his followers (although here, too, much depends on how one interprets "definitional extension"—a deceptively tricky question which I will raise in more detail at the end of this paper).

I chose the present method partly for dramatic effect and partly because I am not really sure how clearly the two theses can be untangled from one another—how much the philosophical motivation behind a given form of the semantical thesis infects the thesis itself.

First of all, of course, Frege was no empiricist. True, one of the philosophical aims of the *Grundlagen* was to refute Kant's doctrine that arithmetic consisted of synthetic a priori propositions. But Frege readily conceded what no empiricist would concede—that Euclidean geometry was synthetic a priori. He says:

> In calling the truths of geometry synthetic and a priori, he [Kant] revealed their true nature. And this is still worth repeating, since even today it is often not recognized. If Kant was wrong about arithmetic, that does not seriously detract, in my opinion, from the value of his work. His point was, that there are such things as synthetic judgments a priori; whether they are to be found in geometry only, or in arithmetic as well, is of less importance (*Grundlagen* § 89).

So for him, establishing the analyticity of arithmetical judgments is not a way of defending empiricism against Kantian attack. It has another purpose, one which I hope to uncover by examining how he introduces and defends his views.

Frege opens the *Grundlagen* by deploring the fact that no one seems to have given a satisfactory answer to the question "What is the number one?":

> ... is it not a scandal that our science should be so unclear about the first and foremost among its objects, and one which is apparently so simple? Small hope, then, that we shall be able to say what number is. *If a concept fundamental to a mighty science gives rise to difficulties, then it is surely an imperative task to investigate it more closely until those difficulties are overcome* ... (*Grundlagen* xiv [my italics]).

This sets the stage. It is to be an inquiry into the foundations of arithmetic with a view to overcoming, the "difficulties" to which its fundamental concepts give rise. It is tempting to think that Frege is speaking tongue-in-cheek, that he does not really regard our inability to give a satisfactory account of the concept of number as a genuine difficulty *within that science*. To be sure, it is a philosophical worry—one appropriate for philosophers—but not a difficulty *internal to the science of number itself*. But this would be a mistake. Frege emphasizes that it is a matter with which mathematicians themselves *qua* mathematicians must be concerned, even though the inquiry will of necessity contain a substantial philosophical component:

> I realize that ... I have been led to pursue arguments more philosophical than many mathematicians may approve; but any thorough examination of the concept of number is bound always to turn out rather philosophical. It is a task which is common to mathematics and philosophy (*Grundlagen* xvii).

In urging that a proof is incomplete unless the definitions have been thoroughly justified, he says:

> Yet ... the rigor of the proof remains an illusion ... so long as the definitions are justified only as an afterthought, by our failing to come across any contradiction. By these methods, we shall, at bottom, never have achieved more than an empirical certainty, and we must really face the possibility that we may still in the end encounter a contradiction which brings the whole edifice down in ruins. For this reason I have felt bound to go back rather further into the logical foundations of our science than perhaps most mathematicians will consider necessary (*Grundlagen* xxiv).

We should take him at his word. It is a concern with the foundations of arithmetic that motivates his study.

Small wonder, given his title.

But such a concern might be interpreted in two different ways, corresponding to the interests of a philosopher and to those of a mathematician. Typically, the philosopher takes a body of knowledge as given and concerns himself with epistemological and metaphysical questions that arise in accounting for that body of knowledge, fitting it into a general account of knowledge and the world. That is Kant's stance. He studies the nature of mathematical knowledge in the context of an investigation of knowledge as a whole. And that was the positivist's stance, though they reached quite different conclusions.

But a mathematician's interest in what might be called "foundations" is importantly different. *Qua* mathematician, he is concerned with substantive questions about the truth of the propositions in question, as well as slightly more "philosophical" issues concerning how such propositions are properly established. The interests of the two groups are not disjoint—nor can these questions be sharply separated. But the differences are significant, and it is important to keep them in mind as we approach Frege. I claim that the Frege of the *Grundlagen* has the mathematician's motivation; that where he appears to deal directly with the more typically "philosophical" issues (Are the propositions of arithmetic analytic or synthetic? A priori or a posteriori?), *it is because he has restructured those questions and posed them in such a form that the answers they require will answer the substantive mathematical questions which are his principal concern*. Thus, if logicism is the complex of philosophical views I described in the first part of this paper, Frege was no logicist.

So, on his view, unless we do as he urges, "we must really face the possibility that we may still in the end encounter a contradiction which brings the whole edifice down in ruins."[9] What needs to be done? Quite starkly, this. The propositions of arithmetic stand in need of *proof*. We cannot simply take them for granted, on intuition, or accept them because they have proved useful in their many applications. " ... in mathematics, a mere moral conviction supported by a mass of successful application, is not good enough" (*Grundlagen* § 1). It is quite the same situation as if, in some more advanced branch of mathematics, a body of "knowledge" had arisen but never been adequately justified. " ... it is in the nature of mathematics always to prefer proof, where proof is possible ... " (*Grundlagen* § 2).

In § 1, Frege explains the *general* need for rigor and proof in mathematics. In § 2, he defends his search for proofs of such propositions as $7 + 5 = 12$ or the Associative Law of Addition, by the last remark I quoted and by likening the matter to the case of Euclid's proofs of "many things which anyone would concede him without question" (*Grundlagen* § 2). He then moves on to what I take to be the heart of his view when he explains the aim of proof as follows:

Philosophical motives *too* have prompted me to enquiries of

---

[9]It is of course the bitterest irony that Frege had to face that possibility when his system led to contradictions—and that he would not have had to face it had he not pursued his foundational investigations.

this kind. The answers to the questions raised about the nature of arithmetic truths—are they a priori or a posteriori? analytic or synthetic?—must lie in this same direction. For even though the concepts concerned may themselves belong to philosophy, yet, as I believe, no decision on these questions can be reached without assistance from mathematics—though this depends of course on the sense in which we understand them (*Grundlagen* § 3 [my emphasis]).

As the "too" I emphasized indicates, the motives discussed so far have been mathematical, not philosophical. Only now is he turning to what he feels may be considered the philosophical aspect of his work. And he serves notice that he will so construe the "philosophical" questions of the a priori and analytic character of arithmetical truths that they shall have mathematical answers. No doubt it will be somewhat controversial to what extent Frege's redefinitions of these concepts are simple clarifications and to what extent they are important reconstruals. This will depend on what we take Kant's and Leibniz's intentions to have been. What concerns me more is the contrast between these concepts as defined by Frege and the corresponding notions that are woven into the texture of the philosophical views I have called "logicism" and which I outlined in the first part of this paper.

The best way to seek out these answers is to follow § 3 paragraph by paragraph, adding what interpretive commentary seems appropriate. The section is short, but rich in the substance of Frege's thought.

It not uncommonly happens that we first discover the content of a proposition, and only later give the rigorous proof of it, on other and more difficult lines; and often this same proof also reveals more precisely the conditions restricting the validity of the original proposition. In general, therefore, the question of how we arrive at the content of a judgement should be kept distinct from the other question, "Whence do we derive the justification for its assertion?" (*Grundlagen* § 3).

This is a seemingly innocent distinction, pointing out that we often form propositions in our minds—indeed come to believe them—and only later (or perhaps never) arrive at proofs of those propositions (or of suitably restricted versions of them). The point of this remark is to try to separate the notion of the content of a judgment from that of the justification for the judgment—in the sense of justification introduced in the previous section: namely, the "support" of the judgment; the propositions on which it "depends" for its truth. Frege's attempt to divorce these two ideas (content and justification) will be crucial to his critique of Kant, a central aspect of his redefinition of analyticity, and a pivotal point of difference with later "logicists."

So we must recognize this distinction as the first leg of an attack on Kant. The reason is transparent. For Kant, the distinction between analytic and synthetic propositions was primarily a distinction in the *content* of the propositions. And the epistemological *point* was that this distinction in content had, for the analytic propositions, the immediate consequence that they were a priori—that they were knowable independently of experience just on the basis of a consideration of their *content*. For it was a fact about the content of an analytic proposition that it was possible to notice that in merely *entertaining* such a proposition one could not think the concept of its subject term without thinking along with it in the appropriate way the concept of its predicate term. Thus a major problem of the *Critique* was establishing the very possibility of a priori *synthetic* judgments: it seemed obvious why analytic judgments were a priori; but an elaborate theory had to be developed to account for the a priori character of judgments that did not pass the simple test that certified them as analytic and hence obviously a priori. Twentieth-century "logicists," following Kant in this respect, accorded a priori status to an enlarged class of analytic propositions *on the basis of their content*—for truth-in-virtue-of-meanings is simply an extension of Kant's distinction and of the epistemological analysis that went along with it. (I should add, parenthetically, that once the class of propositions has been enlarged beyond the subject-predicate propositions to which Kant limited his attention, the easy route to the a priori from the analytic is no longer available.) On this revisionist view, Kant was wrong because, hobbled by an inadequate notion of content owing to a primitive logical and semantical theory, he failed to appreciate the fact that arithmetical propositions were also true for the same kind of reason—merely in virtue of their content. Thus, on the received view of Frege's work and of its relation to this tradition, one would have expected him to claim precisely what I attributed to the "logicists"—that where Kant had gone wrong was in his analysis of the *content* of arithmetical propositions. And indeed, as we see above, Frege criticized Kant's distinction between analytic and synthetic judgments for not being exhaustive (*Grundlagen* § 88). Although it would be tempting to construe this as Frege's criticism of Kant's analysis of the *content* of arithmetical judgments, a construal which would tend to locate Frege in the epistemological tradition that runs through Kant to the contemporary logicist, it would be misleading to do so. For Frege shows in the very next paragraph that he has set up his distinction between content and justification for the express purpose of eschewing further talk of content. The paragraph is full of oblique references to Kant's discussion of analyticity and has a footnote in which he claims to be following Kant. I quote both the paragraph and the footnote:

> Now these distinctions between a priori and a posteriori, synthetic and analytic, concern, as I see it,* not the content of the judgment but the justification for making the judgment .... When a proposition is called a posteriori or analytic in my sense, this is not a judgment about the conditions, psychological, physiological and physical, which have made it possible to form the content of the proposition in our consciousness; nor is it a judgment about the way in which some other man has come, perhaps erroneously, to believe it true; rather, it is a judgment about the ultimate ground upon which rests the justification for holding it to be true.

———————————

   * By this I do not, of course, mean to assign a new sense to these terms, but only to state accurately what earlier writers, Kant in particular, have meant by them (*Grundlagen* § 3).

   The point of introducing the content/justification distinction is to place *both* the a priori/a posteriori *and* analytic/synthetic distinctions squarely on the side of justification, something he will carry through to its conclusion when he explicitly defines all four concepts in the following and concluding paragraph of § 3. As for the above paragraph, the oblique references to Kant are the denials that in calling a proposition analytic or a priori we are concerned in any way with the conditions that have made it possible to form the content of the judgment or, by implication, what in fact happens when one forms the judgment in our minds. This is Kantian language. It has a psychologistic flavor, and Frege wants none of it. What he particularly wishes to avoid is treating the analyticity of propositions in terms of what happens in the mind when one entertains the proposition. Just such a discussion quite properly provided the link for Kant between the analyticity of the proposition and its a priori character. Frege's *reasons* for wishing to avoid such talk in general, and the difficulties to which (in my opinion) he was ultimately led by his particular brand of anti-psychologism, are fascinating questions in their own right, but the subject of another paper. I mention it only to bring into sharper focus the contrast Frege is drawing between his own position and Kant's, footnote disclaimers to the contrary.

So, to resume the argument, Frege sees *both* the question of the analyticity of the judgment and of its a priori character as concerning the *justification* of the judgment. Accordingly, he will provide definitions for these concepts (analytic/synthetic, a priori/a posteriori) that reflect this view. Since arithmetical propositions are at issue, the question of their justification is properly a matter for mathematics. Therefore, the concepts will be so defined as to make it a properly *mathematical* question whether some arithmetical judgment is analytic or synthetic, a priori or a posteriori. This will be in full accord with his remarks at the end of the first paragraph of § 3, which I quote again here for the sake of convenience:

> For even though the concepts concerned [analytic, synthetic; a priori, a posteriori] may themselves belong to philosophy, yet, as I believe, no decision on these questions can be reached without assistance from mathematics—though this depends of course on the sense in which we understand them (*Grundlagen* § 3).

The sense in which Frege will understand them will be one that attempts to give some content to the notion of "the ultimate ground upon which rests the justification for holding ... a judgment to be true." For this is the metaphysical notion on which his view depends. I say "metaphysical" to contrast the dependence to which he is alluding with epistemic dependence. There may be a hierarchical structure to our beliefs, with the hierarchy representing the relation of foundation or justification that a person's beliefs may bear to one another: the relation of dependence that *actually* obtains and which may vary from person to person even though the related beliefs might themselves be close to identical. On some (e.g., foundationalist) views, beliefs do form such a structure; on others (e.g., holist), they do not. Frege is not concerned with such a relation, but with relations of dependence *among the propositions themselves*, whether or not they are believed and however those beliefs may be related to one another in the epistemic world of any individual. To prove a proposition involves (at least) deducing it from the propositions on which it "depends" in this metaphysical sense. It involves tracing its ancestral lines of dependence back to propositions that are themselves "fundamental" or "primitive" and *have* no proofs—which cannot be reduced to more fundamental propositions.[10] I will now quote the balance of section 3, in which Frege gives his definitions and thereby fixes the sense of the questions: Are the propositions of arithmetic synthetic or analytic? A priori or a posteriori? I will devote the balance of my paper to a commentary on this paragraph.

> This means that the question is removed from the sphere of psychology, and assigned, if the truth concerned is a mathematical one, to the sphere of mathematics. The problem be-

----

[10] It is interesting to contrast Frege's attitude toward the relation between logical axioms and mathematical theorems with that expressed by Russell and Whitehead in the following passage drawn from the Preface to the second edition of *Principia matbematica*: " ... the chief reason in favour of any theory on the principles of mathematics must be chiefly inductive, i.e. it must lie in the fact that the theory in question enables us to deduce ordinary mathematics. In mathematics, the greatest degree of self-evidence is usually not to be found quite at the beginning, but at some later point; hence the early deductions, until they reach this point, give reasons rather for believing the premises because true consequences follow from them, than for believing the consequences because they follow from the premises." A. N. Whitehead, B. Russell, *Principia mathematica*, 2nd ed., Cambridge: Cambridge University Press, 1925, Vol. I, p. v.

comes, in fact, that of finding the proof of the proposition, and of following it up right back to the primitive truths. If, in carrying out this process, we come only on general logical laws and on definitions, then the truth is an analytic one, bearing in mind that we must take account also of all propositions upon which the admissibility of any of the definitions depends. If, however, it is impossible to give the proof without making use of truths which are not of a general logical nature, but belong to the sphere of some special science, then the proposition is a synthetic one. For a truth to be a posteriori, it must be impossible to construct a proof of it without including an appeal to facts, i.e., to truths which cannot be proved and are not general, since they contain assertions about particular objects. But if, on the contrary, its proof can be derived exclusively from general laws, which themselves neither need nor admit of proof, then the truth is a priori (*Grundlagen* § 3).

To determine whether a proposition is analytic, look for a proof of it in which the basic propositions are "primitive truths"—propositions which themselves have no proofs. If there exists such a proof (one in which appeal is made only to definitions and to "primitive truths") and the primitive truths evoked include only laws of logic, the proposition in question is analytic. If not, it is synthetic. So, an analytic proposition is one that can be proved from logical axioms alone plus definitions. At least two aspects of this definition deserve comment.

First, Frege includes among the relevant propositions on which a given proposition depends "all propositions upon which the admissibility of any of the definitions depends." This is a consequence of his view that definitions must not simply be introduced in a proof; the proof is not complete unless they are justified as well. (Recall *Grundlagen* xxiv.) Many kinds of questions enter into the judgment of the admissibility of a definition, and it would be too difficult to review them all here. Frege discusses at least these two: (a) Will the introduction of this definition lead to contradictions? and (b) Will the introduction of this definition prove fruitful—i.e., can we prove things with it that we could not have proved without it (*Grundlagen* § 70)?

The inclusion of this element in the definition of analyticity introduces a peculiar problem for Frege when he is discussing the analyticity of the laws of arithmetic. It is this: Normally, a negative answer to the first question (Will the introduction of this definition lead to contradictions?) or any positive answer to the second (Will it prove fruitful?) will require a proof involving some sort of induction, perhaps up to ω, ω^2, or even ϵ_0. If the proposition under consideration *is itself the relevant induction principle*, then either (1) no definitions are involved in its proof, in which case it is irreducibly arithmetical and not analytic; or (2) definitions *are* involved and once more induction is itself one of the principles on which it depends, since some appeal to induction would be required to demonstrate the admissibility of those definitions.[11] Unfortunately, Frege does not consider the question and leaves the notion of dependence insufficiently determinate to resolve this problem. For, whether *on Frege's definition* arithmetical truths are in fact analytic would de-

[11]If the definitions are *explicit* definitions, then the requirement Frege imposes in the *Grundgesetze*, of establishing the existence and uniqueness of the defined entity, would suffice to guarantee that the system including the definition was a conservative extension of the original system, and hence consistent if the system was consistent prior to the introduction of the definitions. So, at best, whether a given law is analytic depends on whether the laws required in the proof of the existence and uniqueness of each of the defined entities employed in its proof are themselves laws of logic. I say "at best" for two reasons: (a) Definitions that are not explicit, but perhaps contextual, might have to be treated as new axioms whose justification requires at least whatever apparatus is needed to prove the consistency of the enlarged system; second, (b) even in the simple case of explicit definitions, although existence and uniqueness suffice to guarantee relative consistency, I am not sure that Frege's requirement that definitions be fully justified does not impose the further condition *that it be proved that existence and uniqueness suffice to guarantee relative consistency*.

This tangle exists because, from the syntactical viewpoint at least, the justification of definitions involves the proof of straightforward combinatorial theorems, something which Gödel showed us long ago was often equivalent to very difficult arithmetical questions, and worse. Consequently, if the basic laws on which a theorem depends include the laws on which the justifications of the definitions are based, it might well be that the theorem is not analytic *in Frege's sense*. In one sense, this is a quibble: he could omit this troublesome condition. But the issue is an important one to Frege. Rigor in mathematics is one of his most powerful motives, and his insistence on not employing definitions without providing them with the proper justification is a theme that runs through all his work.

pend on whether the "logical" principle of induction—i.e., induction in primitive logical notation—is sufficient to establish the admissibility of the definitions introduced in the proof of the *mathematical* principle of induction. If not, then arithmetic is not analytic *on Frege's definition*. But this is a very complicated issue which cannot be explored more fully here; I mention it as an interesting and relevant aspect of Frege's definition of analyticity.

The other matter on which I must comment, also inconclusively, I fear, also has to do with definitions. If we accept the view that I have been urging—that the problem of the *Grundlagen* is to argue that it is probable that one can find proofs of heretofore unchallenged but unproven arithmetical propositions—and if we take seriously Frege's view that finding such proofs is a *mathematical* problem like any other, then we must also regard the definitions that should be employed in these proofs as mathematical definitions like other mathematical definitions. In the *Grundlagen*, Frege does not tell us explicitly what semantical conditions these definitions must meet. (He does say a good deal in the *Grundgesetze*). Nor is this the place to provide a positive account of my own—either of the nature of mathematical definitions or of what Frege's positive view on this might have been. But the things he does say, although leaving it open what positive account he would offer, do render certain accounts unlikely.

Definitions are not *simply* conventions of abbreviation; for if they were, the requirement of fruitfulness cited above would make little sense. The fruitfulness would be a matter only of psychological heuristic and not something to which Frege would attach much importance. So, even if, formally, the definiens must serve at least as an "abbreviation" for the definiendum, the importance and principal role of the definition must lie elsewhere than in this function. Viz., Cantor's definitions of transfinite numbers, which Frege himself cites and praises (with reservations).

Similarly, mathematical definitions do not standardly reflect preexisting synonymies. The reasons are many. Quite apart from the uncertain status of the concept of synonymy, often a new term is introduced in the definition and there is consequently no question of preexisting synonymy. But, more important, typical and important cases of mathematical definition, of precisely the kind that

Frege has in mind, just do not fit that model. To return to one example on which Frege himself comments, consider Cantor's Theory of Transfinite Numbers. Frege praises the theory as extending our knowledge but takes Cantor gently to task for having appealed to "the rather mysterious 'inner intuition'" (*Grundlagen* § 86) in developing the theory "where he ought to have made an effort to find, and indeed could actually have found, a proof from definitions" (*Grundlagen* § 86). Frege then goes on to add, "For I think I can anticipate how his two concepts [following in the succession, and Number] could have been defined" (*Grundlagen* § 86). Surely, whatever Frege may be claiming here, he is not claiming that Cantor overlooked an appeal to preexisting synonymies which he, Frege, thinks he can produce. The analysis of this case—one which closely parallels the case of Number—is complicated. But whatever the correct answer, it does not seem as if it will be in terms of either preexisting synonymies or conventions of abbreviations.

If the two cases I have just mentioned exhaust the kinds of definitions that preserve sense or meaning, it remains an open question whether definitions of the kind employed in *Grundlagen* and in its formal counterpart, *Grundgesetze*, if they are adequate, must even preserve *reference*. I have myself argued elsewhere that this need not be so.[12] What did Frege think? I should like to point to two passages in which it seems clear that, for arithmetic at least, Frege did *not* expect *even reference* to be preserved by his definitions. The two passages I have in mind both concern the definition of Number. The first is a footnote to the definition of "the Number which belongs to the concept *F*" as "the extension of the concept 'equal to the concept *F*'" (*Grundlagen* § 69). The footnote, keyed to the word "extension," reads as follows:

> I believe that for "extension of the concept" we could write simply "concept." But this would be open to two objections:
>
> > 1. that this contradicts my earlier statement that individual numbers are objects, as is indicated by the use of the definite article in expressions like "the number two" and by the impossibility of speaking of ones, twos, etc.

---

[12] Benacerraf, "What numbers could not be."

in the plural, as also by the fact that the number constitutes only an element in the predicate of a statement of
number;

2. that concepts can have identical extensions without
themselves coinciding.

I am, as it happens, convinced that both these objections
can be met; but to do this would take us too far afield for
present purposes. I assume that it is known what the extension
of a concept is (*Grundlagen* § 69).

This is fairly conclusive, unless his way out of the second objection
consists in arguing that for number concepts, concepts with identical
extensions are not only identical with one another *but also identical
with their extensions*, an unlikely course for Frege, given his views
about the distinction between concepts and objects: concepts cannot
be identical with anything. Identity is a relation reserved for objects.

The second passage occurs in the conclusion, as he comments on
the same definition:

This way of getting over the difficulty cannot be expected to
meet with universal approval, and many will prefer other methods of removing the doubt in question. I attach no decisive
importance to bringing in the extensions of concepts at all
(*Grundlagen* § 107).

It may help to recall what "the difficulty" in question was. After
giving a *contextual* definition of the expression "the number which
belongs to the concept F" only for the context of an identity in which
both sides have the same form—e.g., "the number which belongs to
the concept F is identical with the number which belongs to the concept G"—Frege noted that for his definition to be logically complete,
it must fix the sense of all contexts containing that phrase. For example, an adequate definition should determine the truth-value of
"the Number which belongs to the concept 'moons of Jupiter' is
identical with Agamemnon." However, the definitions provided up
to that point are not up to this task, and a further specification was
needed. Frege chose the definition I cited. Thus, in precisely this
context—the one most critical for determining whether he required
definitions to preserve reference—Frege backs off and allows that

different definitions, providing different referents (not "bringing in
the extensions of concepts at all") might have done as well. It is
as if the mathematical job of the definitions had already been done
and all that remained was some logical tidying up, important, but
of no mathematical consequence and for all that mattered to mathematics, something which would be done equally well in a number of
different ways. The moral is inescapable. Not even reference needs
to be preserved.

More needs to be said. It might be objected that at the time
of *Grundlagen* Frege had not developed the concepts of sense and
reference to a sufficient extent to imbue the questions I am raising
with sense, and hence that they should not be raised. Although it
is beyond the scope of the present paper to present the case in full
detail, I believe that the very same questions can be raised about
Frege's accomplishments in *Grundgesetze*,[13] which is certainly late
enough. I will sketch my reasons.

In *Grundgesetze* Frege actually carries out the constructions that
he only promises in *Grundlagen*. He constructs a system, formal in
the technical sense, whose fundamental principles are those that
he takes to be the basic laws of logic and from which he derives,
by the introduction of definitions, the principles he had previously
identified as the fundamental laws of arithmetic. In the course of
that construction, several points appear at which "arbitrary" choices
must be made—arbitrary in the sense that they are not determined
by what has gone before, but which must nevertheless be made, if
only for the sake of completeness. An example will serve to illustrate.

Frege introduces what he calls "courses-of-values" to represent
the extensions of concepts. He stipulates that two functions have
the same *course-of-values* if they have the same value for every
argument. If the function is one

> whose value is always a truth-value, one may accordingly say,
> instead of "course-of-values of the function," rather "extension
> of the concept"; and it seems appropriate to call directly a con-
> cept a function whose value is always a truth-value (*Grundge-*

---

[13] All quotations from *Grundgesetze* are from the translation of Montgomery
Furth, *The basic laws of arithmetic: exposition of the system*, Berkeley and Los
Angeles: University of California Press, 1964.

setze I, § 3).

So far, so good. The only defining condition he has imposed on courses-of-values is the contextual one that the courses-of-values of two functions shall be equal if they have the same value for every argument. He then notes that nothing that he has said bears on whether the two truth-values, the True and the False, are themselves courses-of-values, and if so, which ones. He summarizes this position:

> Thus without contradicting ... [here he repeats the contextual definition] ... it is always possible to stipulate that an arbitrary course-of-values is to be the True and another the False (*Grundgesetze* I, § 10).

He then picks a particular one and stipulates that it is to be the True and another the False. The problem and its solution have exactly the same form as in the case of numbers and the extensions of concepts. And the philosophical consequences are also the same. If we call the one he picked "George," then "George = the True" lacked a truth-value before he did the picking, and acquired the True as its value from the pick. But had Frege not picked George but something else instead, "George = the True" would have been false. Since George then figures in every course-of-values, he figures in the extension of every (non-empty) concept. Had he not been the lucky one chosen, the extension of every concept would have been different.

Of course it does not make any mathematical difference. But *that* it makes no mathematical difference is an important philosophical point concerning what we must construe definitions such as Frege's to accomplish. Although I cannot pursue the matter further here, I hope that these examples make it clear that a straight-forwardly "realist" construal of Frege's intentions or accomplishments will fail to do justice to his practice.

As I promised, the conclusion is unsatisfying. It seems clear that definitions for Frege are not a number of things we might have thought they might be. But it remains unclear what he thinks they are. This makes his notion of analyticity correspondingly unclear, or at least unspecified. If we accept the view that he is simply requiring that proofs be given for the arithmetical propositions that we have heretofore accepted without proof, then the notion is no worse than that of mathematical proof itself: it is hard to say what one is, but mathematicians produce and recognize them daily. Of course, that is not good enough for Frege, who wanted to remove the concept of mathematical proof from the realm of intuition and reduce it to a small number of precisely stated formal rules of logic. What we have learned from this discussion is that he will be unsuccessful in this task until he does the same for his concept of definition.

This brings me finally to Frege's concept of a priori.

> ... if ... the proof of a proposition can be derived entirely from general laws, which themselves neither need nor admit of proof, then the truth is a priori (*Grundlagen* § 3).

First, for other writers, a priori had a definite direct connection with knowledge. Not so for Frege, since nothing in the above definition suggests that any a priori propositions are knowable at all—unless it is the reference to the fact that the ultimate truths from which a priori propositions can be proved do not themselves stand in *need* of proof. But this is rather empty since nowhere in the *Grundlagen* does Frege suggest an account of what it is to stand in need of proof. He asserts that arithmetical propositions do, but the grounds seem to be principally his conviction that they are *susceptible* of proof.

Second, I should like to note that the idea of propositions that do not admit of proof derives from the rationalist conception I attributed to Frege of a hierarchy of propositions, some of which are absolutely basic and form the foundation on which all the others "rest." He pays further homage to this conception when he agrees with Hankel's criticism of Kant's doctrine that numerical identities constitute an infinite set of unprovable and self-evident propositions. The reader will undoubtedly recall that Hankel had criticized Kant for supposing that numerical identities were all self-evident and yet unprovable.

> Hankel justifiably calls this conception of infinitely numerous unprovable primitive truths incongruous and paradoxical. The fact is that it conflicts with one of the requirements of reason, which must be able to embrace all first principles in a survey (*Grundlagen* § 5).

There must be only finitely (or manageably) many first principles, from which all other a priori truths can be deduced. Their
surveyability is a matter to which Frege pays no further attention,
because, I think, to do so would require him to give an account of
how we can and do know what we know—an account that would
force him into a discussion of the conditions under which our beliefs
constitute knowledge, a topic which he correctly perceived would
involve certain psychological issues but which he (wrongly, I think)
sweeps out with his anti-psychologistic broom. But, as I said above,
that is the topic of another paper.

I will close my discussion of § 3 with an amusing sidelight: On
Frege's definitions, are analytic truths all a priori? Presumably so,
since an analytic truth is one whose proof involves only first principles of *logic* (and definitions), and an a priori truth is one that
can be proved exclusively from *general laws* which neither need nor
admit of proof. To all appearances, it remains only to verify that
the first principles of logic are themselves general laws which neither
need nor admit of proof. Clearly, Frege believed that they were, *par
excellence*. But, just as clearly, Frege believed that he had shown
that all arithmetic truths were analytic. This creates a problem, for
it implies that there is a set of logical first principles from which all
arithmetic truths may be deduced, using only definitions and principles of logical inference. Frege's characterizations of the nature of
logical proof make it clear that the notion of proof he has in mind
is an "effective" one, in the technical sense. Thus, if all arithmetic
truths are analytic, there is a set of logical truths from which all
arithmetic truths are effectively derivable. But this implies that if
logic is recursively axiomatizable, so is arithmetic. And we know
from Gödel's first incompleteness theorem that arithmetic is not.
It follows that logic is not, whatever you take as logic, so long as
it is adequate for the derivation of arithmetic. But if logic is not
even recursively axiomatizable, its first principles constitute a class
of "infinitely numerous unprovable primitive truths" and it is therefore "incongruous and paradoxical" and thus "conflicts with one of
the requirements of reason." So—either

(1) not all arithmetical truths are analytic;

or

(2) not all logical truths are a priori (though all are trivially analytic);

or

(3) perhaps the conception of infinitely numerous unprovable primitive truths is not incongruous and paradoxical after all.

None of the above is a comfortable settling place for Frege, for I think he is quite serious about all three views. Indeed, I think it is their conjunction that forms for him much of the philosophical motivation for the *Grundlagen*. I have been arguing that his attempt to establish the analyticity of arithmetic was not to be construed as an attempt to enter an ongoing philosophical debate between Kant and the empiricists, and indeed that his very construal of the question took it out of that arena. It was rather an attempt to prove propositions that had yet to be proved, that he believed *could* be proved, and that he believed *should* be proved. Surely, much of the rationale for making this attempt is provided by his general view of proof, of the role of logic in proof, and of the hierarchical structure of all a priori propositions.

III. CONCLUSION

In this paper I have not touched upon the most exciting and important parts of the *Grundlagen*—Frege's actual discussion of the concepts of arithmetic. I have concentrated rather on trying to place that discussion in the philosophical context in which I think it belongs. The *Grundlagen* was written much more as a work of mathematics than is usually conceded. Or, to retreat to autobiography, than I had been brought up to believe. So, on my view, not only is the *Grundlagen* not a work in the Kantian/empiricist tradition, having as its principal purpose the refutation or establishment of disputed philosophical doctrines, Frege considered it only incidentally a philosophical work. In discussing the "philosophical" issues, he had to redefine certain philosophical concepts so that questions framed in terms of them had mathematical answers. Frege construed the enterprise of the *Grundlagen* as first and foremost a mathematical one, with its problem central to mathematics; and he considered the argument of the *Grundlagen* as merely *the sketch* of a substantive answer to that problem: to prove the heretofore unproven arithmetical propositions. In successfully completing that task, he would incidentally have answered what seems to be a philosophical question: Are the truths of arithmetic analytic or synthetic? But only after having reconstrued that question to suit his own purposes. The *Grundlagen* contains only a sketch, because in the *Grundlagen* he does not give rigorous proofs. That is left for later, but it must be done.

> The demand is not to be denied: every jump must be barred from our deductions (*Grundlagen* § 90).

Philosophy comes in as a convenient vehicle for bolstering his claim that arithmetical propositions must be *proved*. So we see him beginning § 4 with the conclusion he thinks should be drawn from his discussion of § 3, over which we have labored at such length

> Starting from these philosophical questions, we are led to formulate the same demand as that which had arisen independently in the sphere of mathematics, namely that the fundamental propositions of arithmetic should be proved, if in any way possible, with the utmost rigour ... (*Grundlagen* § 4).

3

Frege and the Rigorization
of Analysis

This paper has three goals: (i) To show that the foundational program begun in the *Begriffsschrift*, and carried forward in the *Grundlagen*, represented Frege's attempt to establish the autonomy of
arithmetic from geometry and kinematics; the cogency and coherence of 'intuitive' reasoning were not in question. (ii) To place
Frege's logicism in the context of the nineteenth century tradition in
mathematical analysis, and, in particular, to show how the modern
concept of a function made it possible for Frege to pursue the goal of
autonomy within the framework of the system of second-order logic
of the *Begriffsschrift*. (iii) To address certain criticisms of Frege by
Parsons and Boolos, and thereby to clarify what was and was not
achieved by the development, in Part III of the *Begriffsschrift*, of a
fragment of the theory of relations.

Reprinted from *Journal of philosophical logic*, **23** (1994) 225–46.

I wish to thank Robert Butts, John Corcoran, Michael Dummett, Michael
Friedman, Michael Hallett, and especially, Mark Wilson, for their considered
comments on earlier drafts, and their continued encouragement. I am indebted
to The Social Sciences and Humanities Research Council of Canada for financial
support.

I. RIGOR AND SKEPTICISM

Paul Benacerraf, in his important paper, "Frege: the last logicist,"[1] raises a number of questions about the intellectual motivations for Frege's logicism. Benacerraf makes the correct point that when Frege's foundational interests are viewed in their mathematical context, they stand in sharp contrast with the logical empiricists' attempts to show the analyticity of arithmetic and more generally, of all a priori knowledge. Any applications of Frege's logicism to such an empiricist theory of the a priori is at best an unintended by-product of his foundational concerns. Benacerraf argues that assimilating Frege to a much later empiricism distorts his purpose to such an extent that if this were the point of logicism, we should not look to Frege for its intellectual origins. Frege's logicism comes at the end of another tradition, and that tradition is primarily a mathematical one. On Benacerraf's account, Frege's work is the culmination of the process of making rigorous the calculus and the theory of the reals. Frege sought to do for arithmetic what Cauchy, Bolzano, Weierstrass, Cantor and Dedekind did for analysis: viz., to secure for it a 'rigorous foundation.' But what exactly was the goal of rigorization?

For Benacerraf, Frege's interest in rigor is driven by the problem of providing a proper justification for believing the truth of the propositions of arithmetic: 'The propositions of arithmetic stand in need of *proof*' (p. 51) is Benacerraf's gloss on Frege's remark that ' ... in mathematics a mere moral conviction, supported by a mass of successful applications, is not good enough.' (*Grundlagen*[2] § 1, quoted by Benacerraf, p. 51.) According to Benacerraf, both Frege

---

[1] Chapter 2, above. Page references to Benacerraf are to this paper.

[2] All references to the *Grundlagen* (*Die Grundlagen der Arithmetik*, Berlin: George Olms Verlagsbuchandlung, 1961, reprint of the 1884 edition) are to the translation by J. L. Austin, *The foundations of arithmetic: a logico-mathematical enquiry into the concept of number*, Evanston: Northwestern University Press, 1980. Quotations are also from Austin's translation. For ease of reference, I have, in the case of the *Grundlagen* and wherever else this was possible, referenced works by section number. In cases where only reference by page number proved possible, I have inserted the page numbers of the German original (or standard German edition) in parentheses, immediately after those of the English edition. When a work is generally known by its German title, I have so referred to it. Throughout this paper I have followed standard translations.

and the nineteenth century analysts viewed the process of rigorization as one of providing clear definitions and mathematical proofs where none had previously been given. This comports well with a number of Frege's statements, for example ' ... it is in the nature of mathematics always to prefer proofs, where proofs are possible ... ' (*Grundlagen* § 2, quoted by Benacerraf, p. 51), and ' ... the rigor of the proof remains an illusion ... so long as the definitions are justified only as an afterthought, by our failing to come across any contradiction.' (*Grundlagen* p. ix, quoted by Benacerraf, p. 50.)

If Benacerraf's account of the goal of rigorization is accepted, then Frege's foundational investigations are epistemological in the standard sense that the questions and metaquestions they address surround the *justification* of mathematical propositions for the purpose of insuring that we will not 'in the end encounter a contradiction which brings the whole edifice down in ruins.' (*Grundlagen* p. ix, quoted by Benacerraf, p. 51.) It should therefore be clear that the question Benacerraf is raising is not[3] whether Frege's foundational interests were mathematical or philosophical. They were obviously *both*. Benacerraf is claiming that our understanding of Frege's foundational interests is invariably distorted when these interests are viewed from the perspective of any particular philosophical school; although I don't address it directly, I believe that what I have to say is supportive of this broad interpretive claim.

It would be foolish to deny that the goals of cogency and consistency were an important part of the nineteenth century interest in rigor. And it is easy to find quotations from Frege which show that he sometimes at least wrote as if he took his task to include securing arithmetic, in the broad sense, which includes both the arithmetic of natural numbers and the theory of the reals, against contradiction. But I think it can be questioned just how far skeptical worries about the consistency or cogency of mathematics, generated perhaps by a certain incompleteness of its arguments, were motivating factors for Frege's logicism, or for the other foundational investigations of the period. This is a large historical issue which I will not attempt to resolve here. Rather, I will restrict myself to identify-

---

[3] As Joan Weiner has suggested in *Frege in perspective*, Ithaca: Cornell University Press, 1990.

ing another, largely neglected, component of Frege's concern with
rigor; this component not only has some intrinsic interest, it also
elucidates his views on the nature and significance of intuition in
mathematical proof, as well as his conception of his mathematical
and foundational accomplishments, especially those of the *Begriffsschrift*. Frege's concern with rigor is closely tied to his rejection
of intuition in arithmetical reasoning. This interconnection has not,
I think, been fully understood; at any rate, its importance for our
appreciation of the scope and justification of his logicism has not
been sufficiently appreciated. Since, on the view to be presented
here, Frege inherited his concern with rigor from the tradition in
analysis initiated by Cauchy and Bolzano, and carried forward by
Weierstrass, Cantor and Dedekind, what I have to say concerning
Frege is intended to apply to their foundational interests as well.
Indeed taken solely as a historical thesis about the rigorization of
analysis, I believe my remarks are neither novel nor contentious,
having been well emphasized by the historians.[4] But to my knowledge, with the recent exception of Michael Dummett[5] no one has
applied these ideas to Frege in the way I propose here.[6]

II. RIGOR AND SPATIO-TEMPORAL INTUITITION

Shortly after the publication of the *Begriffsschrift* Frege wrote a
long study of its relationship to Boole's logical calculus.[7] The paper

[4]Especially A. P. Youschkevitch, "The concept of function up to the middle
of the 19th century," *Archive for history of exact sciences*, **16** (1975/76) 37–85,
Judith Grabiner, *The origins of Cauchy's rigorous calculus*, Cambridge: MIT
Press, 1981, and Ugo Bottazzini, *The higher calculus: a history of real and
complex analysis from Euler to Weierstrass*, Berlin and New York: Springer-
Verlag, 1986.

[5]See his book, *Frege: philosophy of mathematics*, Cambridge: Harvard University Press, 1991. This paper was written before Dummett's book appeared.

[6]In his paper, "Kant, Bolzano and the emergence of logicism" (Chapter
1 of the present volume), Alberto Coffa has stressed Frege's affinity with the
tradition in mathematical analysis, especially with Bolzano. Coffa's perspective
is however somewhat different from mine since on his account, pure intuition
was 'a cancer that had to be extirpated in order to make mathematical progress
possible.' (p. 37, above) I hope to show that whatever its role in the development
of the calculus, this was not a factor in Frege's elimination of intuition from the
proofs of various arithmetical principles concerning the natural numbers.

[7]"Booles rechende Logik und die Begriffsschrift," (1880/81) in Hans Hermes

carries out a detailed proof (in the notation of the *Begriffsschrift*) of the theorem that the sum of two multiples of a number is a multiple of that number. In addition to the laws and the (explicit and implicit) rules of inference of the *Begriffsschrift*, Frege appeals only to the associativity of addition and to the fact that zero is a right identity with respect to addition. He avoids the use of mathematical induction by applying his definition of 'following in a sequence' to the case of the number series; 'following in a sequence' is Frege's expression for the ancestral of the relation which orders the sequence.[8] The paper also includes definitions of a number of elementary concepts of analysis (again in the notation of the *Begriffsschrift*). In the course of this discussion Frege presents an illuminating account of the considerations that shaped his logic.

Speaking of the choice of proposition given a 'step by step' or 'gap-free' proof in the *Begriffsschrift*, Frege writes:

> So as not perhaps to overlook precisely those transformations which are of value in scientific use, I chose the step by step derivation of a sentence which, it seems to me, is indispensable to arithmetic, although it is one that commands little attention, being regarded as self-evident. The sentence in question is the following:
>
> If a series is formed by first applying a many-one operation to an object (which need not belong to arithmetic), and then applying it successively to its own results, and if in this series two objects follow one and the same object, then the first follows the second in the series, or vice versa, or the two objects are identical.

This is the proposition (133) whose proof concludes the *Begriffsschrift*. I sometimes refer (somewhat loosely) to what this proposition establishes as the connectedness of the ancestral. I say 'somewhat loosely' because the condition imposed on 'connected elements,'

et al., eds., Gottlob Frege: *Nachgelassene Schriften*, Hamburg: Felix Meiner Verlag, 1983. Published in English translation as "Boole's logical calculus and the concept-script," in Hans Hermes et al., eds., Gottlob Frege: *Posthumous writings*, Chicago: University of Chicago Press, 1980, Peter White and Roger Hargreaves trs.

[8]Sometimes the translations I quote use 'following in a series.' Frege's definition of the ancestral is reviewed in section III below.

namely, that they follow one and the same element, is not part of the standard characterization of connectedness, and because connectedness, in this extended sense, holds only of the ancestral of *many-one* relations.

Frege continues,

> I proved this sentence from the definitions of the concepts of following in a series, and of many-oneness by means of my primitive laws. In the process I derived the sentence that if in a series one member follows a second, and the second follows a third, then the first follows the third.[9] Apart from a few formulae introduced to [accommodate] Aristotelian modes of inference, I only assumed such as appeared necessary for the proof in question.
>
> These were the principles which guided me in setting up my axioms [*ursprünglischen Sätze*] and in the choice and derivation of other sentences. It was a matter of complete indifference to me whether the formula seemed interesting or to say nothing. That my sentences have enough content, in so far as you can talk of the content of sentences of pure logic at all, follows from the fact that they were adequate for the task.[10]

It seems not to have been noticed, at any rate, it has not been sufficiently emphasized, that neither in this paper nor in the *Begriffsschrift* does Frege suggest that connectedness is not correctly regarded as self-evident, or that without the proof of the *Begriffsschrift* one might reasonably doubt the connectedness of the ancestral, and with it, the justification of all those propositions which require it. The point is rather that without a gap-free proof of this proposition, or of its specialization to the number series, one might be misled into thinking that arithmetical reasonings which depend on connectedness import considerations based on 'intuition.' As

---

[9]This is Proposition 98 of *Begriffsschrift*; it clearly expresses the transitivity of the ancestral.

[10]Gottlob Frege: *Posthumous writings*, p. 38 (42f). I don't know whether it is generally recognized that the first complete axiomatization of first-order logic was determined by reflection on what the proof of the connectedness of the ancestral would require, but it is a remarkable fact that the discovery of a gap-free proof of exactly this proposition should have sufficed to generate the first-order fragment of Frege's logic.

Frege puts the matter in the introductory paragraph to Part III of the *Begriffsschrift*:

> Throughout the present example [i.e. the proof of 133, which occupies the whole of Part III] we see how pure thought, irrespective of any content given by the senses or even by an intuition a priori, can, solely from the content that results from its own constitution, bring forth judgements that at first sight appear to be possible only on the basis of some intuition.[11]

This point is also made in the *Grundlagen* when, near the end of the work (§§ 90–91), Frege comments on the *Begriffsschrift*. Frege is quite clear that the difficulty with gaps or 'jumps' in the usual proofs of arithmetical propositions is not that they may hide an unwarranted or possibly false inference, but that their presence obscures the true character of the reasoning:

> In proofs as we know them, progress is by jumps, which is why the variety of types of inference in mathematics appears to be so excessively rich; ... the correctness of such a transition is immediately self-evident to us; ... whereupon, since it does not obviously conform to any of the recognized types of logical inference, we are prepared to accept its self-evidence forthwith as intuitive, and the conclusion itself as a synthetic truth— and this even when obviously it holds good of much more than merely what can be intuited.
>
> On these lines what is synthetic and based on intuition cannot be sharply separated from what is analytic[.] ...
>
> To minimize these drawbacks, I invented my concept writing. It is designed to produce expressions which are shorter and easier to take in, ... so that no step is permitted which does not conform to the rules which are laid down once and for all. It is impossible, therefore, for any premiss to creep into a proof without being noticed. In this way I have, without borrowing any axiom from intuition, given a proof of a proposition [viz., 133] which might at first sight be taken for synthetic ...

---

[11] *Begriffsschrift* § 23, p. 55, in Jean van Heijenoort, ed., *From Frege to Gödel: a sourcebook in mathematical logic, 1879-1931*, Cambridge: Harvard University Press, 1967, Stefan Bauer-Mengelberg, tr.

There is an understandable tendency to pass over such passages, because of the extreme difficulty of the Kantian concept of an a priori intuition. But I think it is possible to understand Frege's thought without entering into a detailed investigation of this concept. It suffices to recall that for the Kantian mathematical tradition of the period, our a priori intuitions are of space and time, and that the study of space and time falls within the provinces of geometry, kinematics, and perhaps, mechanics. It then follows that the dependence of a basic principle of arithmetic on some a priori intuition would imply that arithmetic lacks the autonomy and generality we associate with it: To establish its basic principles, we would have to appeal to our knowledge of space and time, and then arithmetical principles, like the connectedness of the ancestral and mathematical induction, would ultimately come to depend on our knowledge of spatial and temporal notions for their full justification. Frege's point in the passages quoted is that even if it were possible to justify arithmetical principles in this way, it would be a mistake to suppose that such an external justification is either necessary or appropriate when a justification that is 'internal' to arithmetic is available. The search for proofs, characteristic of a mathematical investigation into foundations of the sort Frege is engaged in, is not motivated by uncertainties concerning basic principles or their justification, but by the absence of an argument which establishes their autonomy from geometrical and kinematical ideas. And autonomy is important since the question of the *generality* of the principles is closely linked to the issue of their autonomy, and particularly their independence, from our knowledge of space and time.[12]

[12]The view presented in the text may perhaps be further clarified if it is contrasted with Philip Kitcher's. (See his paper, "Frege's epistemology," *The philosophical review*, **88** (1979) 235–63.) Kitcher writes: '[The reasoning of mathematicians who rely on spatio-temporal intuition] leaves much to be desired. Transitions from one proposition to another are made in huge jumps so that, even if the premises were known for certain, it might be reasonable to doubt the conclusion. Existing proofs codify the slipshod practices of mathematicians. They should be replaced by genuine proofs ... which proceed by steps invulnerable to doubt' (p. 245). Thus Kitcher, like Benacerraf, holds that a jump is a source of doubt. For both Kitcher and Benacerraf the purpose of rigor—of gap-free proofs—is the removal of such sources of uncertainty. Kitcher therefore agrees with Benacerraf's view of Frege's foundational interests. They differ over

On this 'minimalist' reading of the allusion to Kantian intuition, the rigor of the *Begriffsschrift* is required in order to show that arithmetic has no need of spatial or temporal notions. Indeed, Frege's concern with autonomy and independence suffices to explain the whole of his interest in combating the incursion of Kantian intuition into arithmetic. In this respect, Frege's intellectual motivations echo those of the nineteenth century analysts who sought to free the calculus and the theory of the reals from any dependence on the sciences of geometry and kinematics. Thus as early as 1817 Bolzano wrote: 'the concepts of *time* and *motion* are just as foreign to general mathematics as the concept of space.'[13] And over 50 years later Dedekind was equally emphatic:

> For our immediate purpose, however, another property of the system ℝ [of real numbers] is still more important; it may be expressed by saying that the system ℝ forms a well-arranged domain of one dimension extending to infinity on two opposite sides. What is meant by this is sufficiently indicated by my use of expressions borrowed from geometric ideas; but just for this reason it will be necessary to bring out clearly the corresponding purely arithmetic properties *in order to avoid even the appearance [that] arithmetic [is] in need of ideas foreign to it*.[14]

It must be conceded that the issues raised by the foundations of analysis are rather more varied than they are in the case of the arithmetic of natural numbers; in the case of the real numbers, the usual explanation of the purpose of rigor, as a guarantee of cogency

the broader interpretive claim, Kitcher holding that the epistemological concerns they both attribute to Frege fall within a traditional, quasi-psychologistic, theory of knowledge and mathematical proof. But this claim is highly questionable if, as the discussion in the text shows, Kitcher's (and Benacerraf's) particular account of the goal of rigor is unsupported. To my knowledge, Michael Dummett is the only commentator who has gotten Frege's view of the significance of gaps exactly right. See especially pp. 241f of his paper, "Frege and Kant on geometry," *Inquiry*, **25** (1982) 233–54.

[13]S. B. Russ, "A translation of Bolzano's paper on the intermediate value theorem," *Historia mathematica*, **7** (1980), p. 161.

[14]Dedekind, "Continuity and irrational numbers," in R. Dedekind, *Essays on the theory of numbers*, New York: Dover, 1963, W. W. Beman tr., p. 5, emphasis added.

and a hedge against inconsistency and incoherence, while not a complete account of the matter, is certainly part of the story. But when we turn to Frege's primary concerns about the domain of intuition, namely, the arithmetic of natural numbers, skepticism simply plays no role in any of his arguments against its use. We just don't find Frege rejecting intuition for fear that it is a potentially faulty guide to truth. Those few passages which suggest otherwise are invariably concerned with arithmetic in the broad sense which includes real analysis. When we factor arithmetic by real analysis we also factor out doubts about cogency and consistency. What is left is a concern with autonomy and independence of the sort expressed by Bolzano and by Dedekind in the passages just quoted. Thus while Benacerraf is certainly right to emphasize the mathematical context of Frege's foundational investigations, he has, I think, missed a distinctive feature of those investigations.[15]

We may briefly summarize our discussion of the historical situation as follows. The interest in rigor has both philosophical and mathematical aspects. To begin with, we require proofs 'internal' to arithmetic. Unfortunately, this idea is never given much of a positive characterization, but is always presented as a prohibition—a prohibition against the incursion of spatial and temporal notions into demonstrations of the propositions of arithmetic. Finding proofs which are free of such notions is the mathematical aspect of rigorization. The motivation underlying the foundational interest in rigor has an epistemological component. It is not, as Benacerraf has correctly observed, the program of showing that the truths of arithmetic are truths in virtue of meaning, or conventions of language.

---

[15] In the case of real analysis, Frege emphasized the importance of establishing its independence from spatio-temporal intuition as early as his 1874 qualifying essay: 'there is ... a noteworthy difference between geometry and arithmetic in the way in which their fundamental principles are grounded. The elements of all geometrical constructions are intuitions and geometry refers to intuitions as the source of its axioms. Since the object of arithmetic does not have an intuitive character, its fundamental propositions cannot stem from intuitions either.' See Frege's "Dissertation for the *Venia docendi* in the Philosophical Faculty of Jena" (Jena: Friedrich Frommann, 1874), reprinted in English translation as "Methods of calculation based on an extension of the concept of quantity," in B. McGuinness, ed., Gottlob Frege: *Collected papers in mathematics, logic and philosophy*, Oxford: Basil Blackwell, 1984, Hans Kaal, tr., pp. 56f (1).

Benacerraf is also right to insist that it is not usefully viewed from the perspective of any particular philosophical doctrine. But the foundational interest in rigor is not a concern with justification in the face of doubt—mathematical or skeptical. Rather, Frege's epistemological interests are broadly architectonic: the philosophical dimension to his foundational program is to establish the independence of our knowledge of arithmetical principles from our knowledge of spatial and temporal notions. This is so basic a feature of the tradition in foundations to which Frege belongs that Dedekind, for example, actually explains his logicism in terms of it:

> In speaking of arithmetic (algebra, analysis) as part of logic
> I mean to imply that I consider the number concept entirely
> independent of the notions or intuition of space and time, that
> I consider it an immediate result from the laws of thought.[16]

III. SPATIO-TEMPORAL INTUITION AND ABSTRACTION

While it is clear how, for example, the problem of characterizing continuity might suggest the necessity of spatio-temporal ideas in the theory of the reals, it is far from obvious how, in the case of the arithmetic of natural numbers, an arithmetical principle might be thought to require spatio-temporal notions. A suggestion of how such a view might arise, and even come to seem inevitable, is contained in Frege's discussion of attempts by Thomae, and others, to connect the possibility of counting with position in space and time. The grouping to which this section belongs is entitled 'Attempts to overcome the difficulty.' The difficulty referred to is formulated in the previous section (§ 39) as an objection to the Euclidian idea that a natural number is an agglomeration, either of units or of objects, arrived at by a process of abstraction. The problem seems to be this: If we abstract from the individuality of the elements of a collection, we will not have several objects, but will have instead only

---

[16]From the first page of the first edition Preface to *Was sind und was sollen die Zahlen?* See R. Dedekind *Essays on the theory of numbers*, p. 31; the parenthetical clarification of 'arithmetic' is Dedekind's. For an extended discussion of Dedekind and Cantor along related lines, see Michael Hallett, "Physicalism, reductionism and Hilbert," in A. D. Irvine, ed., *Physicalism in mathematics*, Dordrecht and Boston: Kluwer, 1989, pp. 228ff.

one. So in this case no collection can be identified with a number greater than one. But if we do not abstract over their individual differences, each collection of (say) four objects will be a different number four.[17]

For someone committed to the idea that the numbers are 'formed' by abstraction, it is necessary to seek some constraint on how far the process of abstraction can be carried. This is evidently the idea behind Thomae's suggestion that numbers are associated with the positions of a spatio-temporal series:

> If we consider a set of individuals or units in space and number them one after the other, for which time is necessary, then, abstract as we will, there remains always as discriminating marks of the units their different positions in space and in the order of succession in time. (Quoted by Frege, *Grundlagen* § 40.)

Thus one source of the persistence of the idea, that our knowledge of the arithmetic of natural numbers depends on our concepts of space and time, is an internal difficulty in the abstractionist theory of the formation of simple numerical concepts like counting. The theory is committed to the process of 'subtracting' properties by its account of the difference between numbers and ordinary collections: a collection 'becomes a number' when, by the 'removal' of properties from its members, it becomes a collection of 'units.' But without some restriction on how far such subtractions of properties may be carried, the abstractionist theory of number is threatened with the absurdity that there is only one number. Spatio-temporal position provides the necessary constraint.

---

[17]Cp. *Grundlagen* § 39: 'If we try to produce the number by putting together different distinct objects, the result is an agglomeration in which the objects contained remain still in possession of precisely those properties which serve to distinguish them from one another; and that is not the number. But if we try to do it in the other way, by putting together identicals, the result runs perpetually together into one and we never reach a plurality.' For an extended discussion of Cantor's abstractionism, see Chapter 3 of Michael Hallett's, *Cantorian set theory and limitation of size*, Oxford: Oxford University Press, 1989. For a concise discussion of Husserl's abstractionism, see Michael Dummett's essay, "Frege and the paradox of analysis," in his collection, *Frege and other philosophers*, Oxford: Oxford University Press, 1991.

Notice that when abstraction is used to explain the formation of the ordinal notion of number, what for us is a metaphorical use of 'position,' when we speak, for example, of position in a series, is not at all metaphorical for an abstractionist like Thomae. Lacking a general notion of order, Thomae's understanding of ordinal position is *literally* tied to the example of spatio-temporal series. This observation did not escape Frege's attention, for when, in the *Grundlagen* (§ 80), he recalls the *Begriffsschrift* definition of the ancestral, he is careful to remark on the fact that his definition is not restricted to spatial or temporal series:

> It will not be time wasted to make a few comments on this. First, since the relation [whose ancestral has been defined] has been left indefinite, the series is not necessarily to be conceived in the form of a spatial and temporal arrangement, although these are not excluded.

Frege's account of the ancestral proceeds 'schematically': he takes a fixed, but arbitrary, binary relation f, and shows how to form its ancestral. The account is general ('the relation has been left indefinite'), since, aside from the assumption that f is a binary relation, *no further conditions are imposed on f*. Frege's definition is easily presented in three simple steps.[18]

Suppressing explicit reference to the relation f, we abbreviate 'property F is hereditary in the f-sequence' by $\text{Her}(F)$, and define it by the condition

$$\forall d \forall a (dfa \ \& \ Fd \rightarrow Fa),$$

given by Proposition 69 of the *Begriffsschrift*.

For 'after x, property F is inherited in the f-sequence' we write '$\text{In}(x, F)$,' and define this as

$$\forall a(xfa \rightarrow Fa).$$

For 'y follows x in the f-sequence'—the ancestral of f—we write 'xf^*y.' With the above abbreviations, Frege's definition of the ancestral becomes:

$$xf^*y \equiv \forall F(\text{Her}(F) \ \& \ \text{In}(x, F) \rightarrow Fy).$$

[18] See George Boolos, "Reading the *Begriffsschrift*" (Chapter 6, below), whose exposition I follow.

What is striking about Frege's definition comes out most clearly when it is compared with characterizations to which it turns out to be equivalent. The 'intuitive' or preanalytic characterizations are very heavily imbued with number-theoretic and spatio-temporal terminology, as is apparent when, for example, we say that y follows x in the f-sequence if, beginning with x, y is reached after only finitely many applications of the procedure f. Frege's definition appears to avoid this, and his remark, in the *Grundlagen*, to this effect seems clearly justified.

IV. FREGE'S THEORY OF SEQUENCES

Part III of the *Begriffsschrift* is entitled 'Some topics from a general theory of sequences.' For a modern reader, the work deals with topics from the theory of binary relations. The modern analogue of a sequence in Frege's sense is simply a nonempty set X, together with a binary relation f on X, i.e., $X \times X \supseteq f$. However, we will see that something conceptually important is missed if we identify f with a set of ordered pairs.

The major conceptual innovation of Part III is the definition of the ancestral, and the principal result is the proof of its connectedness. Frege's proof makes extensive use of an implicit substitution rule which allows him to substitute formulas for property variables. It is well known that the substitution rule is equivalent to a comprehension scheme,

$$\exists P \forall x (Px \leftrightarrow \mathcal{A}(x)), \quad \text{where } P \text{ is not free in } \mathcal{A},$$

and that it therefore constitutes an assumption concerning property existence.[19] The properties required by Frege's proofs are expressed by formulas of the form 'mf^*x,' and thus depend on the definition of the ancestral. Since the definition of f^* contains a second-order quantifier, properties defined by such formulas are defined impredicatively. Frege understands the formula, 'mf^*x,' to express the property of following m in the f-sequence. (See for example p. 80, § 31.)

---

[19]See Boolos, p. 171 for a proof.

It has long been known that properties of this form are closely related to Dedekind's notion, *the chain of a*.[20] Recall that for Dedekind, a *chain* is a pair $\langle X, f \rangle$, consisting of a nonempty set X and a one-one function f, such that

$$\mathrm{dom}(f) \supseteq X$$

and

$$X \supseteq f[X].$$

For an element $a \in X$, Dedekind writes a_0 for the *chain of a*, and puts

$$a_0 = \bigcap \{ \, X : \langle X, f \rangle \text{ is a chain } \& \ a \in X \, \}.$$

To state the exact connection between Dedekind's chains and Frege's properties, define the *weak ancestral*,[21] $f^*_=$, of f by

$$x f^*_= y \equiv x f^* y \lor x = y.$$

Then, on the assumption that f is one-one, m_0 is just the extension of the property, $m f^*_= x$. Because of their connection with chains, such properties evidently possess a rich mathematical content.

It is clear that Frege's focus in the *Begriffsschrift* is on principles that the number sequence (i.e., the set of natural numbers under the successor relation) shares with other sequences. This suggests a natural division in his logicism between this early concern with 'principles,' such as the connectedness of the ancestral and the reconstruction of inductive reasoning,[22] and his later concern

---

[20]Presented in sections 37 and 44 of *Was sind und was sollen die Zahlen?* As Dedekind notes in his letter to Keferstein, the notion of a chain does for his analysis of the number sequence what the ancestral does for Frege's general theory of sequences. See Hao Wang, "The axiomatization of arithmetic," *The journal of symbolic logic*, **22** (1957) 145–57, for a translation of portions of the letter together with an informative discussion. The letter to Keferstein is reprinted in its entirety in Jean van Heijenoort, ed., *From Frege to Gödel: a sourcebook in mathematical logic, 1879-1931*, Cambridge: Harvard University Press, 1967, Stefan Bauer-Mengelberg and Hao Wang, trs.

[21]The terminology and notation are taken from Boolos.

[22]In §§ 24 and 26 of the *Begriffsschrift*, Frege proves a generalized form of mathematical induction, which he formulates: 'If x has the property F that is hereditary in the f-sequence, and if y follows x in the f-sequence, then y has the property F' (*Begriffsschrift* p. 62, Proposition 81).

with numbers as 'objects,' the concern in the *Grundlagen* and the *Grundgesetze* with the definition of the natural numbers and the proof of their infinity. How should we assess the success of Frege's account of such principles?

There is an influential line of criticism, originating with Poincaré,[23] but most forcefully presented by Charles Parsons.[24] Parsons is concerned with the Fregean analysis of the predicate, 'is a natural number,' but if his objection is successful, exactly the same difficulties confront Frege's definition of the ancestral. The objection focuses on the use of induction in the proof of the 'adequacy' of the definition of the ancestral, where an adequacy proof is one which establishes that the definition succeeds in defining a concept that is equivalent to the 'intuitive' one. Let 'xf^*y' denote the reconstructed ancestral, and let 'y follows x in the f-sequence' denote the intuitive ancestral. Then to establish adequacy, we proceed inductively by showing that xf^*y if, and only if, y follows x in the f-sequence when y is 1 'f-step' from x, and if the equivalence holds when y is n f-steps from x, it also holds when y is $n + 1$ f-steps from x. We conclude that the equivalence holds however finitely-many f-steps separate x and y. But in proceeding in this way, we rely on our preanalytic understanding of the ancestral and on the *underived* principle of mathematical induction. The claim that the definition of the ancestral yields a *foundation* for inductive reasoning cannot be sustained, Parsons-Poincaré argue, if, in order to establish the adequacy of the definition, we must fall back on intuitive reasoning.

In the 'Postscript' to his original paper, Parsons suggests that this argument is inconclusive since it is open to the Fregean to use the derived principle of induction in the proof of adequacy. Parsons argues that the circularity this seems to introduce is similar to the 'circularity' that accompanies standard metalogical justifications of formal reconstructions, so that such a use of the derived principle of induction may therefore be acceptable. But there is a simpler

---

[23]Who urged it against Couturat; there is no indication that Poincaré ever read Frege. For a discussion and references, see Warren Goldfarb's, "Poincaré against the logicists," in W. Aspray and Phillip Kitcher, eds., *History and philosophy of modern mathematics, Minnesota studies in the philosophy of science*, Vol. XI, Minneapolis: University of Minnesota Press, 1988.

[24] "Frege's theory of number," Chapter 7, below.

and more decisive objection to the Parsons-Poincaré line of criticism. If the cogency of intuitive reasoning is not at issue, then it is a wholly pragmatic question whether we should justify the adequacy of the definition of the ancestral with the intuitive or the derived principle of induction: Evidently, to convince Poincaré, Frege would do well to use underived induction. But this tells us rather more about Poincaré than about Frege's account of the principles. And it certainly leaves intact Frege's claim to have given an account of *following in the f-sequence* that is free of any dependence on intuition, even in the rather vague sense of 'intuition' presupposed by Parsons and Poincaré. The 'Poincaré-type' objections have succeeded in gaining a following only because the foundational program they are directed against has been misrepresented as a naive form of foundationalism.

More recently, George Boolos[25] has raised a question about the mathematical development of Part III, and in particular, about Frege's use of substitution in the proof of 133. Boolos puts the point by having his Kantian interlocutor ask whether it isn't 'an intuition of precisely the kind Frege thinks he has shown unnecessary that licenses [the inference to properties like x follows m in the f-sequence]' (p. 172, below). To crystallize the point, we should distinguish the general theory—the case which 'leaves f indefinite'— from its application to the natural number sequence. While both cases suggest questions about the bearing of intuition on assertions of property existence, the questions are rather different. Once the coherence of investigating an arbitrary but fixed f is admitted, it is difficult to see how a question can be raised about the intuitive— i.e. spatio-temporal—character of the properties of the elements it relates, or about the spatio-temporal character of the knowledge on which the inference to such properties rests. This of course leaves open the possibility that in some *other* sense intuition may have a role to play in the general case. Nevertheless, the *Kantian* can make his point only by denying the coherence of a standard practice. I think it is clear that Frege can safely ignore such a Kantian, and assume that the notion of a property of elements of an arbitrary but fixed sequence is not necessarily tied to spatio-temporal intuition.

[25] "Reading the *Begriffsschrift*."

What of the Kantian who allows the general theory outlined in Part III, but questions whether its application to the sequence of natural numbers is capable of proceeding without spatio-temporal intuition? This is a reasonable concern, and it is evident that addressing it will require considerations that go beyond what is explicitly presented in the *Begriffsschrift*.

In Frege's later work, most explicitly in *Function and concept* which appeared twelve years after the *Begriffsschrift*, properties are assimilated to 'concepts,' and concepts are then treated as a special type of function. But the assimilation of properties to functions is just an elaboration of the *Begriffsschrift*'s rejection of the subject/predicate analysis of sentences in favor of function/argument (or function/variable) notation. And it is clear that both developments were the result of reflection on the function concept in mathematical analysis.[26] Frege in effect recognized that a traditional notion of logic, namely, a *property* of numbers, could be subsumed under a slight generalization of the mathematical concept of an arithmetical function, that the arithmetical properties required for the development of arithmetic are contained in the class of characteris-

---

[26] *Function and concept* is of course replete with allusions to analysis. And the subtitle of the *Begriffsschrift*—*a formula language, modelled upon that of arithmetic, for pure thought*—is an allusion to Frege's appropriation of the function/variable notation of analysis, something that is easily missed if 'arithmetic' is understood too narrowly. The influence of analysis is explicitly acknowledged in § 10, where Frege writes: ' ... the concept of function in analysis, *which in general I used as a guide*, is far more restricted than the one developed here.' (p. 24, emphasis added.) His point is that the *Begriffsschrift* characterization of a functional expression is general enough to subsume *both* the usual notion of a functional expression for a function of first level *and* that of a functional expression for a function of second level, so that the *Begriffsschrift* notion of a functional expression—and therefore, its concept of a function—is less 're-stricted' than what one finds in analysis. The fact that Frege's celebrated explanation of a function is actually a characterization of a functional expression is sometimes interpreted as showing that the *Begriffsschrift* confuses functions with functional expressions. (For two recent examples see David Bell, "How 'Russellian' was Frege?," *Mind*, **99** (1990) 267–77, and Peter Simons, "Functional operations in Frege's *Begriffsschrift*," *History and philosophy of logic*, **9** (1988) 35–42.) But this is just a mistake. The functions of the later work are termed 'single-valued procedures' (*eindeutige Verfahren*) in the *Begriffsschrift*. And Frege certainly recognizes the distinction between single-valued procedures and their linguistic expression.

tic functions, and that no *other* notion of property is needed for the development of arithmetic. An important consequence of the rigorization of analysis was the vindication of the idea that a real-valued function is simply a many-one correspondence between real numbers. Fairly early in the century Dirichlet (in 1837) and Lobatchevsky (in 1834) presented clear and unambiguous statements of the modern notion of a *continuous* function from this point of view. But the idea that the general notion of an arithmetical function could only be captured by the concept of an arbitrary correspondence—and that it might be mathematically useful so to capture it—is probably best credited to Riemann's work of 1854.[27] A decisive step in this development was taken when it was recognized that the concept of a real-valued function is not constrained by the notion of a possible trajectory of a classical particle, or that of a piece-wise continuous curve—steps that clearly removed the concept from its earlier dependence on geometry and kinematics. Since dependence on geometry and kinematics pretty much exhausts what for Frege is meant by dependence on a priori intuition, it is reasonable to assume, on the basis of this development, that the notion of an arithmetical property, whether of real or of natural numbers, owes nothing to Kantian intuition. It follows that the techniques developed in the *Begriffsschrift* yield successful derivations of mathematical induction and the structural properties of the ancestral of the successor relation, and that all this is achieved without appealing to 'intuition,' in the then

[27]Thus Bottazzini remarks that 'if for Dirichlet ... the generality of the definition [of a real-valued function] did not go along with a consistent practice in the study of equally "general" functions, the opposite is true for Riemann. In his thesis (*Habilitationsschrift*) of 1854, ... he revealed to the mathematical world a universe extraordinarily rich in "pathological" functions.' (*The higher calculus: a history of real and complex analysis from Euler to Weierstrass*, p. 217.) Bottazzini goes on to remark that when one looks at the historical development, 'the modern concept of a function of a real variable, in its full generality, began to emerge in mathematical practice only towards the end of the 1860s as Riemann's writings became more widely known.' (p. 217.) The later questioning of the notion of an arbitrary correspondence by the French school of quasi-intuitionists, towards the end of the nineteenth century, shows that the story is an intricate one. An illuminating discussion of the importance of the development of the function concept for philosophy of science is presented by Mark Wilson in his paper, "Honorable intensions," in S. Wagner and R. Warner, eds., *Naturalism*, South Bend: Notre Dame University Press, 1993.

accepted understanding of what such an appeal would mean.

The strategy behind this defence of Frege's account of the 'principles,' is slightly unusual: We have exploited a feature of the *mathematical* development of the notion of an arithmetical property in order to defend a *logicist* thesis regarding the arithmetic of natural numbers. The thesis defended is the relatively weak claim that certain principles do not depend on spatio-temporal intuition, something that the mathematical development sustains. But there is considerably more to the mathematical development of the function concept and of the notion of an arithmetical property than their removal from the domain of spatio-temporal intuition. Our defence is successful to the extent that it does not depend on the positive characterization of a function as a many-one correspondence, or that of an arithmetical property as a *set* of numbers. In order to subsume the positive characterization of these notions, Frege must appeal to the value ranges of functions and the extensions of concepts. To what extent does the *Begriffsschrift*'s account of the principles avoid Frege's theory of extensions and value ranges together with the difficulties associated with their introduction?

It is well known that, taken by itself, Frege's theory of concepts is consistent, concepts being naturally stratified; it is only the combination, concepts plus extensions, that leads to paradox. Moreover, the stratification of concepts is naturally extended to the properties of the *Begriffsschrift*, yielding a consistent interpretation of that work, since there is nothing in the formal apparatus of the *Begriffsschrift* which associates properties with extensions. This is an important point of difference between the *Begriffsschrift* and the *Grundgesetze*; extensions are nowhere mentioned in the earlier work. In any case, they appear to play no fundamental role.[28] It is only when Frege sets out to prove the infinity of the natural number

---

[28] An impression that is borne out by a remark in the correspondence where, in a letter to Jourdain (dated 23.ix.1902) which discusses the *Begriffsschrift*, Frege writes: 'This distinction between a heap (aggregate, system) and a class, which had not perhaps been drawn sharply before me, I owe, I believe, to my conceptual notation, although you will not perhaps discover any trace of it in reading my little work. Some of the reasonings that occurred in writing it may well have left no trace in print.' The letter is collected in Brian McGuinness, ed., Gottlob Frege: *Philosophical and mathematical correspondence*, Chicago: University of Chicago Press, 1980, p. 73.

sequence that extensions become prominent. However these considerations do not show that the connection between properties and their extensions may be ignored. Without an account of this connection, our grasp of the properties and relations of the *Begriffsschrift* is incomplete. For example, what are we to say of the connection between the properties $mf_{\leq}^{*}x$ and Dedekind's chains? To express the connection, we of course would appeal to an axiomatic development of the set concept. But for Frege this is highly problematic, since it brings his whole program to rest on yet another 'special science.' If Frege did not consider this consequence, this was because he assumed Basic Law V of the *Grundgesetze* would provide the required 'representation' of properties and relations by their extensions. Once again, what strikes one as the insurmountable obstacle to Frege's logicism, even to the implementation of the modest program Frege set himself in the *Begriffsschrift*, is its failure to derive extensions from properties: the sorry history of Basic Law V.

JOHN P. BURGESS

4

Frege and Arbitrary Functions

I. QUESTION

The orthodox notion of a function in present-day real analysis is a broad one, involving no restriction to differentiable or definable functions, and admitting "arbitrary" functions from real numbers to real numbers. Since the process by which this notion became established was a long one, whose description occupies considerable space in general histories of mathematics, and ended only in the twentieth century, one may ask of an essentially nineteenth century

The present paper is a revised version of the author's "Hintikka *et* Sandu *versus* Frege *in re* arbitrary functions," *Philosophia mathematica* (Series III), **1** (1993) 50–65, and appears by the kind permission of the editor of that journal, Robert Thomas.

The paper in its original form was a response to Jaakko Hintikka and Gabriel Sandu, "The skeleton in Frege's cupboard: the standard versus nonstandard distinction," *The journal of philosophy*, **89** (1992) 290–315. The main changes have been (i) incorporation into the body of the paper of some remarks made in private communications by Michael Dummett, Carl Hempel, and Mark Wilson, to all of whom I am most grateful, and (ii) omission of polemical discussion of Hintikka and Sandu. The anti-Fregean conclusions of Hintikka and Sandu are also challenged by William Demopoulos and John L. Bell, "Frege's theory of concepts and objects and the interpretation of second-order logic," *Philosophia mathematica* (Series III), **1** (1993) 139–56, and by Richard Heck, Jr. and Jason Stanley, "Frege and second-order logic," *The journal of philosophy*, **90** (1993) 416–24.

figure like Frege whether his notion was as broad. One may ask whether Frege's notion of function involved (A) definability restrictions, or even perhaps (B) differentiability restrictions.

The answers to such questions will have implications for the overall interpretation of Frege's philosophy. Or rather, they will if the questions can be answered without *presupposing* some overall interpretation of Frege's philosophy, on the evidence of Fregean texts addressing more or less directly the issue of the breadth of the notion of function. Unfortunately, the available texts—those anyone can quickly find in obvious sources, plus those scholars so far have found in unobvious sources—are few.

§ II and § III of the present note summarize some relevant historical and methodological background. § IV and § V examine the texts relevant to questions B and A respectively. It is concluded that question B can be answered in the negative, but question A cannot be answered on the basis of these texts.

II. HISTORY

Seventeenth century mathematicians considered algebraic and elementary transcendental functions, and eighteenth century mathematicians began to consider the so-called special functions, definable as integrals of functions previously admitted, a special form of the method of defining a function as the limit of a sequence of functions previously defined. Euler, who was in advance of his time, considered also functions obtainable by splicing together different functions on different intervals, a special form of the method of defining a function by cases. All these functions are, except perhaps at a finite number of points, continuous and differentiable. In the early nineteenth century, inspired partly by the applied work of Fourier and partly by the foundational concerns of Cauchy, more pathological functions began to be considered.

Towards mid-century, Dirichlet used a more general form of definition by cases to introduce the function that takes the value unity or zero according as the argument is rational or not, which is discontinuous and hence non-differentiable at all points. Riemann and Weierstrass, using more general forms of definition as a limit, introduced functions that are continuous at all points but non-differentiable at infinitely many points or even at all points. The introduction of such examples was an aspect of these mathematicians' work towards realizing the goals of rigor and generality, of always acknowledging among the hypotheses of a theorem any special properties of the functions involved that are necessary for the proof, and of always seeking the weakest such special hypotheses that are sufficient for the proof.

Great progress was made by such mathematicians in foundations of *real* analysis, despite the fact that work in this area was for them only one of several activities. Another was the development of *multivariate* analysis, which considers functions from and to *pairs*, *triples*, and so on, of real numbers. Still more important was the development, originating in work of Gauss and Cauchy earlier in the century, of developing a *complex* analysis. In a sense, complex numbers are just pairs of real numbers, but the notion of *derivative* in complex analysis is not just the notion from multivariate analysis as it applies to the case $n = 2$. In particular, complex differentiability implies the much stronger property of *analyticity*.

The passage from real to complex numbers is but an instance of the generalization of the notions of "quantity" and of "sum" and "product" to new objects and operations on them. In particular, a function may be considered a kind of quantity, and composition of functions as a kind of operation, giving rise to a new kind of algebra. Also higher-order functions, functions whose arguments are themselves real or complex functions, may be considered, giving rise to a new kind of analysis, though the development of this *functional* analysis took place mainly in the twentieth century, and largely presupposes the late nineteenth century development of set theory. In set theory itself the notion of function is applied in contexts where the arguments and values may be objects of whatever sort.

Set theory itself was an outgrowth of work in foundations of real analysis. Continuity, differentiability, and the like are properties that may hold at some arguments and fail at others. The study of sets of arguments at which such properties hold or fail was the origin of Cantor's special theory of sets of points on the real number line, which was in turn the origin of his general theory of sets. Cantor hoped to show that certain theorems that hold of a function provided the set of points at which it is ill-behaved is finite also hold even if the set is infinite, provided that it is in some sense "small."

Central to Cantor's set theory is the distinction of smaller and larger among infinite numbers. (One sense in which an infinite set may be "small" is to have a small number of elements.) The smallest infinite number or *transfinite cardinal* is that of the set of natural numbers, called *countable*. This is also the cardinal number of the set of rational numbers, as a corollary to what may be called the *lesser cardinality theorem*, that the number of finite sequences from a countable set is countable. The cardinal of the set of real numbers is larger, as a corollary to what may be called the *greater cardinality theorem*, that the number of subsets of (or functions from and to) a set is greater than the number of elements of the set. The cardinal of the set of real numbers is called *continuum*, and is also the cardinal of the set of complex numbers. By the same theorem, the cardinal of the set of real functions is larger still, and may be called *supercontinuum*. Continuing in this way, one obtains an infinity of larger and larger cardinals. Indeed, their number is uncountable and superuncountable and even what Cantor calls *absolutely infinite*.

While many of these developments of the general theory of sets lay to one side of the main line of work in foundations of analysis, certain developments in the special theory of point sets remained central. Such was the work done on defining a notion of *measure* for sets in the line and plane generalizing the notion of length and area for intervals and rectangles. (One sense in which an infinite set may be "small" is to have measure zero.) Such work was central to the solution of the main problem bequeathed by the nineteenth century real analysts to their turn-of-the-century successors, that of generalizing notions and results pertaining to integration.

The solution to this problem is due to the Paris School of Baire, Borel, and Lebesgue. Baire and Borel introduced a very extensive hierarchy of functions from and to the reals and a parallel one of sets of reals, in which the most pathological examples specifically considered in the nineteenth century, the Dirichlet function and the Cantor set, occur already at the bottom few levels. Borel also defined by a "bottom-up" method a notion of integral applicable to all (but only) functions in this hierarchy. Lebesgue defined by a "top-down" method another notion of integral also applicable to (at least) all functions in this hierarchy, called "L–" or "Lebesgue–" measurability in contrast to "B–" or "Borel–" measurability.

Left open were the questions whether there are any functions that are L-measurable but not B-measurable, and any functions that are not L-measurable. An affirmative answer to the former question is implied by cardinality estimates, since it can be shown that the number of B-measurable functions is continuum, while the number of L-measurable functions is supercontinuum. An affirmative answer to the latter question is implied by the choice axiom, published by Zermelo in 1904. A peculiarity of both proofs is that they show that there exists a function with a certain property, but provide no *definition* of such a function.

This peculiarity gave rise to controversy, notably an exchange in 1905 of five letters[1] among leading French analysts of the day, Hadamard (pro-choice, two letters) and the Paris School (anti-choice, one letter each from of Baire, Borel, Lebesgue). Hadamard (second letter) observes that the position of the Paris School, on which all mathematical objects must be definable, is incompatible with Cantor's set theory, on the tacit assumption that what is meant is definability in some language L of the usual *finitary* kind, in which every expression is a finite sequence of symbols from a finite alphabet. For by the lesser cardinality theorem, there will then be only countably many expressions to serve as definitions for mathematical objects, whereas by the greater cardinality theorem there are supposed to be uncountably many mathematical objects (continuum-many real numbers, supercontinuum many real functions, and so on).

The same observation is involved in the semantic paradoxes of König and Richard, which also date from 1905. Paradox results from the further observation that Cantor's diagonal argument actually enables one to *define* an enumeration of all possible definitions (in L) and then *define* a mathematical object not definable by any of these definitions (in L), by a definition itself involving the notion of "definability (in L)." Analysis of these semantic paradoxes and related set-theoretic paradoxes led Poincaré to adumbrate a stricter version of definabilism than that of the Paris School, called *predicativism*, in which certain kinds of circularities (such as the definition in terms of definability) are banned as vicious. Predicativism was

---

[1]René Baire, et al., "Cinq lettres sur la théorie des ensembles," *Bulletin de la Société Mathématique de France*, (1905) 261–74.

also, somewhat later, advocated by Weyl, but neither this nor any other version of definabilism won many strict adherents.

Partial sympathy with definabilism did motivate some work in the theory of point sets by the Moscow and Warsaw schools in the 1920's. Out of the latter emerged the *model theory* of Tarski in the 1930's. On Tarski's view, the notion of definability for a (formal or symbolic) language L can be legitimately mathematically defined, but not within the language L itself, only within a richer *metalanguage* L'. This view has influenced not only the orthodox but also many dissidents: Hierarchies of ever richer languages also figure in later characterizations of predicativism. By Tarski's time, however, definabilism in any version had ceased to be a live issue for working analysts.

III. METHODOLOGY

Methodological problems arise already in connection with questions about whether some mathematician M of the past incorporated a continuity or differentiability restriction into the notion of function. Most important is the problem whether one should give most weight to what M *says* in definitions, or to what M *does* in proofs.

One finds in the works of many mathematicians of the past definitions of the type:

> One variable quantity y is called a function of another variable quantity x if to each of the values that x successively assumes, there is associated a unique value that y thereupon assumes.

with glosses of the type:

> The notion of function involves no restriction on the *mode* of dependence of the values of y on those of x, of which a broad variety are possible.

Such definitions-*cum*-glosses appear as early as the middle eighteenth century, and as late as the early twentieth century. The actual wording may differ from writer to writer, "variable quantity" corresponding to one writer's "changing quantity" and to another writer's "indefinite quantity," and "mode of dependence" above to one's "method of determination" and another's "law of correlation."

And if M offers such a gloss, this seems to indicate that M is rejecting traditional restrictions on the notion of function.

However, if M habitually uses special properties of functions such as continuity and differentiability in the *proofs* of theorems without acknowledging them as special hypotheses in the *statements* of the theorems, this seems to indicate on the contrary that M has incorporated these special properties into the very notion of function. And unfortunately, many an M who has offered a definition-*cum*-gloss like that above has also used unacknowledged special hypotheses in proofs, especially prior to the rigorization of analysis in the mid-to-late nineteenth century.

The broadening of the notion of function in the mid-to-late nineteenth century led to certain ambiguities in the usage of the word "function" which create another methodological problem. Wilson observes that the same author will use "function" to mean "arbitrary function" in real analysis, and to mean "analytic function" in complex analysis. Moreover, in functional analysis and set theory, the word "function" is often avoided in favor of "functional," "operator," or "mapping."

Methodological problems become more difficult in connection with questions about whether some M incorporated a definability restriction into the notion of function, since *neither* of the two kinds of indications just considered may be reliable. On the one hand, given how vague such words as "method" or "law" are, even if M *does* impose a restriction of definability, M may well take this restriction to be *already* implicit in the requirement that there be a "method" or "law," and not some *further* requirement on the *kinds* of "methods" or "laws" allowable. Hence M may well be willing to endorse a gloss like that above. On the other hand, since definability is not—or at least, was not at the relevant period, long before Tarski—a straightforwardly mathematical property like continuity or differentiability, even if M *does* impose a restriction of definability, M might well find no way to *use* this restriction in a mathematical proof.

M's imposing such a restriction might well be indicated in the *opposite* way, not by accepting proofs others reject, but rather by rejecting proofs others accept. For one thing, if M imposes such a restriction and is conscious that a given proof uses the axiom of choice, M may be expected to reject the proof. For another thing, if M imposes a definability restriction and is conscious of the bearing of cardinality estimates on the issue, M may be expected to reject proofs using such cardinality theorems. But unfortunately, prior to the debates that began in 1905, many mathematicians were *un*conscious that certain proofs use the axiom of choice, and *un*conscious of the bearing of cardinality estimates, of the tension between a transfinitistic mathematics and a definabilist philosophy. It would not be easy to guess from their pre-1904 writings alone that Baire, Borel, and Lebesgue would support definabilism and renounce transfinitism in the post-1904 debates. Moreover there is a tension even in the work of *Cantor*.

Cantor's strictly mathematical work unconsciously used the choice axiom and famously originated cardinality theory, but it is with some justice that Lebesgue in his letter describes himself as holding "with Cantor" that a set or function must have a definition. For this is what Cantor seems to presuppose in his various more philosophical formulations. Jourdain, in his long introduction to his translation[2] of key papers of Cantor collects several of these. One of these defines a set to be (p. 54, my italics):

> ... any multiplicity that can be thought of as one, that is to say, any totality of definite elements which can be bound up into a whole *by means of a law*.

while another requires of a set that (p. 46, my italics):

> ... in consequence of *its definition* and of the logical principle of the excluded middle, it must be considered as intrinsically determined whether any object belonging to this sphere belongs to the aggregate or not ... .

The tension between what Cantor does in theorems and what he says in glosses seems most extreme in the case of cardinal numbers: He proves in one place that there is an absolute infinity of them, and he asks rhetorically in another place (p. 80, Cantor's italics):

> Is not ... [a] cardinal number ... an abstract picture ... in *our* mind?

[2]Georg Cantor, *Contribution to the founding of the theory of transfinite numbers*, New York: Dover, n.d., "unabridged and unaltered reprint of the English translation first published in 1915," P. E. B. Jourdain, trans.

As if there were room in a human mind for an absolute infinity of pictures! (A similar tension can be found in Dedekind.) If Zermelo's and Hadamard's view, and not Lebesgue's or Borel's, is now the orthodoxy, this is doubtless because, as van Heijenoort in an introduction to his translation of a work of Zermelo[3] observes (p. 199), "Cantor's definition of set [has] had hardly more to do with the development of set theory than Euclid's definition of point with that of geometry."

IV. DIFFERENTIABILITY

That functions are not restricted to be continuous or differentiable was becoming established orthodoxy in the mathematical community, at least in Germany, already before the beginning of Frege's career. Frege did his graduate studies at Göttingen, where Dirichlet and Reimann had taught, and began his teaching and research career at a time when Weierstrass was one of the most influential mathematicians in Germany. These facts alone make it scarcely conceivable that Frege could have ignored or rejected examples of discontinuous and undifferentiable functions. Still, even from his early career, when his work was closer to mainstream mathematics, there is a lack of texts bearing directly on question B, since analysis seems to be an area in which he did only teaching, not research. One of his early works, his dissertation for the *venia docendi*,[4] is largely a contribution to the generalizing tendency in the mathematics of the period. Even this, however, pertains more to algebra than to analysis. There are, nonetheless, in this work two half-sentences (with two *omissions*) which do bear on question B.

The work begins with quasi-philosophical remarks on considering functions (of one or several real or complex variables) as "quantities" and composition as an "arithmetic" operation. It next turns to consider a class of mathematical problems that then arise, of which the simplest is this: Given a function F, to find a function G such

[3]Ernst Zermelo, "Investigations in the foundations of set theory I," in Jean van Heijenoort, ed., *From Frege to Gödel: a source book in mathematical logic, 1879-1931*, Cambridge: Harvard University Press, 1967, 199–215.

[4]Gottlob Frege, "Methods of calculation based on an extension of the concept of quantity" (1874), in Brian McGuiness, et al., eds., Gottlob Frege: *Collected papers in mathematics, logic and philosophy*, Oxford: Blackwell, 1984, 50–92.

that the composition of G with itself will be equal to F. Frege at first considers such problems at the highest level of generality, and attacks the problem by considering the consequences for the functions involved of the relation that is supposed to hold between them. There is little to be said at this high level of generality.

So he soon (p. 60) begins considering:

consequences for the *derivatives* of the functions

He *omits* to add:

if they have derivatives.

Though from the standpoint of Dirichlet, Riemann, and Weierstrass, Frege is at this point descending to a lower level of generality, where F must be a *differentiable* function, he does *not* take formal note of the fact. But indeed, there is not much more to be said at this level of generality either. So he soon makes a further descent, to the level where F must be a *linear* function, of which he *does* take formal note.

He remains at that level throughout the bulk of the paper, which thus belongs to *linear algebra*, a field far from foundations of real analysis. Frege in effect reminds the reader of the level on which he has been proceeding at the very end (p. 92), just before raising the question of the possibility of lifting his results to the *quadratic* level. Referring back thirty-odd pages he writes:

Apart from some general propositions valid for all functions ... , we have been concerned almost exclusively with linear functions.

He omits to add:

... and some further general propositions valid for all differentiable functions ...

What is one to make of Frege's omissions? One can certainly conclude that the need to take account of non-differentiable functions in the foundations of real analysis was not uppermost in his mind when writing his thesis. Given that the work is in another field of mathematics, and given the marginal place of the quoted passages in the work, it would seem imprudent to conclude more before other texts from his mature years have been examined.

In his mature years, Frege's work had two sides, the creative and the critical. On the creative side, he originated a new branch of mathematics, mathematical logic, in his *Begriffsschrift*. Precisely because this and his other major works are so innovative, discussion of the scientific work of mainstream mathematicians, whether in function theory or any other established field, occupies little space in them. Nonetheless, since one of the central undefinable notions of Frege's system is that of "function," an obvious source to examine is the place in his major work the *Grundgesetze* where he introduces that notion. One quickly finds there a brief reference to mainstream developments. Another obvious source to examine is the well-known collection of Geach and Black of minor works from the same period, which turns out to contain two works with the word "function" in the title. In one of these,[5] one quickly finds a parallel passage. It is instructive to set the two passages side-by-side:

Grundgesetze I,[6]§ 2	*Function and concept*
	Now how has the reference of the word 'function' been extended by the progress of science? We can distinguish two directions in which this has happened.
To the fundamental arithmetical operations mathematicians have added, as ways of forming functions, the process of proceeding to the limit as exemplified by infinite series,	In the first place, the field of mathematical operations that serve for constructing functions has been extended. Besides addition, multiplication, exponentiation, and their

[5] *Function and concept* (1892), in Peter Geach and Max Black, trans. and eds., *Translations from the philosophical writings of Gottlob Frege*, Oxford: Blackwell, 1970, 21–41.

[6] "Selections from Volume I. Translated by P. E. B. Jordain and J. Stachelroth," in *Translations from the philosophical writings of Gottlob Frege*, 137–58.

differential quotients and integrals;

and finally the word 'function' has been understood in such a general way that the connexion between value of function and argument is in certain circumstances no longer expressed by signs of mathematical analysis, but can only be signified by words.

Another extension has consisted in admitting complex numbers as arguments, and consequently as values, of functions.

In both directions I have gone still farther.

converses, the various means of transition to the limit have been introduced—to be sure, without people's being always clearly aware that they were thus adopting something essentially new.

People have even gone further still, and have actually be obliged to resort to ordinary language, because the symbolic language of Analysis failed, e.g. when they were speaking of a function whose value is 1 for rational and 0 for irrational arguments.

Secondly, the field of possible arguments and values for functions has been extended by the admission of complex numbers. In conjunction with this, the sense of the expressions 'sum,' 'product,' etc. had to be defined more widely.

In both directions I go still further.

The cryptic allusion in the better-known work to functions only definable colloquially and not symbolically corresponds in the lesser-known work to an explicit citation of the Dirichlet function. This suffices to answer question B in the negative. In the pages that follow these passages, Frege indicates just *how* he "goes still further." One way is in allowing not just real or complex numbers but any objects whatsoever to be arguments and values of a function. The other way is by introducing a new language in which functions previously definable only *colloquially* will become definable *symbolically*. It does seem that for virtually *any* nineteenth-century existence theorem, the (perhaps only colloquial) definition of a specific example or "witness" which is always provided (perhaps only implicitly) in the proof can be explicitly symbolized in (the language of higher-order arithmetic and hence in) *Begriffsschrift*. But this does not suffice to answer question A.

V. DEFINABILITY

On the critical side of the work of his mature years, Frege's objections are mainly directed against nineteenth century psychologism and formalism. His targets include figures as famous as Cantor and as forgotten as Czuber. His arguments include grand ontological principles—a mathematical entity is not to be identified with anything spatial or temporal, and hence not with anything mental or linguistic—and petty idiomatic details. But there is an omission in Frege's arguments.

Frege's arguments bear against views that would *identify* a mathematical object with some mental image or linguistic definition of it, but not against views that would, like twentieth century constructivism or predicativism, make the *warrant* for asserting an existence theorem depend on the provision of some image or definition or other representation of a specific example, without *confusing* the mathematical object in the example with its representation. And the latter kind of view is more relevant than the former kind of view to question A. Arguments based on the choice axiom or cardinality theorems, unlike the kinds of arguments Frege advances, would bear against views of the latter kind as much as against views of the former kind. Now the choice axiom had not yet been published, but the cardinality theorems had, so one needs to explain *why* Frege does not invoke them.

The explanation is *not* that Frege rejected Cantor's results. A sufficient explanation is that Frege (like so many others) was largely *unaware of the bearing* of Cantor's cardinality theorems on the issues that concerned him. If he had pondered that bearing, he would surely have begun by translating Cantorian jargon into Fregean jargon. He would then immediately have seen that the Cantorian greater cardinality theorem says that there are more Fregean "concepts" than Fregean "objects." He would then immediately have seen that this contradicts an axiom of the Fregean system, according to which there is a distinct "object" associated with each "concept," namely, the "class" that is its "extension." He would then surely have gone on to ponder whether or not the Cantorian proof can be reproduced within the Fregean system. He would then surely have seen that it can, and would thus have seen that his system is inconsistent.

Indeed, it was in just this way that Russell, who *did* attempt to collate Cantorian and Fregean notions, discovered his paradox, as he indicates in an autobiographical essay[7] written a half-century after the event (p. 44):

> In June 1901, this period of honeymoon delight came to an end. Cantor had a proof that there is no greatest cardinal; in applying this proof to the universal class, I was led to the contradiction about classes that are not members of themselves. It soon became clear that this is only one of an infinite class of contradictions. I wrote Frege ...

The 1902 letter from Russell to Frege,[8] however, did not give the reasoning leading up to the discovery of his paradox, and Frege remained unaware of the bearing of cardinality estimates in 1903. For after his edifice collapsed in paradox, Frege tinkered for a while with repairs, hastily adding an Appendix to Volume II of the *Grundgesetze*. But Dummett has observed that this can also very easily be seen not to work if one is aware of the bearing of cardinality estimates.

As 1904 and the sixtieth anniversary of Boltzmann's birth approached, Frege had abandoned tinkering with repairs and taken up searching through the rubble for salvageable material for a contribution to a *Festschrift* volume. This occasion piece[9] consists of criticism of certain accounts of what sorts of entities functions are.

---

[7]Bertrand Russell, "My mental development," in R. E. Egner and L. E. Denonn, eds., *The basic writings of Bertrand Russell*, New York: Simon and Shuster, 1961, 37–50.

[8]Dated 16.vi.02. In *From Frege to Gödel: a source book in mathematical logic, 1879-1931*, 124–25.

[9]"What is a Function?" (1904), in *Translations from the philosophical writings of Gottlob Frege*, 107–16.

It identifies in its first paragraph two kinds of definition of function Frege will be criticizing, and correspondingly falls into two parts, though there is no formal break on the page. Frege's criticisms involve the same mixture of objections of grand principle and of petty detail (and the same absence of appeal to cardinality estimates) as in the works of his prime years. There is, nonetheless, in each of the two parts of this work, a half-sentence that does bear on question A.

The kind of definition criticized in the first part is that which, like the definition-*cum*-gloss exhibited earlier, identifies a function with a "variable" quantity, specifically with one such that "depends" on another such. Frege's criticism is directed against the definition proper, and specifically against its mention of "variable quantities." His criticism is partly a matter of ontological principle: The notion of "variation" and hence of time should have no place in pure analysis. (His closest approach to saying anything about Boltzmann's work in thermodynamics comes when he gives an example of what *could* be legitimately called a "variable quantity" in *applied* analysis, the length of a metal rod that expands as it is heated.) It is partly a matter of idiomatic detail: The closing sentence of the first part of the essay notes that if functions were variables, "then elliptic functions would be elliptic variables." (But Weierstrass's ℘, for instance, is always called an "elliptic function," not an "elliptic variable.")

The gloss accompanying the definition is relevant to the issue of the breadth of the notion of function to be adopted in real analysis, but Frege is not interested in it from this perspective. In the first paragraph Frege contrasts the (bad) identification of the function with the "dependent variable" with the (better though perhaps still not quite right) identification of the function with the "mode of dependence." He cites the gloss mainly because it is *only* in the gloss that the "mode of dependence"—or in the terminology of Czuber, his main target, the "law of correspondence"—gets mentioned at all. Nonetheless, it is worth mentioning that Frege does not *criticize* the gloss for what it says about the breadth of the notion of function. In a half-sentence relevant to question A, Frege even offers a gloss on the gloss tending to neutralize any definabilistic connotations attaching to Czuber's use of the word "law":

... *generality* is what the word "law" denotes.

Frege then makes an abrupt transition (p. 112) from his "modes of dependence" or Czuber's "laws of correlation," topic of the first part of the paper, to "equations" and "expressions" and "signs," topic of the second part. He does so in a single sentence asserting that the latter are typically used to express the former:

> Our general way of expressing such a law of correlation is an equation, in which the letter 'y' stands on the left side whereas on the right there appears a mathematical expression consisting of numerals, mathematical signs, and the letter 'x,' e.g.:
> '$y = x^2 + 3x$.'

He then mentions the definition of function he will be criticizing:

> The function has indeed been defined as being such a mathematical expression.

The objections to this formalistic definition that he intends to pursue at length are partly matters of grand principle:

> ... a mathematical expression, as a group of signs, does not belong in arithmetic at all. The formalist theory, which regards signs as the subject-matter of arithmetic, is one that I may well consider to be definitively refuted by my criticism in the second volume of my *Grundgesetze der Arithmetik*. The distinction between sign and thing signified has not always been sharply made ...

and partly matters of petty detail: If one *were* going to consider an expression, the *right* expression to consider would not be the righthand side of an equation like that above, but rather something like:

$$( )^2 + 3( )$$

with blanks instead of letters "x."

But before pursuing such objections, which he takes to admit of no response, he briefly mentions another objection that could be made that he will *not* be pursuing, and one that is clearly relevant to the issue of the breadth of the notion of function:

> In recent times this concept has been found too narrow.

Presumably by way of indicating *why* he will not be pursuing it, he mentions a response it invites, that the difficulty mentioned in the objection is to be avoided by:

> ... introducing new signs into the symbolic language of arithmetic.

This half-sentence, seemingly relevant to question A, requires close examination. A series of questions arise.

(1) Given the convoluted dialectical context, one may ask whether Frege is (a) asserting something as his own considered opinion, or merely (b) not challenging something that might be claimed by a hypothetical opponent. To decide this, one must consider the phrase immediately preceding that just quoted in the original:

> Indessen wäre dieser Übelstand durch ... wohl zu vermeiden.

The translation given by Geach and Black (who did not anticipate the weight that has come to be placed on this passage) is:

> However, this difficulty could easily be avoided by ...

which suggests (a). An alternate translation has been suggested by Hempel:

> However, this difficulty could presumably be avoided by ...

which suggests (b). My own fairly extensive consultations with native speakers of German has tended to confirm that the adverb "wohl" well deserves its reputation as one of the most equivocal in the German language. I have so far found no one prepared to defend its translation in the present context as "easily," and many who have suggested translations akin to "presumably" in that they indicate some subtle degree of guardedness.

(2) Even assuming (1)(a), Dummett has asked whether Frege is asserting (a) that there is a language in which for every function, there is an expression, or merely (b) that for any function there is a language in which there is an expression.

(3) Even assuming (2)(a), one may ask whether Frege is (a) cutting back the notion of "function" to fit a language, or merely

(b) stretching the notion of "language" to accommodate the functions. In orthodox model theory since Tarski theorists generally have considered uncountable sized "languages," and in developing his model for non-standard analysis, Robinson specifically considered the supercontinuum-sized "language" obtained by adding to the symbolic language of arithmetic a new symbol for each real function there happens to be. To anyone aware of the bearing of cardinality estimates, such "languages" are clearly essentially different from humanly *comprehensible* finitary languages like German or *Begriffsschrift*. But might not Frege have had the simple-minded thought that one could add a new symbol for each function there happens to be, "to be sure, without being clearly aware that he was thus adopting something essentially new?"

(4) Even assuming (3)(a), even assuming Frege is asserting as his own opinion that there is a *finitary* symbolic language L in which for every function there is an expression, one must ask what were his *grounds* for this assertion. They cannot be *purely* conceptual: It cannot be built into his very *concept* of function that a function must be definable in a *symbolic* language L, if he accepted the Dirichlet example at a time when it was only defined in *colloquial* language. Presumably his grounds must be at least *partly* inductive. Presumably either (a) definability in *some language or other*, colloquial or symbolic, is built into Frege's notion of function, and by something like induction from his success in providing definitions in *Begriffsschrift* for functions defined in German he has convinced himself that a language L of the kind required could be developed; or else (b) definability as such is *not* built into Frege's notion of function, but by something like induction from his success in providing definitions in *Begriffsschrift* for "witnesses" for nineteenth century existence theorems—whose proofs all do provide at least implicitly and at least in colloquial language definitions of "witnesses"—he has convinced himself that a language L of the kind required could be developed. But which is it?

From an orthodox, classical standpoint, in either case his induction rests on too narrow a base, since he does not consider twentieth century existence theorems—whose proofs do not provide even implicitly even in colloquial language definitions of witnesses. Moreover, his conclusion is false, since presumably there are among the supercontinuum many real functions some that admit no definition in humanly comprehensible language at all. But equally from a classical or a predicativist standpoint, his induction rests on too narrow a base, since he has not considered functions definable in terms of the notion of definability-in-L. Moreover, his conclusion is false, since there are functions definable in the metalanguage L' that do not admit a definition in L. Even under the far from indubitable assumptions (1)(a), (2)(a), (3)(a), the text only shows that Frege has, from *any* standpoint, orthodox or heterodox, made the (very common) mistake of arriving by induction from too narrow a base at a false conjecture. It does not show which standpoint was Frege's own. It does not answer question A.

Frege soon gave up searching the rubble at the disaster site, and left the area of his previous research altogether. In his last years he became absorbed in a program for a geometrical foundation for mathematics. He worked on his program in comparative isolation. If he failed to give what many would consider due attention to the works of Poincaré on relativity and *geometry*, the field in which he was then working, it will be hardly surprising if he failed also to give such attention to the works of Poincaré on predicativity and *analysis*, and the field he had by then abandoned. It was only after the debates among the French and other mathematicians that began in 1905 had run their course that it became more or less established orthodoxy in the mathematical community that functions are not restricted to be definable. To ask about which side Frege *would have* supported if he *had* participated in those debates risks anachronism: Those debates presupposed awareness of the paradoxes; and we *know* what Frege's reaction to the discovery of the paradoxes was: It was to abandon the field altogether.

5

Frege: The Royal Road
from Geometry

I would like to submit this mode of thinking to philosophers,
who otherwise often limit themselves to mathematical trivial-
ities, for a consideration of its principles.

Felix Klein, *The development of mathematics
in the nineteenth century.*

I

Ludwig Wittgenstein posed the question, "What grounds can be
given in favor of our presumption that the proper continuation of the
'even number' series 2, 4, 6, 8 is 10, rather than, e.g., 62?" The tenor
of Wittgenstein's discussion suggests that, however he expected the
worries raised by his examples to be dispersed, their proper "philosophical" solution was not expected to affect the practice of the
working mathematician, any more than the definitive resolution of
skeptical doubts about the external world will force new methodology upon subatomic physics. Within more sophisticated contexts,

From *Noûs*, **26** (1992) 149–80. Reprinted by kind permission of Blackwell
Publishers and the author.

I would like to thank Tom Ricketts, John Bell, Burton Dreben and, especially,
Bill Demopoulos and Ken Manders for helpful comments. Bob Batterman and
Keith Humphrey drew the figures.

analogous worries about the proper continuation of a mathematical practice can assume quite practical dimensions. Such cases occur when questions of "the proper setting" for a mathematical question arise. For example, Leonhard Euler and other early mathematicians claimed that the "sum" of the series $1 - 1 + 1 - 1 + \ldots$ ought to be $1/2$, a "howler" for which they are still roundly mocked by writers of the E. T. Bell school. It is true that the familiar calculus level approach to infinite summation pioneered by A. L. Cauchy does not grant the Euler series a sum, yet it turns out that a variety of generalized notions of summation exist that fully ratify the answer "$1/2$." Viewed retrospectively, these generalized treatments appear to be "better settings" for the mathematical contexts in which Euler originally employed his series manipulations. It is true that Euler never *thought* of these generalized conceptions, but, by the same lights, he never thought of Cauchy's account either. The Cauchy approach is nowadays "central" in that it serves as the springboard for the introduction of the generalized notions, but this consideration alone should not force us to evaluate Euler retrospectively as *making mistakes* in the summation of series (by the lights of the generalized notions, Euler rarely made errors).

The fact that "better settings" for old mathematical problems can be uncovered after a long period of mathematical digestion raises unsettling doubts about mathematical certainty. The mathematician Phillip Davis, in an interesting article that asks whether *any* proposition can be definitively established as "mathematically impossible," writes:

When placed within abstract deductive mathematical structures, impossibility statements are simply definitions, axioms or theorems . . . . [But] in mathematics there is a long and vitally important record of impossibilities being broken by the introduction of structural changes. Meaning in mathematics derives not from naked symbols but from the relationship between these symbols and the exterior world . . . . Insofar as structures are added to primitive ideas to make them precise, flexibility is lost in the process. In a number of ways, then, the

closer one comes to an assertion of an absolute "no," the less
is the meaning that can be assigned to this "no."[1]

That is, although a particular approach to summation can be rendered precise within, say, *Principia mathematica*, the final arbitration of the *proper* truth value for "$1 - 1 + 1 + \ldots$ has a sum" will not rest simply with the "naked symbols" of the selected formalization, but will instead turn upon the question of *which* formal setting best suits the "meaning" of the original mathematical ideas. The discovery of the "best account" of original meaning may not emerge until long after the fact. Davis's worry about the permanence of mathematical certainty (or impossibility) derives from the fact that, although the truth values of many statements transfer unscathed into the revised setting, the truth value of many others do not. It seems impossible to predict the eventual fate of any claim in advance, no matter how unshakable it may now seem. Wittgenstein's worries about the future continuation of "$2, 4, 6, \ldots$" may look as if they may hinge upon questions whether human beings can unambiguously internalize algorithms possessing infinitely many applications, but Davis's worries depend solely upon semantical concerns: when can a formalization be established as "optimally respecting the inner nature" of a certain class of mathematical notions?

"Better settings" for old problems came into their own during the nineteenth century (although the earlier introduction of the negative and complex numbers can be viewed in this light also). It was discovered that a wide variety of traditional mathematical endeavors could be greatly illuminated if their basic questions were reset within richer domains of objects than had previously seemed appropriate. For example, mathematicians had long struggled with the rather intractable formulas for arc length along an ellipse or lemniscate. Niels Abel recognized that it was advantageous to study such functions (or, more properly, their inverses) over the complex numbers, even though the "arc length" of a path stretching between 0 and $2i$ prima facie sounds like a ridiculous notion. Bernhard Riemann then saw that these so-called "elliptic functions" behave in a quite simple

---

[1]Phillip Davis, "When mathematics says no," in Phillip J. Davis and David Park, eds., *No way—the nature of the impossible*, New York: W. H. Freeman, 1987, pp. 176–77.

manner if they are viewed as fundamentally defined upon a doughnut of complex values rather than over the usual complex plane.
Their behavior had originally seemed complicated because the core
functions had been "projected" out of their natural domain into the
"artificial" setting of real values only. It is considerations of this
sort that led Hermann Weyl to proclaim that "Riemann surfaces"
(such as the doughnut) represented the "native land, the only soil
in which the functions grow and thrive."[2]

The claim here is that the relevant mathematics achieves a simple and harmonious closure only when viewed in the proper—often
extended—setting. One should compare such recastings—and the
analogy was frequently drawn by many nineteenth century mathematicians—to the completion of physical explanations by appeal to
unseen molecular structures. When a branch of physics or mathematics is "reset" through an expansion of its ontological territory,
the older explications of technical vocabulary frequently come under review and are modified to suit their reconstituted surroundings.
We noted that the new explications often reverse the truth-values
of apparently well established propositions. Such reversals do not
trouble the physicist much, but they are worrisome to our sense of
the permanence of mathematics. Davis's concern derives from this
source. When he asserts that "meaning does not derive from naked
symbols," he is maintaining, quite correctly, that the "proper" definition of a mathematical term should not rest upon the brute fact
that earlier mathematicians had decided that it should be explicated
in such-and-such manner, but upon whether the definition suits the
realm in which the relevant objects optimally "grow and thrive." In
this respect, Davis's views about mathematical meaning resemble
some of the doctrines about the proper meaning of physical terms
such as "water" or "gold" expounded lately by Saul Kripke and
Hilary Putnam. The doctrine that an unsuspected setting secretly
determines the "true meaning" of certain mathematical vocabulary
might well be labeled *hidden essentialism*, in tribute to its similar-
ities to essentialist claims about "natural kind terms."

Insofar as I can tell, most working mathematicians subscribe,
in their practice, to some form of "hidden essentialism" (although

---

[2] *The concept of a Riemann surface*, Reading: Addison-Wesley, 1955.

they might be reluctant to accept this thesis explicitly). Mathematicians are frequently quite scathing about colleagues who "don't work in the proper setting." The discovery of an optimal mathematical placement can represent the highest form of achievement—I have recently heard it said that, whereas a complete proof of Fermat's theorem might prove mathematically rather insignificant, the older partial results on this problem (such as E. E. Kummer's) are "deep" in that they reveal hidden aspects of the integers that become manifest only when they are imbedded within richer realms of "ideal numbers."

Despite this *de facto* insistence upon the "proper" setting of mathematical notions, the semi–"official" philosophy prevailing among most mathematicians seems insufficient to ground the distinction between "better" and "worse" settings. The "official" position, dominant since the start of this century, maintains that any self-consistent domain is equally worthy of mathematical investigation; preference for a given domain is justified only by aesthetic considerations, personal whim or its potential physical applications. Upon such grounds, how can the mathematician who uncovers the "proper setting" for a traditional problem be distinguished from the mathematician who simply changes the topic? How can one give teeth to the conceit that certain mathematical concepts (e.g., "elliptic function") cry out for *particular* settings in which "they grow and thrive?" With only the tools of the "official" philosophy, preference for "better settings" simply devolves to becoming a matter of "I like it better" or "Isn't this pretty?"

What one would ideally like, as an excellent pair of papers by Kenneth Manders[3] has made clear, is a deeper understanding of mathematical concepts that relates the "meanings" of mathematical concepts to the hidden factors that *drive* disciplines into reconstituted arrangements, regroupings that in some richer way "better respect" the meanings of the concepts involved. Although a richer account of mathematics' hidden essentialism seems desirable, the bare mention of the term "meaning," with all of its attendant vagaries, readily

---

[3]See his "Logic and conceptual relationships in mathematics," in The Paris Logic Group, eds., *Logic colloquium '85*, Holland: Elsevier, 1987, 193–211, and "Domain extension and the philosophy of mathematics," *The journal of philosophy*, **86** (1989) 553–62.

explains why mathematicians, when pressed about their essentialist claims, frequently beat a hasty retreat to the limited verities
of their official "any self consistent formalism is equally worthy of
study" philosophy. Yet, lacking a richer account of continuity of
meaning through change of setting, the very idea of "mathematical
progress" falls prey to Davis's worry—is *every* mathematical claim
potentially subject to some reconstitution of domain that might reverse our current estimation of its truth-value? Let us now turn to
what was probably the most notorious case of "domain extension"
arising within the nineteenth century.

II

After 1820, geometry under the lead of Jean-Victor Poncelet grew
into a much *stranger* subject than could be expected simply from an
acquaintance with schoolbook Euclid. By most standards of comparison, the world of the so-called "projective geometer" is considerably more bizarre[4] than non-Euclidean geometry per se. Why?
The projective school decided that space should contain many more
points and lines than usual. These extra "extension elements" stem
from two basic sources: To any plane, one adds extra points located along a line at infinity and also a full complement of "imaginary points" whose coordinates are allowed to be *complex* numbers.[5]

---

[4]Good accounts of the projective revolution are: J. L. Coolidge, *A history
of geometrical methods*, Oxford: Clarendon Press, 1940, John Theodore Merz,
A history of European scientific thought in the nineteenth century, Vol. II,
New York: Dover, 1965, and Ernest Nagel, "The formation of modern conceptions of formal logic in the development of geometry," in *Teleology revisited*,
New York: Columbia University Press, 1971. Interesting evaluations by Frege's
contemporaries can be found in Felix Klein, *The development of mathematics
in the nineteenth century*, Brookline: Mathematical Sciences Press (no date;
originally published in 1925), M. Ackerman, tr., and the famous "Presidential
address to the British Association, September 1883" of Arthur Cayley (in his
Collected papers, XIII, Cambridge: Cambridge University Press, 1889–97).
        By "projective geometry," I mean "geometry of the school founded by Pon-
celet," a sense in which the metrical attributes of the expanded geometrical realm
are not excluded. A possibly less misleading term would be the old-fashioned
"geometry from a higher point of view," a phrase that properly suggests that Euclidean geometry is not being overthrown or diminished, but simply reorganized.
        [5]The two processes, complexification and projective completion, must be per-
formed concurrently. To avoid undue technicalities, I shall sometimes illustrate

Most of these imaginary points don't even have the grace to hide away at infinity; some of them hover at rather peculiar nearby distances.

The addition of these extra elements would be unproblematic if they merely served as technical devices to streamline proofs pertaining to the original Euclidean realm; many philosophers have characterized the extension elements in such a fashion. Paul Benacerraf writes that:

> [The points "at infinity"] are introduced as a convenience to make simpler and more elegant the theory of the things you really care about.[6]

But the characterization "things one really cares about" is wrong; the extended space *displaces* Euclidean space as the primary focus of interest in complexified projective geometry. For example, one of its great theorems tells us that on any cubic surface exactly twenty-seven lines can be inscribed, but the theorem does not determine which of these lines are "real."

Rather than thinking of the extra points as "conveniences," the projective geometers saw the additions as revealing the "true world" in which geometrical figures live. Familiar figures such as circles and spheres have parts that extend into the unseen portions of six dimensional complex space, so that when we see a Euclidean circle, we perceive only a portion of the full figure (nineteenth century geometers liked to claim that we see the full shape of geometrical figures in the manner of the shadows in Plato's cave). This way of proceeding is still pretty much the norm in algebraic geometry today.

This recasting of geometrical ontology represents one of the most striking manifestations of "hidden essentialism" to be found in the nineteenth century. The projective school regarded these changes

procedures within the affine part of a projective plane.

[6] "Mathematical Truth," *The journal of philosophy*, **70** (1973) 661–80. The quotation is from p. 406 of its reprinting in Paul Benacerraf and Hilary Putnam, eds., *Readings in the philosophy of mathematics*, second edition, Cambridge: Cambridge University Press, 1983. Benacerraf intends this remark as a gloss on David Hilbert's views, who was unlikely to view the extension elements of geometry in the manner described.

as "growing organically"—a phrase drawn from the geometer Jacob Steiner—out of the original world of Euclidean geometry. The question posed in the previous section was: how might we make sense of the claim that such revisions grow "organically" out of prior practice—how can it be meaningfully claimed that geometrical concepts cry out for the extended, projective setting? For a variety of reasons we have tended to avoid serious engagement with these sorts of question in 20th century philosophy of mathematics. Witness a modern geometry text on the topic of the extension elements:

> Questions probably arise immediately in your mind: "What right have we arbitrarily to add points to space? And, even if we may be permitted to do so, must we not differentiate between the 'ordinary' points, which we see as dots on the paper, and the 'ideal points,' which are constructs of our imagination?"

So far so good. Here is the answer provided:

> Planes, straight lines, and points are the entities about which certain postulates are made but which are otherwise undefined. Hence, we can give the name "point" to anything which conforms to our postulates about points and which suits our convenience to designate as "point." In short, *all* our geometry (indeed, *all* our mathematics) deals with constructs of our imagination; and an "ideal" point is on precisely the same footing, from this point of view, as an "ordinary" point.[7]

But it is hard to see from this response why the projective additions are especially "natural" to geometry; why shouldn't we instead add

---

[7]Robert Rosenbaum, *Introduction to projective geometry and modern algebra*, Reading: Addison-Wesley, 1963, p. 16. Historically, "formalist" responses like Rosenbaum's became prevalent among geometers after the publication of David Hilbert's *Foundations of geometry*. One wonders whether some of the cranky animus that Frege displays towards Hilbert in their interchanges may trace to a sense that the philosophical concerns common to Poncelet and von Staudt is likely to be plowed under in Hilbert's approach. Insofar as I am aware, however, this issue is not explicitly raised by either writer. For an interesting later evaluation by a great geometer of the Italian school, cf. Fredrigo Enriques, *The historical development of logic*, New York: Holt, Rinehart and Winston, 1929.

inaccessible cardinals or pineapples to Euclid? Such a weak apology for the extension elements leaves one wide open to Phillip Davis's kind of worry. One might dress up Rosenbaum's answer somewhat, but most modern defenses of the extension elements come down to being simply a matter of "the projective changes are particularly pretty" or the like.

I don't pretend to have a better answer to such worries myself, but many nineteenth century mathematicians thought that they did. A variety of rather sophisticated attempts to rationalize the changes in geometry were set forth. For various historical reasons, most of us today are unaware of these discussions, their importance having been pushed out of view by the later debates over non-Euclidean geometry. But the controversy over the projective revisions is philosophically just as interesting as the non-Euclidean challenge, although completely different in character. The question was not whether established Euclidean theorems of geometry could be reversed by an empirically confirmable non-Euclidean competitor, but whether internal factors might drive Euclidean geometry to reconstitute itself within a richer domain of objects in which various older theorems fail. The projective threat to Euclidean certainty came from within, rather than from without.

The central purpose of this paper is to show that Gottlob Frege's celebrated logicism grew up in the shadow of these debates and that the more obscure parts of his philosophy—his context principle, for example—can be better understood when viewed in this geometrical perspective. Frege's context principle, by way of reminder, is:

> [Never] ask for the meaning of a word in isolation, but only in the context of a proposition . . . . It is enough if the proposition taken as a whole has a sense; it is this that confers on its parts also their content. (*Grundlagen*[8] § 106)

This is one of those slogans that have deeply influenced a lot of famous later philosophers, although few of them seem to agree upon what it means!

---

[8] All quotations from the *Grundlagen* are in J. L. Austin's translation: *The foundations of arithmetic: a logico-mathematical enquiry into the concept of number*, Evanston: Northwestern University Press, 1980.

In any case, the linkage between Frege and geometry is scarcely made obvious within Frege's better known writings. However, Frege did much work in complexified projective geometry and was undoubtedly familiar with the doctrines outlined here. I'll attempt to tell a story here that links the philosophical traditions in geometry to Frege's published opinions.

III

Let us first ask what factors led geometers to place Euclidean geometry in its strange projective setting? We will concentrate upon the *complex* extension points (which are the most interesting). The original impetus for adding these came from the successes of analytic geometry in the fashion of Descartes. Here "analytic geometry" means the treatment of geometrical relations through the use of algebraic manipulations upon equations based upon numerical coordinates. At first blush, the method looks unpleasantly roundabout. Suppose line L cuts a coplanar circle C. The Cartesian method asks that our circle be assigned some equation, say, "$x^2 + (y-1)^2 = 26$." This requires C to be calibrated with two arbitrarily selected coordinate lines X and Y. Likewise, line L must also be tied to X and Y by an equation such as "$y = 0$." The *geometrical* relationship of "intersection" is replaced, in the Cartesian scheme, by the *algebraic* operation of "substitution." Thus if "$y = 0$" is substituted into the equation for the circle and the result is solved, we find the intersection points $\langle -5, 0 \rangle$ and $\langle 5, 0 \rangle$. This algebraic treatment of "intersection" seems cumbersome for two reasons: (i) The procedure relies upon additional figures (the lines X and Y) that seem completely *extraneous*, from a geometrical point of view, to the original figures C and L. (ii) The Cartesian treatment of any iterated series of geometrical relationships is apt to culminate in rather messy algebra.

So much for disadvantages of analytic geometry; what are its *advantages*? The answer is that a *single* Cartesian calculation can often cover a multitude of cases whose traditional (often called "synthetic") treatment would require separate proofs. Let us survey quickly two cases that display this Cartesian economy and, at the same time, introduce some concepts we shall need later.

CASE 1: Let us begin with an intersecting circle C and line L, as above. We have already found the two points $\langle -5, 0 \rangle$ and $\langle 5, 0 \rangle$ that represent L's intersections with C. Now set up the map Φ from $x \rightarrow x'$ along the line defined by the relation $xx' = (\pm 5)^2$. In

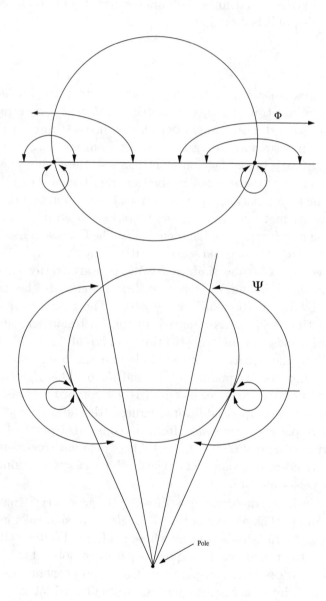

traditional geometrical terms, Φ represents a "harmonic division" of the points $\langle x, 0 \rangle$ and $\langle x', 0 \rangle$ by the intersections $\langle -5, 0 \rangle$ and $\langle 5, 0 \rangle$. A map like Φ is called an *involution* because it has the property that it returns to itself after one cycle. Note how Φ "nests" around the two intersection points, which represent Φ's sole *fixed points*, i.e., places where a point maps to itself under Φ. By a further operation (projecting from the top of the circle), this involution along the line transfers naturally to a second involution Ψ intertwining the points around the circle (the algebra that constructs Ψ from Φ is rather messy and won't be detailed here). The circle map Ψ turns out to have a very simple construction: x will map to x' just in case x' lies at the other end of x's cord in some bundle of lines emerging from a common focus c (in the jargon, c is the "pole" of the line $y = 0$). c has many pretty properties, among them this: Find, by the usual calculus rules, the equations for the circle's tangents at the fixed points. The intersection point of these two tangents will, in fact, prove to be our "pole" c.

CASE 2: In the foregoing procedure, we tacitly assumed that the substitution of the line equation into the circle equation gives real roots, such as the ± 5 of our example. But suppose our line does *not* intersect the circle, as happens if we select the line $y = -5$. Here the Cartesian substitution procedure gives the roots $\pm i\sqrt{10}$, suggesting the ridiculous "points of intersection" $\langle -i\sqrt{10}, -5 \rangle$ and $\langle i\sqrt{10}, -5 \rangle$. But let us retain the courage of our calculations and continue our construction of involution maps. The line map Φ, defined by $xx' = -10$, still represents an involution along the line, but now an *overlapping* map with no apparent fixed points. Along the circle, the induced map Ψ is also an involution, but now constructed from a pole *inside* the circle. The pole and line retain the same pretty properties as in case 1, including the fact that the "pole" is still the intersection of tangent lines to the circle (albeit "imaginary" tangents). Somehow blind algebraic perseverance has led us to sound geometric constructions similar to those in the first case, despite the fact that our intermediate calculations now route through the absurd "intersection points," $\langle \pm i\sqrt{10}, -5 \rangle$. This is one of those cases where it seems, to paraphrase Euler, that our pencil has surpassed our geometrical understanding in intelligence.

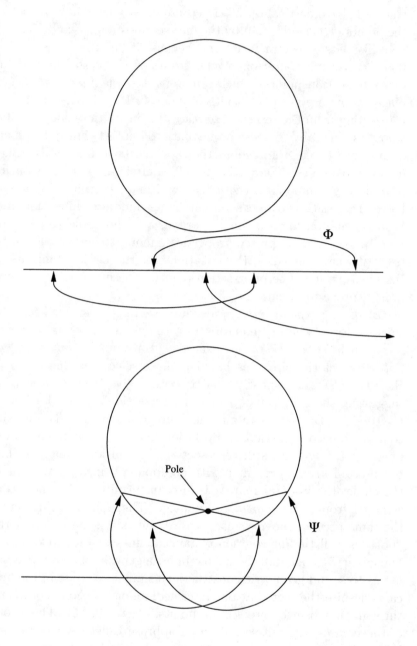

Our Cartesian calculations cover, in a unified setup, two situations that, from a traditional Euclidean point of view, look completely different (and would require completely different proofs to establish). And this advantage maintains itself quite generally—the algebraic techniques unify Euclidean situations that were traditionally regarded as distinct.

Now there were some mathematicians (e.g. Peacock and Boole) who believed that the rules of complex algebra possessed some sort of universal validity as "rules of thought"—these calculations always lead to sound results, in whatever field they are applied—even if the formulas generated by the rules seem completely uninterpretable. On this view, the complex points look as if they were, in Hermann Hankel's phrase, "gifts from algebra." I might remark that one of the best philosophers of my acquaintance was driven to philosophy after having received this kind of "justification" when he encountered complex numbers in his engineering courses!

In any event, this answer was commonly rejected as resting upon a confusion. Charlotte Angas Scott wrote:

> In teaching the elements of analytical geometry we are practically forced to allow, even to encourage, a slipshod identification of the field of geometry with the field of algebra.[9]

And

> It is far too much the custom now to rely on the analogy of algebra to justify the introduction of imaginaries into geometry. Analogy, however, is no justification unless we first prove the exact correspondence of the fields of investigation. In analytical geometry the identification of the two fields is permissible, and is easily explained; but in pure geometry any reference to algebra, expressed or implied, is irrelevant and misleading.[10]

[9]See p. 163 of Charlotte Angas Scott, "The status of imaginaries in pure geometry," *Bulletin of the American Mathematical Society*, **6** (1900) 163–68. See also her "On von Staudt's *Geometrie der Lage*," *Mathematical Gazette*, **5** (1900) 307–331 and 363–70. This is one of the best surveys of von Staudt's thought and, unlike the notoriously turgid original, it attempts to articulate the philosophy behind the method.

[10] "On von Staudt's *Geometrie der Lage*," p. 307.

That is, formulas like "$x^2 + (y-1)^2 = 26$" carry two distinct interpretations: (i) they are "about" circles and (ii) they are "about" the
relationship of various number pairs $\langle x, y \rangle$. The steps in our Cartesian calculation are quite unexceptionable interpreted over the field
of complex numbers, so if geometrical interpretation is eschewed,
there is no danger of unsoundness. It is only under the geometrical
interpretation that the reasoning appears unsound. Scott's complaint is that these two interpretations were often confused.[11]

IV

During the nineteenth century, several noteworthy attempts were
made to explain the successes of the Cartesian calculations in exclusively geometrical terms. The first was due to the great origina-
tor of projective geometry, Poncelet. He agreed that the Cartesian
achievements revealed a deep unity in our two cases but denied that
this unity traced to any "universality of algebra." Rather he gave
reasons for supposing that *hidden elements* existed in Euclidean
geometry—namely, the extension elements. Poncelet pictured the
relationship of cases (1) and (2) in cinemagraphic terms as follows:
Push the line of the diagram of case 1 downward, through the de-
generate middle case, to its final position in the case 2 diagram. As
the line moves downward, its "pole" is pulled inside the circle and
becomes crossed. The induced involutions on both the line and circle gradually change in character from that of being *nested*, as in
the first diagram, to being *overlapping*, as in the last diagram. The
intersection fixed points move together along the line until they fuse
and vanish.

It certainly looks as if some permanent interconnection between
circle and line, rather like a mechanical linkage, persists through
these modifications. In sympathy with this feeling, rather than regarding the fixed points as absent in the last diagram, why not
regard them as simply *invisible*,[12] having been pulled up into a

---

[11]It is possible, of course, to maintain that geometry simply *is* a portion of
complex analysis, a piece, in George Birkhoff's phrase, "dressed up in glitter-
ing intuitional trappings." An early advocate of this point of view was Julius
Plücker.

[12]"But when L moves still further from the center, the points A and B

higher plane by the transformations leading to case 2? The intersection points should be regarded as *always* governing the interactions

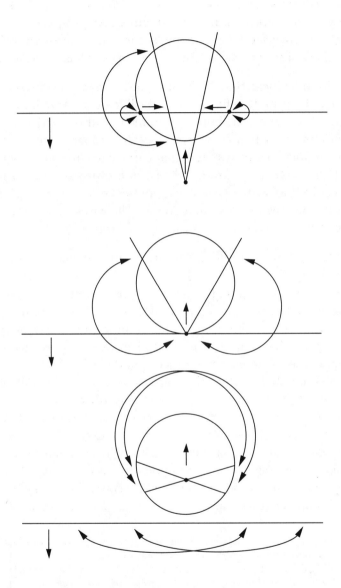

become invisible; yet, for the sake of continuity, we say they still exist, but are invisible or *imaginary*" (J. W. Russell, *An elementary treatise on pure geometry*, Oxford: Clarendon Press, 1893, p. 23).

between circle and line, but they can no longer be "seen" when they become imaginary (although they can always be brought into view by dragging the figure back to case 1). Poncelet argued that in cases when a form persisted, but the accompanying objects vanished, one ought to postulate new elements according to a general principle of so-called continuity or persistence of form. Thus Poncelet:

> Is it not evident that if, keeping the same given things, one can vary the primitive figure by insensible degrees by imposing on certain parts of the figure a continuous but otherwise arbitrary movement; is it not evident that the properties and relations found for the first system remain applicable to successive states of the system ... [even] when those objects cease to have a positive, absolute existence, a physical existence? ... Now this principle, regarded as an axiom by the wisest mathematicians, one can call the *principle* or *law of continuity*.[13]

Obviously it will be difficult to articulate any such a principle in a precise way.

But once the required objects are postulated by the "law of persistence of form," the original success of the algebraic Cartesian methods can be seen as stemming from a kind of *accident*, viz., it so happens that a partial isomorphism exists between the proper, expanded set of geometrical objects and algebraic formula involving complex numbers. But the isomorphism is not completely exact, for the algebraic formulas permit many distinctions that have no inherent geometric significance whatsoever. Poncelet felt that if one directly reasoned geometrically about the extension elements, without taking an unnecessary detour through the realm of the complex numbers, one might improve upon the successes of the Cartesian techniques while avoiding the clumsy crutch of the algebraic methods.[14] Labeling the "hidden elements" of genuine geometry as "gifts

---

[13] Jean-Victor Poncelet, *Traite des proprietes projectives des figures*, in *The history of mathematics—a reader*, John Fauvel and J. J. Gray, eds., Houndsmill: MacMillan Press, 1987, J. J. Gray, trans., p. 543 (order rearranged).

[14] Indeed, in my description of the circle and line cases above, I largely refrained from presenting the gory details of the algebra and only supplied the geometric "effect" that the algebraic manipulations will give (some of these "effects," of course, representing "imaginary" points and lines). Poncelet's "syn-

from algebra" is an injustice akin to crediting Christian, rather than Cyrano, with the authorship of the letters to Roxana. Poncelet's idea was that geometry should be pursued *synthetically*, with the imaginary points postulated following the law of continuity in the same spirit as the existence of parallels is posited in Euclid. The accidental tie to numerical realms exploited in *analytic* geometry should be avoided as an unnecessary diversion.

Note that the "principle of continuity" supplies an answer to the dilemma discussed at the start of this paper. Whatever its shortcomings in terms of clarity, the "continuity" principle represents the new projective elements as the natural outgrowth of the domain of traditional geometry. This "organic" connection through "continuity" gives one the right, thought Poncelet, to claim that the new extension elements represent the *correct* extension of Euclidean geometry. The new elements are added to complete the mechanism that makes Euclidean geometry work. Accordingly, the "ideal" elements are "just as real" as the original elements of geometry, even though we can form no visual picture of what the imaginary elements look like (of course, nineteenth century writers like Frege would have expressed this claim as: "we can form no intuition of the imaginary elements").

V

So much for Poncelet. Between 1847 and 1860, the German geometer G. K. C. von Staudt brought a new perspective to the dilemma of the extension elements.[15] Rather than postulating new entities in Poncelet's manner, von Staudt elected to define the extension elements in terms of accepted Euclidean ideas, so that complexified projective geometry would stand simply as an extension by definitions of the

---

thetic" viewpoint maintains that my algebra-free description of "effects" is the proper way to handle the geometry, whereas the full-fledged Cartesian treatment of our example merely introduces messy complications that trace to the avoidable entanglement of our figures with the "extraneous" coordinate lines X and Y. Selection of fiducial lines X and Y is needed to map geometry into the complex numbers, but Poncelet's contention is that such number assignments are unnecessary.

[15] G. K. C. von Staudt, *Die Geometrie der Lage*, Erlangen, 1847 and *Beitrage zur Geometrie der Lage*, Nuremberg, 1856.

old field. This idea alone would not be novel; von Staudt's originality lay in the particular strategy he followed in achieving this end.
Since there are no regular Euclidean objects that can adequately
substitute for the complex points, von Staudt was forced to rely
upon some unexpected materials. In fact, he selected *abstract concepts* to supply the missing material. His reasoning went roughly as
follows:

"We are familiar with cases where assertions about concepts or
relations are accorded with an object-like status, as when we say
'Personhood is possessed by Socrates,' rather than 'Socrates is a
person.' Consider 'personhood' to be a *concept-object* derived from
the concept 'is a person.' From this point of view, consider the
innocent-looking claim: 'Line L has the overlapping involution (defined by $xx' = -10$) on it.' Once again, we have converted what is
essentially a relation ('x maps to x' by the rule $xx' = -10$') into a
'concept-object' '*the* involution on L defined by $xx' = -10$.' Recall
that Poncelet postulated new points to correspond to the missing
'fixed points' of such a relation on the ground that the involution
relation 'persists' along the line although its normal fixed points
vanish. But rather than following Poncelet and viewing this process
as one of *postulation*, why not simply *define* the desired imaginary
points as the concept-objects denoted by phrases like 'the involution
on L'?"

This is von Staudt's basic idea but, strictly speaking, it doesn't
quite work. We actually need to construct *two* imaginary points
from every involution along a line (each involution has two fixed
points, albeit occasionally "double" ones). Von Staudt evades this
difficulty by correlating each involution with two distinct "senses"—
that is, the two "arrow" directions in which the involution can be
taken to act as a function. Such "involutions with a direction"
can still be regarded as "concept-objects," but concept-objects constructed by more complicated means than simple "personhood" or
"involution."

I have adopted the funny hyphenated term "concept-object" to
represent whatever object it is that we plan to extract logically from
an originating concept. Frege, in his practice, invariably utilizes the
concepts' extensions—that is, sets—as the selected concept-objects.
But he also notes that more intensional "concept-objects" might

have been used in their stead, a topic to which we will return. I believe that Frege is best understood if the actual jump into set theory is postponed until the last possible moment.

In an allied manner, rather than positing that new points exist where parallel lines meet, von Staudt proposed defining a "point at infinity" directly in terms of a concept-object derivable from the relation "x is parallel to L." In this case, however, too many concept-objects will be generated if the criterion of identity for concept-objects is held to follow the usual identification conditions for concepts, for such a procedure will create a distinct concept-object for each concept of the form "parallel to ... ," even when substituents L and L' represent lines parallel to one another. Only *one* concept-object ought to stem from L and L'; otherwise unwanted points at infinity will be produced. If we follow Frege in utilizing the extension of "parallel to L" as the appropriate concept-object, the problem of an overabundance of points at infinity vanishes. But there is no necessity in this choice; a variety of alternative strategies work with "concept-objects" more intensional than raw sets, yet still produce a proper complement of points at infinity. Insofar as I can determine, von Staudt's own discussion is silent upon such questions of intensionality—he simply assumes that a common concept-object can be extracted from the "form" common to all parallel lines. In this case, the desired concept-object possesses a familiar name, viz., "the direction of L," even if someone might wish to debate the question of its exact "intensional" status further (I'll return to these issues of "intensionality" later).

Prima facie, the reconstruction of a thought into "concept-object" format seems as if it ought to be a process falling within the domain of logic. Certainly, many logic texts of the period included sections on "abstraction" that read as if they ought to embrace conversions of this sort. But nobody before von Staudt had tried to define "concrete" objects like points in terms of these abstract entities. Whereas Poncelet believed that new elements should be postulated to meet the demands of "persistence of form," von Staudt saw that the persisting forms themselves, reconstituted in concept-object guise, could serve as the desired extension elements. Complexified geometry is now viewed as a *logical* outgrowth of traditional geometry. Accordingly, we have obtained a second explanation of the "naturalness" of the projective extensions.

In order to establish the claim that projective geometry is a non-creative reorganization of Euclidean geometry, von Staudt needed to supply more than this initial set of definitions. All of the notions utilized in projective geometry must be redefined and its theorems shown to be transcriptions, under the definitions, of regular Euclidean statements (albeit involving concept-objects). Thus von Staudt had to redefine what a line is (to allow for the imaginary ones) and then to redefine "lies upon." Armed with such definitions, claims like "two points always lie upon some line" needed to be proven, even though, in the original Euclidean realm, this proposition is an axiom.

VI

Many 19th century mathematicians felt that von Staudt's explication provided the better rationalization of the extension elements, due to its precision and non-creative character. Frege was almost certainly among their number. I'll postpone my own appraisal of the approach until section XI.

As with Poncelet, these results were *not* taken to show that the ideal points were "fictions" or any less "real" than the regular points. The "real" and the "ideal" form interconnected parts of the hidden structure that lies behind the facts of Euclidean geometry. It is an accidental feature of the way we happen to learn about Euclidean geometry that some of its parts appear to us clothed in intuitive garb and some do not. As a result, we originally detect the presence of the ideal elements only through the rigid relationships they induce upon the visible objects. But another class of beings might directly intuit the imaginaries, rather than the reals.

Such claims have a long tradition in projective geometry, stretching back before von Staudt's time. It was recognized that if one wanted a mental picture of the extension elements, one could do so by shifting the "sensuous appearance" of familiar elements over to the invisible ones.[16] For example, one can use line-like representations to paint a useful picture of the real points at infinity. Although

---

[16]Laguerre was an early proponent of such methods—see J. L. Coolidge, *The geometry of the complex domain*, Oxford: Clarendon Press, 1924.

there is no single visual representation wherein all of the elements of geometry appear simultaneously in intuitive garb, by accepting suitable compromises, any desired element can be pictured at the expense of others. Von Staudt's innovation was to recognize that, as long as complex points were to be "invisible" anyway, why not allow them to be *abstract*? Or, more exactly, why not hold that the distinction between "abstract" and "concrete" object is somewhat conventional, depending upon our epistemological point of entry into the relevant domain of knowledge?

It is important to recognize that Frege frequently endorsed theses of this sort. In § 26 of the *Grundlagen*, Frege claims that two mathematicians might fully agree on the content of projective theorems, despite the fact that one "intuits" geometrical points as "looking like points" and the other as "looking like lines."[17] In his doctoral dissertation, Frege recounted some experiments, in the style of Laguerre, that transferred the intuitive appearance of lines to selected imaginary points.

Accordingly, one should not be mislead by Frege's frequently repeated claim that "the only source of geometrical truth lies in geometrical intuition." By this, Frege means only that our entire body of geometric knowledge stems from facts reported to us by "intuition." In agreement with Poncelet, Frege holds that geometrical data bear no essential relation to any facts about numbers except through "accidental" isomorphisms. On the other hand, although all of the *facts* of geometry must be reported to us in intuition, it does not follow that all of the *objects* proper to geometry must appear in intuitive garb—in particular, the "invisible" complex points will not. Although the corpus of geometrical facts is fixed, their contents may be carved up to reveal unexpected objects such as the imaginary points, which logic is able to extract from geometrical fact in the guise of von Staudt's concept-objects.

Recognizing that Frege views geometry in this fashion gives a deeper reading to assertions like:

---

[17]Compare: "[S]tereometric ideas can be correctly comprehended only when they are contemplated purely by the inner power of the imagination, without any means of illustration whatever"—Jacob Steiner, quoted in Theodore Reye, *Lectures on the geometry of position*, New York: MacMillan, 1898, T. F. Holgate, trans., p. xiii. Echoes of these claims can be found in Descartes.

Time and time again we are led by our thought beyond the scope of our imagination, without thereby forfeiting the support we need for our inferences. Even if, as seems to be the case, it is impossible for men such as we to think without ideas, it is still possible for their connection with what we are thinking of to be entirely superficial, arbitrary and conventional. (*Grundlagen* § 60)

I suggest that the manner in which the "unimaginable" complex elements "support" the otherwise uninterpretable steps in the Cartesian reasoning pattern discussed above provides a nice exemplar of what Frege has in mind.[18]

On such a reading, Frege intends to reject not only "psychologism" in the usual sense, but the deeper presumption that definitional practice in mathematics must respect the "intuitive trappings" of a mathematical object, even in a field such as geometry. Although Frege had not drawn the tone/sense/reference distinction at the time of the *Grundlagen*, these geometrical examples suggest that much of what we might naively regard as the "sense" of a mathematical assertion would in fact have been assigned by Frege to "tone" (or some other "non-objective" category). Likewise, although the "sense" of a proposition can always be computed as a function of the "senses" of its component phrases, such a decomposition should not be regarded as unique—witness the constant propositional content shared by an orthodox Euclidean statement and its reparsing in terms of extension elements. Thus I agree with the commentators who feel Frege had a "holistic" conception of propositional content, rather than one structurally constructed from the senses of its component parts.

Another aspect of von Staudt's approach had long standing antecedents in geometry—viz., the tendency to conceive of objects in their concept-like aspects and vice versa. Thus Julius Plücker noted the following: in suitable (homogeneous) coordinates, the equation for a line looks like $Ax + By + Cz = 0$. If a point (a, b, c) lies upon the line, it satisfies the equation, i.e., $Aa + Bb + Cc = 0$. We normally regard this as a case of the *function* (frequently called a

---

[18]Von Staudt's discovery, sketched below, that Cartesian reasoning can prove valid in the absence of a metric illustrates a further implicature of this quotation.

"form") determined by the coefficients A, B and C acting upon the point-object (a, b, c). Plücker realized that the action could be also seen in reverse: a, b and c are viewed as coefficients in a function now operating on the *line*-object (A, B, C). That is, the common thought "$Aa + Bb + Cc = 0$" can be carved up into distinct functions and objects, depending upon one's point of view.[19] The terms A, B and C represent either a function or a line depending upon how the thought is carved up—in Frege's terminology, whether "(A, B, C)" is regarded in an "unsaturated" or "saturated" way. This insight of Plucker's proved very important in unraveling the mysteries of so-called "duality" in projective geometry (I might mention that the purely synthetic work of Jacob Steiner, in rather different ways, also contributed to the tendency to look for the function-like qualities of geometrical objects such as points, lines and circles).

In general terms, nineteenth century geometers tended to conceptualize mathematical entities in terms of the *roles* they play with respect to their surroundings. The intersections of a circle and line form the basis around a wide variety of other geometrical constructions hover (viz., circular and linear involutions, pole and polar pairs, etc.). When the same clustering of behaviors occur in the apparent absence of the "fixed points" that generate the constructions, it is proper, claimed Poncelet, to postulate the missing "intersections." Von Staudt would have said: The fixed points *evaluate* the behavior of the other constructions in their neighborhood. If we continue to make identical evaluations in the absence of regular points, it is permissible to regard the evaluating concept itself as a new "object," for it serves the same basic purposes that regular objects do. Frege's own discussions of "objecthood" often carry a similar "role playing" emphasis.[20]

---

[19] The tangent and cotangent bundles of a manifold exhibit this dual behavior nicely. Such examples, I believe, are very helpful in unraveling what Frege intended by "the multiple analyzability of a thought." For work in Plücker's vein, see Frege's "Lecture on the geometry of pairs of points in the plane," (1884), in Brian McGuinness, ed., Gottlob Frege: *Collected papers in mathematics, logic and philosophy*, Oxford: Basil Blackwell, 1984.

[20] It is worth mentioning that least one of Frege's stock examples of a non-material object shares some of this evaluative character, viz. the axis of the earth (*Grundlagen* § 26). A theorem of rigid body mechanics says that, at each moment, the earth's movement can be decomposed into a translation and an

While on the subject of von Staudt's contributions, he also worked out an isomorphism that aligned a system of coordinate names such as $\langle 6 + i, 9, -2i \rangle$ with his expanded set of geometrical points. Here von Staudt achieved a remarkable gain in clarity—he showed that a mapping to so-called "projective" coordinates can be set up *without* relying upon the metrical attributes of the underlying geometry. In so doing, he was able to not only account for the successes of analytic geometry (when the complex calculations mirror genuine geometrical relations), but also for its failures (e.g., why the principle of duality fails when metrical relations are involved). The realization that coordinate numbers can be meaningfully assigned to geometry without recourse to any notion of "magnitude" (= metric) was, I believe, an important stimulus to Frege's views of number.

VII

These geometrical considerations give us some sense of the freedom to rework preexisting mathematics that Frege wished to defend. This freedom supplies a way of legitimating the projective revisions without destroying geometry's claims to permanence and certainty. Let's now turn to Frege's own work on the classical number systems.

Frege probably came to his logicism by wondering how one might determine whether algebraic calculations involving complex quantities will supply correct results within a particular field of endeavor—remember our perplexed philosopher/engineer! Rather than seeing the rules of complex algebra as unquestionable "leading principles," capable of extracting sound results from any subject matter whatsoever, Frege viewed the general status of complex calculations in the terms that Poncelet and von Staudt advocated for geometry—the manipulations work only if the field of complex numbers isomorphically mirrors structures pre-existent in the subject under inves-

---

instantaneous rotation about an axis, but this axis will usually wander with respect both to the surface of the earth and to an inertial frame. In short, the laws of mechanics define an "evaluator" that, at each instant, isolates a spatial line to serve as the present position of the abstract continuant, "axis of the earth." Frege's holistic views of "inertial frame" are interesting in this respect. See his "On the law of inertia" (1885), reprinted in English translation in Gottlob Frege: *Collected papers in mathematics, logic and philosophy*.

tigation.[21] The question becomes: exactly what sort of structure is required for the valid application of reasoning involving complex numbers?

In Frege's time, it was frequently presumed that the notion of *real number* is somehow "abstracted" from the notion of distance along a geometrical line. Likewise, the notion of *complex quantity* "abstracts" from vector magnitude within a two-dimensional Gaussian array. The trouble with "abstraction," as is familiar from the stock problems of delineating a unique concept by merely ostending a group of objects, is that it is often unclear which features are retained and which abandoned after the "abstraction." The stock answer presumed that some bare notion of "magnitude" was the vital residue left after the abstraction. As we noted, von Staudt showed that useful calculations with complex numbers work within a geometry that contains no notion of magnitude whatsoever—so something had to be wrong in the "abstraction" view.

Frege concluded, in agreement with several other figures of his time, that real and complex numbers serve only as a means to mark the places that given relations occupy within larger families of allied relations, families structured by relations of "addition" or "composition" and containing selected unit relations. (By a family of relations, I mean something like the collection; "x is 1 ft from y," "x is 2 ft from y," etc.) Two relations within two different fields—within the fields of *distance* and *angle*,[22] say—should be assigned the same

---

[21] Frege published several papers that demonstrated how a variety of unlikely mathematical domains could be coded into complex numbers in such a way that standard analytic techniques prove sound over the encoded domain. See "Methods of calculation based on an extension of the concept of quantity" (1874) and "Lecture on a way of conceiving the shape of a triangle as a complex quantity" (1878), both reprinted in English translation in Gottlob Frege: *Collected papers in mathematics, logic and philosophy*.

[22] See "Methods of calculation based on an extension of the concept of quantity." Note also the following passage: "According to the old conception, length appears as something material which fills the straight line between its end points .... [But on a modern conception] all that matters is the point of origin and the end point; whether there is a continuous line between them ... appears to make no difference whatsoever; the idea of filling space has been completely lost .... The concept [of quantity] has thus gradually freed itself from intuition and made itself independent. This is quite unobjectionable, especially since its earlier intuitive character was at bottom mere appearance .... If, as we have shown, we

real number only if the two relations correlate under the unique isomorphism between the fields that preserves "addition" and aligns the two unit relations. The important project, then, is to isolate what sort of structure a field of relations must have in order that, e.g., the real numbers can be mapped onto them.

Accordingly, rather than regarding r as "abstracted" from concrete distances, why not view the construction in the crisp manner of von Staudt—i.e., r is the concept-object derived from the concept of two relations falling at the same place within their respective families? Or, alternatively, a real or complex number serves as an "evaluator" of position within a family of relations. Such a story gives a cleaner account of the intellectual processes involved in constructing number systems than is suggested by vague appeals to "abstraction." We noted that contemporaneous logic books usually treated this as a process within their ken; Frege instead saw "abstraction" as a hazy perception of the precise logical process involved in converting a thought involving a concept into a thought involving a concept-object (Frege's codification of this logical move became his notorious Axiom V—the axiom that leads to the paradoxes). In my opinion, Frege's logicism starts (although it doesn't conclude) with this assimilation of number to von Staudt's treatment of the extension objects within geometry. Numbers are the concept-objects logically induced by the comparative study of families of relations.

It is always important, I think, to keep Frege's uniform approach to the higher number systems firmly in mind, for it is only within this wider context that delimiting the proper range of application of a number system seems a genuinely urgent project. Few of Frege's contemporaries would have puzzled whether reasoning involving integers alone was likely to prove sound, but it was quite otherwise with the complex numbers. We noted the rather mysterious successes of complex reasoning in analytic geometry; such inexplicable triumphs (coupled with some equally mysterious failures) occurred across the entire face of nineteenth century science. We tend to miss

do not find the concept of quantity in intuition but create it ourselves, then we are justified in trying to formulate its definition so as to permit as manifold an application as possible" (p. 56 of the English reprinting, p. 1 of the original).

On Frege's approach to the reals, cf. the chapters by Peter Simons and Michael Dummett (13 and 14, respectively) below.

the *urgency* of the general application problem if we only read the *Grundlagen* (as opposed to the earlier writings where an especial concern with complex reasoning is more marked). Such considerations lead me to presume that Frege must have roughed out his basic approach to the higher numbers relatively early in his development.

VIII

If these suppositions are correct, then a variety of otherwise mysterious aspects of Frege's philosophical remarks in the *Grundlagen* can be clarified. Many commentators have found the discussion leading up to the definition of "natural number" to be puzzling. Hans Sluga writes:

> But there is something surprising and disturbing about the definition of number in terms of extensions of concepts in the general context of Frege's thought. He had originally reasoned that numbers as logical objects had to be defined contextually [at the start of part IV] .... But the conclusion of that section was that the attempted definition which fulfilled that condition could not legitimately be adopted.[23]

But if numbers are concept-objects induced in the von Staudt manner over appropriate families of relations, this confusion about "contextual definition" can be sorted out. That Frege has the von Staudt analogy in mind is suggested by the statement he provides of his overall objectives:

> Our aim is to construct the content of a judgement which can be taken as an identity such that each side of it is a number. We are therefore proposing not to define identity specially for this case, but to use the concept of identity, taken as already known, as a means for arriving at that which is to be regarded as being identical. Admittedly, this seems to be a very odd kind of definition, to which logicians have not yet paid enough attention; but that it is not altogether unheard of, may be shown by a few examples. (*Grundlagen* § 63)

But what are the "not altogether unheard of" examples that Frege brings forth? In fact, his chief example turns out to be von Staudt's

---

[23] *Gottlob Frege*, London: Routledge and Kegan Paul, 1980, p. 127.

definition of the points at infinity in terms of the relation "x is parallel to y." Unfortunately, Frege chooses to characterize the account as providing a definition for the object "the direction of line L" rather than "the point at infinity corresponding to L." There are reasons why Frege wrote in this fashion, but the unsuspecting reader is scarcely alerted to the geometrical significance of the "direction" concept-object.

In general, Frege often makes puzzling allusions to the radical nature of his "logical" procedures, as when he writes:

> Pure thought, irrespective of any content given by the senses or even by an intuition *a priori*, can, solely from the content that results from its own constitution, bring forth judgments that at first sight appear to be possible only on the basis of some intuition. This can be compared with condensation, through which it is possible to transform the air that to a child's consciousness appears as nothing into a visible fluid that forms drops.[24]

What does this have to do with his account of numbers? In von Staudt's treatment of geometry, our logical ability to reconstitute judgments in terms of concept-objects really does give the appearance of pulling invisible points out of empty air. Assimilating Frege's treatment of number to von Staudt's approach, Frege's radical metaphors begin to seem more appropriate. Certainly, von Staudt's trick of granting abstract concept-objects such as "the direction of a line" with the same mathematical "reality" as points at infinity was widely regarded, in the mid-nineteenth century, as a mathematical gambit "bordering upon impudence."[25] Frege liked the gambit, and used it for his numbers, but in the numerical context the shock value of the move was greatly diminished. Only by recognizing that Frege's "directions" are really "points at infinity" can one regain

---

[24] *Begriffsschrift* § 23, p. 55, in Jean van Heijenoort, ed., *From Frege to Gödel: a sourcebook in mathematical logic, 1879-1931*, Cambridge: Harvard University Press, 1967, Stefan Bauer-Mengelberg, tr.

[25] See Hans Freudenthal, "The impact of von Staudt's foundations of geometry," in P. Plaumann and K. Strambach, eds., *Geometry—von Staudt's point of view*, Dordrecht: D. Reidel, 1981. This excellent article is highly recommended to interested readers.

an appreciation of the radical work that the context principle was meant to accomplish within mathematics.

The apparently conflicting strands of approval and disapproval of "contextual definitions" found in Frege's thought trace to this: Any clear-cut property can serve as the grounds for extracting a definite concept-object. As such, the concept-object can be expected to possess the full range of properties natural to abstract objects—it must represent a "self subsistent object, subject to the usual laws of identity" and the other features that Frege so often emphasized. On the other hand, these abstract objects do *not* stand to regular points and lines in the spatial relations needed in projective geometry. As long as we adhere to the original meaning of "lies upon," concept-objects like "the direction of L" or "the involution Φ" will lie upon no lines whatsoever—abstract objects do not possess spatial locations. Yet we saw that phrases like "p lies upon L" can be *redefined* so that all the claims of complexified projective geometry are validated for concept-objects. But these further definitions must be done piecemeal—it is this piecemeal quality that replaces what used to be regarded as "contextual definition."

In sum, Frege's puzzling ambivalence about "contextual definition" should be understood as a recognition that von Staudt's procedures do justice to the insights displayed in standard applications of "contextual definition," while simultaneously avoiding the false illusion they engender of unbridled creativity over the original domain.

On this reading, Frege's "context principle" can be seen as serving two purposes: (1) to support the claim that conversion of a thought to one involving a concept-object does not change the underlying thought, despite the different picture of the thought that the reorganization may convey; (2) to emphasize that concept-objects should be regarded as possessing the same "objective" status as any other mathematical object. One gets a better feel for the radical import of these claims within von Staudt's original framework, where the "intuitive contents" of geometrical propositions seem to be disregarded in a surprisingly high-handed way. Numbers being rather remote objects to begin with, Frege's applications of von Staudt's methods do not trample upon the claims of expected arithmetical content in so shameless a fashion.

IX

These observations are not sufficient to carry us completely to Frege's logicism. Although the bare creation of numerical concept-objects is sanctioned by pure logic, this does not entail that their non-trivial behavior—e.g., how the "addition" relation acts—is determined by logic alone. Such facts will be determined by the behavior of the basic objects that make up the extensions of the relevant concept-objects. In von Staudt's setting, the non-trivial behavior of the points at infinity—e.g., which lines they lie upon—is determined by the behavior of the objects—the parallel lines—lying in the extensions associated with the points. But notice this: the behavior of the points at infinity is completely dictated by the behavior of a much smaller *subset* of objects than are found in the complete extensions. Suppose we want to know whether three points at infinity a, b, and c lie upon a common line. Rather than worrying about *all* of the parallel lines definitionally associated with a, select the single line L_a passing through the origin as representative. We can likewise concentrate upon the representatives L_b and L_c. a, b, and c can now be determined to be collinear or not by examining the relationships among L_a, L_b and L_c. In short, our points at infinity will behave properly if the selected "representatives" do; we can ignore what happens with the rest.

Consider any of the standard systems of numbers, say the natural numbers. Any system of relations that spawns these numbers as concept-objects can be considered as a *representative* of the collection of numbers. Frege had the brilliant idea that one might be able to construct "representatives" for each of the classical number systems within the ontology of pure logic alone. For the natural numbers, his "representative structure" is well-known; his approach to the reals (and, presumably, the unpublished treatment of the complex numbers) follows a similar pattern. By the observations just made, the behavior of these logical representatives is sufficient to induce the correct behavior on the numerical concept-objects—we don't have to look at *all* systems of relations that count as "representatives" of our numbers.

From this point of view, geometry, rather than logic, could have been selected as the basis for inducing the correct behavior upon

Frege's numbers. Epistemologically, however, it is preferable to utilize a logical representative, because logic alone fixes how such objects act, whereas *both* logic and geometrical intuition are needed to determine how geometrical entities behave. The existence of logical exemplars for each of the classical number systems means that the facts about these numerical concept-objects can be settled by appeal to the truths of logic alone—no higher source of knowledge is needed to get the numbers behaving properly. Frege discussed what might happen if the imaginary number *i* were defined in a deviant manner in terms of some concrete object such as the moon or a second of time:

> Against the particular sense we have proposed to assign to "*i*," many objections can of course be brought. By it, we are importing into arithmetic something quite foreign to it, namely time. The second stands in absolutely no intrinsic relation to the real numbers. Propositions proved by the aid of real numbers would become ... synthetic judgments, unless we could find some other proof for them or some other sense for *i*. We must at least first make the attempt to show that all propositions of arithmetic are analytic. (*Grundlagen* § 103)

Note that the main objection is that, if an interval of time is selected as a referent for "*i*," synthetic facts about time must be cited in order to show that the complex numbers behave properly. This specialized reliance on synthetic facts ought to make one worry whether complex reasoning could serve as adequate "leading principles" in domains of inquiry that possess no analogue of the required synthetic facts. Why should one think that Dirchlet's results on the distribution of primes, obtained using reasoning over the complex numbers, have any validity, given that no "time" is found among the integers? Or that Clebsch's complex techniques in elasticity have any validity, given that this branch of physics involves no temporal processes either? Frege put the point this way:

> All we need to know [with respect to a system of numbers] is how to handle logically the content as made sensible in the symbols and, if we wish to apply our calculus to physics, how to effect the transition to the phenomena. It is, however, a

mistake to see in such applications the real sense of the propositions; in any application a large part of their generality is always lost, and a particular element enters in, which in other applications is replaced by other particular elements. (*Grundlagen* § 16)

X

On this reconstruction of Frege's logicism, it is not part of the argument that the natural numbers, under analysis, should turn out secretly to represent *unique* logical objects. There is little suggestion in Frege of the quasi-Russellian[26] claim that the number 3 *must* be identified with the collection of all 3-membered concepts in order to embrace all of 3's potential applications. Indeed, this claim looks plausible only if one is tempted to suppose that 3 represents an "abstraction" from 3-membered sets, as Frege assuredly did not. More generally, the Frege/von Staudt approach to definitional practice is rather sophisticated and does not obviously succumb to the familiar difficulties scouted by Benacerraf.[27]

This latitude in the treatment of number may surprise many of Frege's readers, due to such passages as:

When we speak of "the number one," we indicate by means of the definite article a definite and unique object of scientific study. There are not divers numbers one, but only one. (*Grundlagen* § 38)

But Frege's point is simply that in a reconstruction of arithmetic only a single number 3 ought to be set up, not that a wide variety of objects couldn't have been selected as suitable. Looking back at von Staudt, one recognizes that imaginary points could have been equally well defined in terms of involutions around circles rather than along lines, a fact that Frege explicitly notes.[28] But one should not

[26] Despite tradition, Russell didn't seem to claim uniqueness either, cf. Chapter XI of *The principles of mathematics* (1903), New York: Norton, 1964.

[27] "What numbers could not be," *The philosophical review*, **74** (1965) 47–73, reprinted in Paul Benacerraf and Hilary Putnam, eds., *Readings in the philosophy of mathematics*, second edition.

[28] "On a geometrical representation of imaginary forms in the plane," (1873) in Gottlob Frege: *Collected papers in mathematics, logic and philosophy*, p. 2

define imaginary points simultaneously in terms of *both* kinds of involution, for too many imaginary points will then be produced.

I stress these facts because it is easy to become confused about two different methodological strands that are potentially involved in constructions like Frege's. On the one hand, we have the von Staudt idea that the work of "concrete" mathematical entities can be duplicated by more abstract "concept-objects." On the other, there is the doctrine that set theory can be used to free particular mathematical objects from the harness of particular "representational forms." For example, there is a sense that, under Gauss's approach to congruence, the objects "3 mod 7" and "10 mod 7"[29] are the "same," albeit tied to two different numerical representations. Richard Dedekind proposed that Gauss's "congruence numbers" could be rendered "representation free" by defining them as equivalence classes such as $\{3, 10, 17, \dots\}$. This practice is ubiquitous in modern mathematics– the standard definition of "manifold" liberates geometrical objects from the fetters of particular coordinate choices by constructing the manifold in terms of a complete "atlas" of coordinate charts. However laudable the desire to construct mathematical objects in "representation free" terms may be, it is not the same impulse that motivates von Staudt's practice. A classic case where the two motivations run together can be found in Dedekind's 1872 recasting of Kummer's "ideal numbers" in terms of the sets algebraists still call "ideals." Kummer merely gave rules in "implicit definition" fashion for calculating with his "ideal numbers"; one hoped that a generalized notion might apply in more general contexts. Dedekind found the generalization by recognizing that the "persisting forms" of Kummer's method could be treated as "concept-objects" in von Staudt's manner. But Dedekind also sought to make his "ideals" representation free, which forced them to become infinitary in a way they otherwise didn't need to be. A rather bitter debate ensued between Kronecker and Dedekind over the virtues and demerits of this "representation free" infinitude.[30]

(p. 4 of the German original). The alternative approach to imaginary points that Frege cites is that of circles and lines combined in pairs, rather than a second form of involution.

[29]That is, 3 and 10 leave the same remainder on division by 7.

[30]For an excellent discussion of the criticisms of Dedekind, see Harold M.

The standard approach to the points at infinity constructs them as extensions consisting of infinitely many lines. This infinitude can be avoided if specific "representations" of each direction are selected instead, in the manner discussed several paragraphs back. That is, points at infinity are now defined in terms of the concept-objects derived from "parallel to L," where the range of L is restricted to the lines running through some selected origin. Such a "representation bound" treatment would lead to considerable inconvenience in the later definitions of concepts such as "lies upon," but at least the technique avoids appeal to infinite totalities. If we wish, the concept-object "the direction of L" could be credited with whatever intensionality we might like without necessarily engendering too many points at infinity. It is important to observe that, in the *Grundlagen*, Frege frequently expressed some ambivalence about his employment of extensions:

Edwards, "Dedekind's invention of ideals," *Bulletin of the London Mathematical Society*, **15** (1983) 8–17, and "The genesis of ideal theory," *Archive for history of exact sciences*, **23** (1980) 321–78. In this essay, I have distinguished von Staudt's general contention that abstract "concept-objects" might serve as adequate replacements from more concrete seeming entities from the more specific proposal that equivalence classes should serve as the "concept-objects." Insofar as I can determine, von Staudt is unspecific about the exact nature of the entities he utilizes; the earliest explicit use of equivalence classes seems to be Dedekind's 1881 reconstruction of Kummer's ideal numbers. One would be curious whether Frege was aware of Dedekind's work at the time of the *Grundlagen*, although Charles Peirce claimed that the equivalence class idea was in the mathematical air long before Dedekind (cf. his *The new elements of mathematics*, Vol. III/1, The Hague: Mouton, p. 130). Much work needs to be done on how the chains of influence might have run in this period (for valuable speculations, see the paper by Freudenthal cited earlier). Not all uses of equivalence classes, by the way, seem as philosophically radical as the von Staudt applications; for example, if one regards the Gaussian congruence classes as carving new "modular arithmetics" out the regular integers. In this kind of case, there is no pressing need to tell a philosophical story *à la* von Staudt that the new domains constructed "have the same reality" as the basic set of objects over which the construction proceeds. Dedekind clearly conceived all appeals to equivalence classes, even in von Staudt-like contexts, in the benign light of the Gaussian paradigm (which, historically, he was among the first to advance). In essence, the thesis of this paper is that many of the philosophical concerns in the *Grundlagen* have been misunderstood because we now tend to view *all* uses of equivalence classes from an uncritical point of view rather like Dedekind's.

In [my] definition[s] the sense of the expression "extension of a concept" is assumed to be known. This way of getting over the difficulty cannot be expected to meet with universal approval, and many will prefer other methods of removing the doubt in question. I attach no decisive importance even to bringing in the extensions of concepts at all. (*Grundlagen* § 107)

Insofar as I can determine, Frege was willing to concede that logic might afford other ways of extracting concept-objects from concepts that result in objects more "intensional" than extensions. Unfortunately, the usual vagaries in the notion of "intension" make their proper criteria of identity less than clear. Frege apparently selected sets as the concept-objects utilized in his von Staudt-like constructions because the conditions of set existence (i.e., Axiom V in the *Grundgesetze*) seemed relatively uncontroversial, even if, from a logical point of view, extensions constitute objects of secondary importance. His views about this topic may have changed as his views about the extensionality of mathematical language[31] developed on route to "On sense and reference."

In the recent philosophical literature (e.g., Stein[32]), Frege is sometimes chastised as an example of a mathematician who, driven by

---

[31] To be precise, Frege might have doubted whether "function" could be properly treated in the set-like way familiar from modern courses on "functions of a real variable." In the early 1880's it was more common to presume that functions, while extensional in their identity conditions, enjoy a richer, non-set-like character such as they are supplied in the context of "functions of a complex variable." By the time of the *Grundgesetze*, "function of a real variable" was well on its way to being widely accepted, so Frege may have relaxed his qualms about sets (of course, many logicians in Frege's time would continue to object to the prevailing "extensionality" among mathematicians). For a good history of the analysis side of these issues, see Ugo Bottazzini, *The higher calculus: a history of real and complex analysis from Euler to Weierstrass*, Berlin and New York: Springer-Verlag, 1986, W. V. Egmond, trans. For a contemporaneous expression of attitude, cf. Arthur Cayley, "Functions," in his *Collected papers*, XIII Cambridge: Cambridge University Press, 1889–97. William Demopoulos is currently investigating these difficult matters; see his Chapter 3, above, as well as John Burgess's Chapter 4.

[32] Howard Stein, "Logic, logos, and logistiké," in W. Aspray, and Phillip Kitcher, eds., *History and philosophy of modern mathematics, Minnesota studies in the philosophy of science*, Vol. XI, Minneapolis: University of Minnesota Press, 1988.

"bad" philosophical motives, became committed to an impossible search for the "true essence" of number, whereas the "good" Dedekind recognized that mathematicians should be concerned only with the "structures" of number systems. There are a variety of ways in which this appraisal misrepresents the positions of the two writers, but, at a minimum, it should be recognized that Dedekind, in his demand for "representation free" accounts of mathematical objects, was somewhat more "absolutist" about mathematical constructions than the historical Frege seems to have been. I imagine that Frege would have found that, at first blush, "representation free" constructions count as aesthetically more pleasing than their competitors, but that this virtue might pale if the resulting constructions become overly complicated (as would happen if the "representation free" demand forces geometers to include both line and circle involutions within the definition of imaginary point).

While on this topic, we might examine an objection to Frege's procedure that George Boolos makes in an important article on Frege's approach to numbers:

> The well-known comparison that Frege draws ... between "the direction of line L" and "the number belonging to concept F" is therefore seriously misleading. We do not suspect that lines are made up of directions, that directions are some of the ingredients of lines. ... The principle that directions of lines are identical just in case the lines are parallel looks, and is, trivial only because we suppose that directions are ... distinct from ... the things of which lines are made. (p. 248 of Chapter 9, below)

In point of fact, in the von Staudt universe some lines—e.g., the lines at infinity—are "made up of directions." Even worse, it is assumed that concepts can apply to the very concept-objects manufactured from them. For example, the original motivation for constructing relation-objects to correspond to the involutary relation "$xx' = -10$" was to supply two "points" that become mapped into themselves by the original relation. In the nineteenth century, concepts (or functions) were not regarded as intrinsically tied to the objects in the domain in which they are first encountered; otherwise it makes no sense to maintain that an elliptic function's "true home" is on a Riemann surface. In this sense, one might wish to resist the wholly extensional point of view that makes the domain of a function constitutive of its very identity.

On the other hand, this point of view seems prima facie at odds with the recognition that the meanings of "line" and "lies upon" must genuinely alter when the shift to the extended domain is made. As is well known, Frege thought the practice of using a common symbol "." to designate the two "multiplication" operations over the reals and the complex numbers was a mistake, encouraging mathematicians to believe that they had proved facts for the complex realm which had really been proved only for the reals. However, as long as the meaning of the "persisting form" "$xx' = -10$" is treated in sufficiently abstract fashion, the two viewpoints can be reconciled. Frege's strictures on definability do *not* entail that one cannot meaningfully inquire whether the triple of objects ⟨number, number, operation⟩ satisfy the form "$x\zeta x' = -10$" in *both* the real and complex domains.[33] Although the specific "multiplication" operations differ in the two domains, we can still study how a particular "form" behaves when we move from the reals to the complex numbers. Indeed, it would be disastrous for Frege's philosophy of mathematics if it could not construct some sense in which the behavior of elliptic functions (or involutions) could be compared over different domains. After all, much of the genius of nineteenth century mathematics—and the primary motivation for the philosophies of Poncelet and von Staudt—lay in the realization that the behaviors of certain mathematical forms become more comprehensible when placed in an extended setting. In short, the assumption[34] of which Boolos complains is deeply imbedded in much intuitive mathematical thought, not simply Frege's.

XI

In summation: I have claimed that Frege's more radical philosophical claims, especially those expressed in the *Grundlagen*, come into

---

[33]Strictly speaking, the "10" and the "$-$" also shift referents in the move to the complex realm, so they should be reassigned to the "tuple" side of this question.

[34]Indeed, the sometimes obnoxious restrictions on construction that Boolos's strictures require still represent a bone of contention among category theorists.

sharper focus when viewed from the perspective of the prior discussions of the extension element problem within geometry. There are two primary reasons for the increased vividness of the geometrical setting: (1) Von Staudt's techniques in geometry run roughshod over the expected "sensuous content" of geometrical propositions, to the extent of harnessing erstwhile concepts to the work of concrete objects. Frege's philosophical pronouncements on definition and context justify these procedures, but their applications within Frege's own context seem very tame, largely because we lack a robust sense of the apparent "content" of arithmetical propositions. (2) Nineteenth century geometry represented a discipline in internal tension, a tension arising from the need to admit new elements while retaining a permanence in subject matter (we want to let geometry enlarge her living quarters a bit without allowing her to bolt the stable altogether). Von Staudt found an unusual approach to definitions that (temporarily) resolved these conflicting demands upon geometry neatly. Frege's own views on definitional practice walk the same delicate line as von Staudt's, but, since Frege's numerical context is a static one, we scarcely appreciate the methodological pitfalls that he felt constrained to avoid in his philosophy of mathematics.

The "royal road" of this paper's title is intended somewhat whimsically. Although virtually every road in nineteenth century mathematics ran directly through geometry, they also—courtesy of Riemann's grand synthesis—wound through virtually every other area of mathematics as well. Although I have concentrated upon geometry here, it should be recognized that closely related problems of "extension elements" developed elsewhere (e.g., Kummer's "ideal numbers" mentioned above). Frege was clearly aware of these other developments and of the deep ties that bound them. Given Frege's rather unpleasant tendency to not acknowledge positive influences, any reconstruction of his thought processes must *per force* be speculative. For the reasons cited in the previous paragraph, the geometrical case remains the most salient setting for understanding the various philosophical attitudes towards mathematical existence that were popular in Frege's era. Accordingly, I have treated it in isolation here, although Frege would have viewed matters from a richer perspective.

Recognition of such geometrical precedents, of course, should not be seen as minimizing the magnitude of Frege's personal achievements. It is worth stating, however, that certain modern commentators regularly undervalue the philosophical merits of Frege's *rivals* in a misguided attempt to glorify Frege's particular form of genius.

In this respect, we should beware of treating the philosophical problems of arithmetic in the hermetically insulated way that Frege's own writings encourage. Hermann Schubert, whose views on numbers Frege castigated in an unfairly derisive review, had been deeply influenced in his approach to number by Poncelet's philosophy of geometrical existence (i.e., with heavy reliance upon "continuity" or "persistence of form"). Viewed independently of geometry, as Schubert presented the doctrine to his readers,[35] this principle looks absolutely silly. Read in conjunction with geometry, "continuity" assumes a rather profound aspect, especially in light of the astonishing "enumerative calculus" that Schubert developed in algebraic geometry largely following its guide. The real purpose of the "principle of continuity," for both Schubert and Poncelet, was to rationalize the "resettings" of mathematical questions which were vital to their geometrical investigations. Schubert's work in arithmetic represented a carry-over of these philosophical ideas, into a setting where the appeals to "continuity" largely lost their cutting edge.

Indeed, looking at Schubert's own geometrical successes, one begins to feel that Poncelet's explication of the "correctness" of the projective revisions was probably superior to von Staudt's. Although von Staudt's methods were undoubtedly more rigorous than Poncelet's, rigor is not everything. Von Staudt's methods are "copycat" procedures in that one knows which concept-objects to extract only after an authority like Poncelet has located the important "forms" that should persist in the extended domain. In contrast, the

---

[35] Hermann Schubert, *Mathematical essays* (1899), Chicago: Open Court. For an interesting perspective on Schubert, see Steven Kleinman, "Rigorous formulation of Schubert's enumerative calculus," in F. Browder, ed., *Mathematical developments arising from Hilbert's problems*, Providence: American Mathematical Society, 1976. It is worth mentioning that various philosophers attempted to extend the "principle of continuity" into general philosophy of language, e.g. Ernst Cassirer, in *Substance and function*, New York: Dover, 1953, W. C. and M. C. Swabey trs.

"principle of continuity," with all of its vagaries, genuinely affords a *discovery procedure* for uncovering the proper extended domain. When S. S. Chern writes that "figure and number fight for the soul of every geometer,"[36] he is referring to the way that questions natural to visual geometry and questions natural to the algebra of coordinate geometry interact to drive geometry forward. We witnessed an early example of this interaction in the Cartesian successes of section III—the subsequent development of algebraic geometry, in which Schubert played an important role, illustrate the continuing influence of these tensions, leading to abstractions far beyond those found in projective geometry. "Principles of continuity" hew closely to the considerations[37] that force the geometrical changes, whereas von Staudt's method submerges these motivations as an artifact of the "context of discovery." Despite these limitations, von Staudt's approach was widely judged in Frege's time to be superior to Poncelet's (albeit more tedious).

Frege's work is commonly characterized as contributing to the nineteenth century "rigorization of analysis," a phrase that normally encompasses investigations of notions of limits, series, the assumption of lower bounds, notions of derivative and integral, and the like. In fact, Frege's own writings stand at a remarkable distance from concerns of these sorts. Rarely does he comment upon the burning issues within analysis in his time—e.g., the possibilities for evading the use of Dirchlet's principle within Riemann's work. Of course, Frege was quite rigorous in his own projects, but this rigor does not appear harnessed to higher mathematical ends in the manner of Weierstrass and other contemporaneous workers on foundations. On the other hand, the puzzles of reconstitution of domain, so acutely raised within nineteenth century geometry, do not hinge upon worries about "rigor" per se, but upon questions of how unexpected modifications of old definitions and theorems can be permitted in mathematics, while retaining some sense that the discipline's subject

---

[36] "From triangles to manifolds," *The American mathematical monthly*, **86** (1979) 339–49, p. 343.

[37] Kenneth Manders, in unpublished work, has given a very deep analysis of the factors that push the early developments. For geometry's later career, see Jean Dieudonné, *History of algebraic geometry*, Monterey: Wadsworth, 1985, Judith Sally, tr.

matter is permanent and immutable. Even though the mathematical task of defining "natural number" is not, on the face of it, an "extension element" problem, I submit that if we see Frege's work as conceived in the shadow of the methodological concerns arising in geometry, we will better understand the character of his philosophy and the manner in which he conceived his logicism.

Returning once again to the "reconceptualization" issues with which we began, I believe that, even today, we scarcely understand the character of mathematical concepts—how it is that they can "grow and thrive" over time. Too much concern with set theoretic formalization—whose purpose, after all, is simply to grant a concept a temporary foothold on rigor—may have dimmed our eyes to this fact. Perhaps the realization that one of the founding parents of set theoretic rigor was deeply concerned with such issues may serve, as a bit of carbonate of ammonium, to reawaken us to the deep problems of conceptual growth within mathematics, even if Frege's particular answers can no longer be regarded as wholly satisfactory.

POSTSCRIPT: A NOTE ON FREGE'S "METHODS OF CALCULATION"

Frege's qualifying thesis, "Methods of calculation based upon an extension of the notion of 'quantity,'"[38] clearly anticipates his later insistence that the wide ranges of applicability of the classical number systems preclude any narrow grounding in Kantian intuition. But, as William Demopoulos stresses in "Frege and the rigorization of analysis,"[39] such an independence from intuition cannot be plausibly demonstrated unless the logistic alternative is specified in a rigorous, gap-free fashion. In contrast to Frege's more characteristic work, "Methods of calculation" is quite cavalier in respect to matters of rigor. What, then, was the source of Frege's early interest in *generality* and *independence from intuition*, given that a concern with rigor per se lay entirely in his future?

The purpose of this note is to suggest a somewhat incomplete response to this puzzle, allied to certain considerations touched upon in "The royal road." In fact, the second half of the nineteenth

---

[38] In Gottlob Frege: *Collected papers on mathematics, logic, and philosophy*, Hans Kaal, tr.

[39] This volume, Chapter 3.

century witnessed a rather widespread interest in varieties of "generality," of which Frege's particular variant represents a now unfamiliar subspecies. However, the subject called "abstract algebra" very much had its roots in Frege-like concerns.

It is a mistake, I think, to consign the labors of Frege, Dedekind, and Weierstrass too briskly to the task of "rigorizing analysis," if only because the scope of the term "analysis" is nowadays construed in a different manner than it was in the 1870's. Today a standard course in the calculus begins with a hurried construction of the real numbers, a "δ/ε" definition of "limit" and the like. Subsequent courses in "real analysis" provide the topological generalizations that support allied operations within functional analysis and so forth. Histories of mathematics often emphasize this "real analysis" line of development, leaving the impression that the alleged nineteenth century "crisis in analysis" was largely a matter of setting this familiar δ/ε machinery into place—that is, a simple matter of "rigorizing" and extending the loose calculus practices in which mathematicians had been engaging since Barrow and Newton. But this popular portrait of the time's troubles ignores the facets of "analysis" that would have seemed most salient to a Frege or a Dedekind.

In *Function and concept*,[40] Frege mentions two directions in which

> the meaning of the word 'function' has been extended by the progress of science: [(i) where] the field of mathematical operations that serve for constructing functions has been extended ... [including descriptions in ordinary language], e.g., when they were speaking of a function whose value is 1 for rational and 0 for irrational arguments; [(ii) where] the field of possible arguments and values for functions has been extended by the admission of complex numbers.

However, these two developments, at first glance, do *not* point the term "function" along a common path of development. Type (i) considerations, whose salience became prominent after Riemann's

[40]Gottlob Frege: *Collected papers on mathematics, logic, and philosophy*, p. 144 (p. 12 of the German original), Peter Geach, tr.

famous essay on trigonometric series[41] was published, represent an important part of the story of the growth of *"real* analysis" sketched above. It is in this tradition that the familiar definition of a "function" as an arbitrary many-one mapping over a domain grew up (following Riemann's essay, it is now often called the "Dirichlet/Riemann conception"). But in Frege's time the term "analysis," without qualifying riders, would have been normally understood to center upon the rather different considerations of the field we now call *"complex* analysis," following Frege's second manner of extension. But the natural behaviors of most garden variety "complex functions" are seldom "many/one." That is, the "complex functions" that were the primary focus of "analysis" in Frege's day turn out not to be "functions" at all in modern usage (though every modern algebraic geometer persists in calling them "functions" anyway).

Indeed, quite aside from the demands of complex analysis, the original usage of "function" conveyed only the sense of an intertwined dependence, rather than a unique specification. The graph of "$x^4 - x^2 + 2x^2y^2 + y^2 + y^4 = 0$" traces out a figure eight (lemniscate). It is natural to regard x and y as "functions" of one another within this formula, despite the fact that neither dependence satisfies the "many-one" requirement embodied in the Dirichlet/Riemann account.[42] Starting from these humble roots, it was amply

---

[41] "Über die Darstellbarkeit einer Function durch eine trigonometrische Reihe," in G. F. B. Riemann, *Werke*, New York: Dover, 1953.

[42] The reader should be warned that Riemann himself should not be regarded as a single minded advocate of the "many-one" conception of "function"; after all, it is his work on Abelian functions that directly lead to the great flourishing of work (by Frege's teacher Clebsch, *inter alia*) upon "algebraic functions," which are not many-one objects. Riemann, moreover, provided a compelling vision of complex functions as specified not so much through linguistic expressions, but almost "physically," through delineation of their singularities, jumps and boundary behavior. Analogy: build a big bathtub and specify the "sources" and "sinks" where the water enters and leaves the tub. These conditions will completely fix how the water will flow in the rest of the tub. For Riemann, a "complex function" automatically corresponds to the "flow" induced by its singularities, et al., whether or not there is any formula that everywhere matches such a flow (a vivid description of Riemann's point of view is sketched in Felix Klein's great monograph, *On Riemann's theory of algebraic functions and their integrals*, New York: Dover, 1963). This Riemannian conception, as much as any, helped sever the notion of "function" from its earlier association with

recognized by Frege's time that a better understanding of the "special functions" long used in mathematical physics could be obtained if they are viewed as merely the real valued slices of "fatter" (= complex valued) entities. This is because the "off the real axis" behavior often crisply reveals how the functions are likely to behave (or misbehave) in numerical approximations. Likewise, only a direct study of the "fatter" functions will display the family resemblances that link the various "special functions" with one another (e.g., classification theorems of Abel's type can be articulated).

Unfortunately, the road to uncovering the properties of the "fatter functions" in their natural realm is not straightforward. The properties of the "rules"—expressed, say, in a Taylor's expansion—that seem to "define" a function within its real valued appearances can easily turn out to fail once the reach of the function stretches over the full complex realm. To figure out how an elliptic function like $sn(x)$ behaves, Jacobi pieced odd scraps of information together in a very delicate manner: (i) he first looked at the local behavior of the inverses of several elliptic functions over the real line, (ii) used that data to evaluate $sn(x)$ along the purely imaginary axis, (iii) used Euler's addition formula to extrapolate to mixed arguments, and so on. Such piecemeal techniques were basically premised on the hope that *suitably selected* local facts about the function would prevail within its uncharted regions as well, even though many examples showed that such confidence was ill-founded in general. Clearly some better framework for settling the existence or nonexistence of complex valued functions with stated attributes was wanted. The schools of Riemann and Weierstrass proposed rather different resolutions of these problems (in which, to be sure, the "many-one" notion of "function" always became salient, as a means of taming the local behavior of the "complex functions"). The trick in all such approaches—most saliently realized in Riemann's picture—is to locate some preliminary heaven (i.e., a Riemann surface) where functions can live a purer life before they assume the sorrier guise of a normal complex "function."

Accordingly, "the crisis in analysis" in Frege's time concerned itself as much with the existence problems of the *intervening struc-*

mathematical formulas.

tures needed to build up complex analysis as with the explications of "limit," "derivative" and "integral." I have sketched this background in some detail, because several recent evaluations[43] of Frege have wandered astray through neglect of this history. With respect to present purposes, however, we should observe that "complex analysis" (understood broadly enough to include the beginnings of algebraic geometry) *is apt to prompt philosophical concerns of its own, more or less distinct from foundational worries*. Specifically, the introduction of "intervening entities" often reveals hidden "commonalities" that forge unexpected links between superficially distinct

---

[43]E.g., Jaakko Hintikka and Gabriel Sandu, "The skeleton in Frege's cupboard: the standard versus nonstandard distinction," *The journal of philosophy*, **89** (1992) 290–315. These authors berate Frege for not appreciating the full ramifications of the Riemann/Dirichlet conception of function. In truth, many mathematicians in Frege's time believed that the Dirichlet/Riemann account did not capture the essence of what enables a "true function" to behave as it does. They would have instead pointed to the "organic unity" exemplified in one of Riemann's complex functions, as described in the previous footnote. After all, it was the recognition that a "function" could be assigned a deeper articulation of its "unity" that originally persuaded most mathematicians to abandon the old reading of a "function" as a linguistically articulated rule. In the contexts surveyed by Hintikka and Sandu, Frege is attempting to draw an analogy between "function" and "concept." Such an assimilation would have seemed unconvincing to much of Frege's intended audience if the Riemann/Dirichlet reading of "function" were dogmatically assumed, for such "functions" lack the "unity" that makes functions prima facie comparable to concepts.

Such adherents of "organic unity" might have happily conceded that the Riemann/Dirichlet notion is important mathematically; as noted previously, the notion provides an important technical prop needed in any workable account of "complex functions," whether one follows Riemann's or Weierstrass's rather different routes to this end.

Although Frege tends to leave his own views on the nature of "functions" rather nebulous, there is no obvious doctrinal reason why he should have opposed the rise of functions of a real variable. His doctrines on meaning allow that a Riemann/Dirichlet function might serve as the proper reference of a function term, its "unity" being left behind in the *sense*. And his notion of "sense" seems so liberal that a god who can inspect an infinite table of values would thereby grasp a unified "sense" appropriate to the most random function of a real variable.

This is not to say that Hintikka and Sandu's objections are not applicable to many modern accounts of concepts and properties. I drew a similar moral independently in "Honorable intensions," in S. Wagner and R. Warner, eds., *Naturalism*, South Bend: Notre Dame Press, 1993.

branches of mathematics. The problem is: how might mathematics accommodate these surprising commonalities?

For example, Riemann's discoveries indicate that deep analogies must exist between the behaviors of families of Abelian functions, collections of geometrical curves and the numbers within algebraic fields. Some similarity in basic *structure* persists between these areas. Kronecker, Weierstrass and Dedekind, in their respective ways, became intrigued by the fact that if various supplemental entities are added to a collection of algebraic numbers or functions, the old class of objects can then be *factored* into a set of basic "prime" entities that allow useful computations with respect to the original class of objects. Some algebraic commonality must lie hidden within these prima facie different subjects—or, as mathematicians like to put it, some abstract "dictionary" must exist to translate between analysis, geometry and number theory. To flesh out this hunch, two basic tasks are required: (i) isolate the manner in which special sets of numbers, functions or curves collect together into natural families (i.e., what we now call rings, fields or modules) and (ii) show how these natural families can be extended by supplementary entities that permit the desired "factoring." Dedekind often used his so-called "ideals" to add the supplementary entities, built up according to the von Staudt "equivalence class" technique. Dedekind seemed to hold that it was "the pure creative act" sanctioned by "logic" that facilitates the construction of these supplementary entities once an originating set of objects has been set in place.[44] Setting aside the psychologism implied in "creative act," the notion that "logic" serves as a wellspring of satellite objects is very close to the doctrine I credit to Frege in "The royal road." In any case, it can hardly be "intuition" that supports the supplementations, since they spring identically from arithmetical and geometrical sources.

[44]In this regard, it is worth remembering that Dedekind held that geometry, considered in its own right, did not guarantee that points existed on a line to correspond to every real number or that every angle would divide into three equal parts. These conclusions, he maintained, were the result of the "logical faculty" filling in the original domain of geometry (cf. *Essays on the theory of numbers*, New York: Dover, 1963, 37–8). Here he seems influenced by both Galois theory and his own researches on algebraic numbers. A similar opinion seems to motivate G. N. Halsted, *Rational geometry*, New York: MacMillian, 1897.

Dedekind's achievements in isolating commonalities were quite remarkable, specifically in showing (with H. Weber) that a path can be forged to a version of the Riemann-Roch theorem that does not rely upon the specifically analytic techniques of the original proofs. This discovery opens the possibility of applying the theorem in areas of number theory where the notion of integration no longer makes sense. The price, in Dedekind's treatment, is a heavier involvement of mathematics with the notions of set theory.

In sum, although Dedekind contributed importantly to the "rigorization of analysis," his philosophical interests in isolating specific kinds of mathematical structure arose as much from an interest in "mathematical generality" as "rigor" per se; from a concern with how the "deep structures" within mathematics can be profitably articulated and interconnected.

The remarks on "independence from intuition" in Frege's "Methods of calculation," I believe, flow from considerations allied to Dedekind's, although the focus of Frege's particular investigation was rather different. We might begin by examining the final example in Frege's essay.

By mid-nineteenth century, a recognized need arose in mechanics to find a method whereby propositions about machine movement in three dimensions can be expressed in some algebraically perspicuous manner. The old fashioned manner of using Cartesian coordinates blurs the spatial relationships proper to the device with those arising from its extraneous dependence upon the coordinate axes (recall Poncelet's related complaint about Cartesian techniques). Today we use the Gibbs/Heaviside vector calculus for mechanical tasks such as this (although, in truth, this format is not optimal if one primarily intends to study compositions of rotational movement). In the vector calculus, algebraic symbols are directly associated to sundry sorts of geometrical object, e.g., directed line segments. In Frege's time, however, it was usually presumed that the desired description of mechanical movement would utilize objects that are more "number-like" than vectors, on the model of the surprising utility of the regular complex numbers in planar situations. William Hamilton originally hoped his "quaternions" would play the desired role; W. Clifford's "dual numbers" represented another effort in this same direction.

But other mathematicians tried to approach the descriptive problem by using standard numbers as measures of unusual quantities. A chief proponent of variant "dual numbers"—Edward Study—complained in 1890:

> In a number of circles, particularly in Germany, there is a widely held view that systems of [extended] complex numbers or similar algorithms have actually been of hardly any use, with the single exception of the ordinary complex numbers. The reason given as justification for this attitude has been that no results could ever be provided by these systems that could not equally well have been provided without their help.[45]

Working in this "ordinary numbers" line, Sir Robert Ball in 1876 described mechanical movements in terms of primitive transitions he called "screws": "A screw is a straight line with which a definite linear magnitude termed the pitch is associated."[46] Calculations prove much simpler (at least for smallish movements) if a complicated machine action is seen as decomposing directly into a set of "screws" rather than as a pattern of independent translations and rotations.

In so doing, Ball saw his work as following directly in the tradition initiated by Julius Plücker. In "The royal road," I remarked on the surprising manner in which Plücker utilized regular complex numbers in his celebrated "line geometry." When numerical coordinates are assigned to lines rather than points and these newly coordinated objects are treated as the basic building blocks of the subject, regular Euclidean geometry becomes reconfigured as a four-dimensional non-standard geometry. It is the novel application of standard numerical evaluators that articulates the hidden structure.

Much of Frege's teaching and occasional mathematical work followed the Plücker tradition closely. Although "Methods of calculation" is not very explicit about motivation, it is likely that Frege hoped that some of his labors might contribute to a fruitful treat-

---

[45] "Über Systeme complexer Zahlen und ihre Anwendungen in der Theorie der Transformationsgruppen," quoted in H. D. Ebbinghaus et al., *Numbers*, New York: Springer-Verlag, 1990, 182.

[46] R. S. Ball, *A treatise on the theory of screws*, Cambridge: Cambridge University Press, 1890, p. 7.

ment of mechanical movements analogous to Ball's (whose contemporaneous research in Ireland would have then been unknown to Frege). It is worth noting that Frege devoted a good deal of his graduate course work to mechanics.

Frege's basic idea is to take a given function f (or set of functions) and use this fiducial f to grade the degrees of "f-ness" of some restricted range of other functions, a range he calls a "quantitative domain." He sees this family of f's as growing from the identity function by continually increasing their degrees of "f-ness." With respect to screw movements, Frege proposes that we might begin with a set of three f_i's describing a specific spiral motion stretching from point a to b. Let this f-movement have a "quantity of spiral motion $= 1$." Frege's plan is to assign further numerical values to other mechanical movements so that if the f-movement's "quantity of spiral motion" is doubled, we will obtain a new screw, described, say, by a system of $g_i(t)$, that extends farther than b and coils more tightly than the original f-motion. We can thus grade a rather complicated family of movements by regular numerical values and compose certain machine movements by numerical addition.

Frege's scheme for generating these numerical assignments (restricting our attention now to a single function f) begins by taking f itself as representing (definitionally) a quantity $= 1$ of "f-ness." The identity function is assigned quantity $= 0$ of f-ness. If $g(t) = ff(t)$ for all t, g has quantity 2 of f-ness. Conversely, if $f(t) = gg(t)$ for all t, then g has quantity $1/2$ of f-ness. If $g(t) = ff(t)$ for only some t, then g doesn't "belong to the same quantitative domain" as f and so shouldn't be assigned any degree of f-ness (although it may still be possible to relate the distinct quantities of f-ness and g-ness in "quantitative equations"). Frege takes it for granted that we can interpolate irrational degrees of f-ness within an f-series. Such structuring of the f-series will allow many "quantitative domains" of functions to be assigned an "infinitesimal generator" (i.e., the partial derivative of the function $f(n, t)$ at $n = 0$, where the "quantity" n is treated as a variable parameter). Indeed, in developed work like Ball's, it is precisely the "infinitesimal screws" that provide the key to the method's descriptive successes.

This kinematic case seems to be the example of the greatest practical potential suggested in "Methods of calculation." The rest of the essay explores other contexts where Frege's "degrees of f-ness" can be introduced as *intermediary entities*. Instead of simply stating that $y = ax + b$, we can say that the range y is "grown" by an increase in f-ness that simultaneously rotates and transports the identity function. That is, a new quantity representing an appropriate degrees of f-ness now intervenes in the linear relationship between x and y.

Although somewhat related ideas—Lie groups, for example—have proved very useful in mathematics, it is hard to see that Frege's specific proposals are likely to bear much fruit. Part of the trouble clearly lies in the adherence to the Plücker tradition that Study decried; rather than inventing new mathematical objects—vectors, dual numbers—especially adapted to the task at hand, Frege hoped to manage with *standard numbers harnessed to non-standard applications*. Such a conservative stratagem has the merit of raising fewer ontological questions, but it proved too inflexible to service mathematics' needs. But we know this only in hindsight; in the 1870's, Frege was not unique in his number-centered orientation.

If the suggested placement of "Methods of calculation" is correct, Frege's interest in issues of applicability would have been born in the hope that a proper explication of number might help mathematicians cast aside the blinders of intuition and see numbers in their unsuspected appearances as simplifying intermediaries. As Frege put it, the hope was "partly to bring to notice a class of problems which might be solved with the help of a further developed theory."[47]

Since these ambitions were never realized, Frege retreated in his later writings to two forms of apology for his investigations: (i) that a rigorous demonstration of arithmetic's independence of intuition can be supplied; (ii) that his logical insights are needed to prevent mathematics from falling into conceptual disaster. Although Frege's specific remarks with respect to (ii) are generally well taken, the over-arching conceit that mathematics runs the risk of a "major calamity"[48] due to its faulty explications of "variable," etc., passes

---

[47]Gottlob Frege: *Collected papers on mathematics, logic, and philosophy*, p. 58 (p. 2 of the original). On this reading, the usual parsings of Frege's "application problem" miss the center of his concern.

[48]"Logical defects in mathematics," in Gottlob Frege: *Posthumous writings*, H. Hermes et al., eds., Chicago: University of Chicago Press, 1979, p. 165.

unsubstantiated. It is hard to see these soundings of deep error as anything other than a pathetic bid for attention on the behalf of work whose own solid virtues have passed unappreciated.

There is no doubt that writings like "Logical defects in mathematics" or the Frege-Hilbert correspondence encourage the picture, rather common nowadays, of Frege as a philosophical gadfly working outside of the mathematical mainstream. The fact that Frege rarely addressed the key questions troubling analysis in his time further contributes to the belief that Frege was not "motivated" in the manner of a typical mathematician—as if the presumption of a "typical motivation" for mathematics is coherent! But we have observed that Dedekind (and, to some extent, even Weierstrass himself) was driven as much by a desire to understand hidden structure as anxiety about rigor. It was simply that Frege's specific hopes for general structure became lost in the course of mathematics' later rambles.[49]

One of the clearest cases where the notion of variable as an "indeterminate object" led to genuine harm in mathematics can be found in the loose appeals to "generic points" in algebraic geometry. It is surprising to me that Frege nowhere addresses this concrete practice, for some of the contemporaneous discussions of the problem were probably known to him.

[49]I would like to thank Bill Demopoulos, Ken Manders and Jamie Tappenden for helpful discussions. Tappenden has two forthcoming papers that survey allied topics in a revealing way: "Geometry and generality in Frege's philosophy of arithmetic," and "Frege on extending knowledge and 'fruitful concepts.'"

PART II

The Mathematical Content of *Begriffsschrift* and *Grundlagen*

6

Reading the *Begriffsschrift*

The aim of the third part of the *Begriffsschrift*, Frege tells us, is:

> to give a general idea of the way in which our ideography is han-
> dled ... Through the present example, moreover, we see how
> pure thought, irrespective of any content given by the senses or
> even by an intuition a priori, can, solely from the content that
> results from its own constitution, bring forth judgements that
> at first sight appear to be possible only on the basis of some
> intuition ... The propositions about sequences developed in
> what follows far surpass in generality all those that can be de-
> rived from any intuition of sequences. If, therefore, one were
> to consider it more appropriate to use an intuitive idea of se-
> quence as a basis, he should not forget that the propositions
> thus obtained, which might perhaps have the same wording
> as those given here, would still state far less than these, since
> they would hold only in the domain of precisely that intuition
> upon which they were based.[1]

From *Mind*, **94** (1985) 331–44. Reprinted by kind permission of The Mind
Association and the author.

I am grateful to Michael Dummett, Robin Gandy, Daniel Isaacson, David
Lewis, and Simon Blackburn for helpful comments. This paper was written while
I was on a Fellowship for Independent Study and Research from the National
Endowment for the Humanities.

[1]Gottlob Frege, *Begriffsschrift, a formula language, modeled upon that of*

He then proceeds to give a definition, proposition 69, on which he comments, 'Hence this proposition is not a judgement, and consequently *not a synthetic judgement* either, to use the Kantian expression. I point this out because Kant considers all judgements of mathematics to be synthetic.'[2]

In the preface to the *Begriffsschrift* he states, 'To prevent anything intuitive [*Anschauliches*] from penetrating here unnoticed, I had to bend every effort to keep the chain of inferences free of gaps.'[3] It is evident from the anti-Kantian tone of these remarks that Frege regards himself as showing the inadequacy of a certain (unspecified) Kantian view of mathematics by supplying examples of judgements that he thinks 'at first sight appear to be possible only on the basis of some intuition,' but which pure thought, 'solely from the content that results from its own constitution,' can bring forth. However an exact statement of the Kantian position under attack might run, the view is one according to which no non-trivial mathematical judgement is 'possible' without 'a priori intuition.'

My principal aim in this paper is to examine Frege's procedure in the third part of the *Begriffsschrift* in order to see how, and how well, a Kantian view of Frege's examples might be defended and to determine to what extent Frege could claim to have shown the truth of a view that may be called *sublogicism*: the claim that there are (many) interesting examples of mathematical truths that can be reduced (in the appropriate sense) to logic. Inevitably, the uncertainties and obscurities attaching to the notions of *intuition* and *logic* will leave these matters somewhat unresolved. I will however argue that a compelling case for Frege's view can be made against a certain sort of defence of Kant.

The issue between Frege and Kant is joined over a certain technical point that arises in connection with the marginal annotations of the derivations of part 3. If we wish to understand the issue, we

---

arithmetic, for pure thought, §23. All references are to the Bauer-Mengelberg translation, found in Jean van Heijenoort, ed., *From Frege to Gödel: a source book in mathematical logic*, Cambridge: Harvard University Press, 1967.

[2]Ibid. The remark that Kant considers all judgements of mathematics to be synthetic seems somewhat intemperate: Kant might of course agree that 69 is no *judgement*, hence no synthetic judgement.

[3]*Begriffsschrift*, Preface, p. 5.

cannot avoid examining the wallpaper. There is a further reason for looking at the formalism of part 3: at least one little-known but major master-stroke is hidden there, and one of the subsidiary aims of this paper is to call attention to it, repellent though the notation in which it is cloaked may be. Another aim of the paper is simply to render part 3 more accessible.

Before we examine Frege's achievement we must review the special notational devices which Frege introduces in part 3. Fortunately, there are only four of them.

The first of these '$\left| \begin{smallmatrix} \delta \\ \\ \alpha \end{smallmatrix} \left(\begin{smallmatrix} F(\alpha) \\ \\ f(\delta,\alpha) \end{smallmatrix} \right. \right.$,' is defined in proposition 69 to mean something that we might notate: $\forall d \forall a (Fd \ \& \ dfa \rightarrow Fa)$. (I have written '$dfa$' in place of Frege's '$f(d,a)$.') Since the relation f—Frege calls it a *procedure*—is fixed throughout part 3, I shall use the abbreviation 'Her(F),' suppressing 'f,' for this notion instead. ('Her' is for 'hereditary.')

The second, '$\underset{\beta}{\gamma} f(x_\gamma, y_\beta)$,' is Frege's abbreviation for the strong ancestral of f, whose celebrated definition is presented in proposition 76. Abbreviating '$\forall a (xfa \rightarrow Fa)$' as 'In($x, F$)' (again suppressing mention of the fixed f), we may give the definition as: $\forall F (\text{Her}(F) \ \& \ \text{In}(x, F) \rightarrow Fy)$. We shall use: xf^*y for this notion.

The third, '$\underset{\beta}{\gamma} f(x_\gamma, z_\beta)$,' is the abbreviation for the weak ancestral, defined in proposition 99 as $xf^*z \lor z = x$. We write this: $xf^*_=z$.

Finally, Frege defines '$\underset{\varepsilon}{I} f(\delta, \varepsilon)$' in proposition 115 to mean: $\forall d \forall e \forall a (dfe \ \& \ dfa \rightarrow a = e)$. We write this: FN (for 'f is a function').

We can now say what the judgements are which Frege thinks can be brought forth by pure thought solely from the content that results from its own constitution—or, as we may say, can be proved by purely logical means—but which, he thinks, appear at first sight to be possible only on the basis of some intuition. We can then take up the question whether the means used to prove them are in fact 'purely logical.'

If we look at the table with which the *Begriffsschrift* ends and which indicates which propositions are immediately involved in the derivations of which others, we find that there are only two propositions in the third part not used in the derivation of any others: number 98 and the last one, number 133. Since these propositions are not used to prove any others, I do not find it too far-fetched to suppose that Frege thought of these as illustrating the falsity of the Kantian view with which he is concerned.

The translation into our notation of 98 is: $xf^*y \ \& \ yf^*z \rightarrow xf^*z$. That of 133 is: $FN \ \& \ xf^*m \ \& \ xf^*y \rightarrow yf^*m \lor mf^*_{=}y$.[4] These state that the (strong) ancestral is transitive and that if the underlying relation f is a function, then the ancestral connects any two elements m and y to which some one element x bears the ancestral. The analogy with the transitivity and connectedness of the less-than relation on the natural numbers, which is the ancestral of the relation *immediately precedes*, will not have escaped the reader's notice, and I dare say it did not escape Frege's.

Although Frege does not explicitly single out 98 and 133 as noteworthy in any way, it is quite reasonable to suppose that he regarded both of them as the sort of proposition that would justify the anti-Kantian viewpoint sketched above. For not only are these two the only propositions in part 3 not used in the demonstration of others, their content can be seen as a generalization of that of familiar and fundamental mathematical principles, for the grasp of whose truth some sort of 'intuition' was often supposed in Frege's time to be required. Moreover, one who attempts to convince himself of the truth of, for example 98, might well hit upon an argument that would seem to make appeal to the sort of intuition which Frege was concerned to show unnecessary. Suppose that y follows x in the f-sequence and z follows y. Then if one starts at x and proceeds along the f-sequence, one can eventually reach y. Ditto for y and z. Thus, by starting at x and proceeding along the f-sequence, one can eventually reach z, first by going to y, and thence to z. Thus z follows x in the f-sequence. Intuition it might be suggested, discloses to us that any two paths from x to y and y to z can be combined into one single path from x to z: intuit them both and then attach in thought the beginning of the second to the end of the first. Or

---

[4]I do not know why Frege chose to use the variable 'm' here instead of (say) 'w.'

some such thing.

The procedure Frege employs in the derivation of 98 is of considerable interest, and we shall look at its final steps. Having arrived at

(84) $$\text{Her}(F) \ \& \ Fx \ \& \ xf^*y \rightarrow Fy$$

and

(96) $$xf^*y \ \& \ yfz \rightarrow sf^*z,$$

Frege generalizes upon z and y in 96 to obtain $\forall d \, \forall a (xf^*d \ \& \ dfa \rightarrow xf^*a)$. He then substitutes $\{ a : xf^*a \}$ for F (as we might put it) in the definition of $\text{Her}(F)$ to obtain 97, which we can write: $\text{Her}(\{ a : xf^*a \})$. He then reletters x and y as y and z in 84, substitutes $\{ a : xf^*a \}$ for F in 84, and discharges $\text{Her}(\{ a : xf^*a \})$ to obtain the desired 98.

Frege appears to regard the substitution of a formula for a relation letter in an already demonstrated formula as on a par with substitution of a formula for a propositional variable or relettering of a variable. Of course, in standard first-order logic, substitution of formulae for relation letters gives rise to no special worries: any formula demonstrable with the help of substitution is demonstrable without it. (Frege performs several such substitutions in part 2, which contains none but first-order notions.) But this is emphatically not the case as regards part 3 of the *Begriffsschrift*. The capacity to substitute formulae for relation letters gives the whole of Frege's system, which is not a system of first-order logic, significantly more power than it would otherwise have.

Although a Kantian opponent could well make an objection at this point to Frege's use of substitution, there is a more pertinent objection to be made: no one can sensibly think that *every* mathematical judgement must be based on some intuition. For certainly there are some trivial mathematical judgements which need not be so based, among them analytic judgements concerned with mathematical matters and others of a *trivial* logical nature, such as '5 + 7 = 5 + 7' or 'if 5 + 7 = 12, then 5 + 7 = 12.' Moreover, among such judgements are those that follow from definitions with only a *small* amount of *elementary* logical manipulation. And one of these is Frege's 98. For, let us face it, Frege's proof of 98 is unnecessarily non-elementary. One needs no rule of substitution at all to prove that if xf^*y and yf^*z, then xf^*z. For suppose xf^*y and yf^*z. We want to show xf^*z, i.e. $\forall F(\mathrm{Her}(F) \ \& \ \mathrm{In}(x, F) \to Fz)$. So suppose $\mathrm{Her}(F)$ and $\mathrm{In}(x, F)$. We want to show Fz. Since yf^*z, we need only show $\mathrm{In}(y, F)$, i.e. $\forall y(yfa \to Fa)$. So suppose yfa. Since xf^*y, $\mathrm{Her}(F)$ and $\mathrm{In}(x, F)$, Fy. And since $\mathrm{Her}(F)$ and yfa, Fa, QED. The trouble with 98, our Kantian might complain, is that although the *above* proof of 98 is certainly a proof by logical means alone, 98 does not *look at first sight as if* it must be based on an intuition.

Frege has not yet laid a glove on the Kantian. 98 *is* a weak example. Of course Frege's rendering of 98, 'If y follows x in the f-sequence and z follows y in the f-sequence, then z follows x in the f-sequence,' might have been a better choice, but the Kantian might then have been in a position to raise questions about the grounds for *reading* 'xf^*y' as 'y follows x in the f-sequence,' plausibly arguing that this reading is itself justified only on an intuition.

No such objection can be raised against 133, FN & xf^*y & $xf^*m \to yf^*m \lor mf^*_=y$, of which an 'intuitive' proof might go as follows. Suppose FN, xf^*y, and xf^*m. Since xf^*y and xf^*m, there is an f-sequence leading from x to y and an f-sequence leading from x to m. And since FN, each thing bears f to at most one thing; thus at no point along the way can either of these paths diverge from the other. Thus the paths coincide up to the point at which any shorter one gives out. Since xf^*m and xf^*y, we eventually reach both m and y; when we have done so, we will evidently have reached y before m, reached m before y, or reached m and y at the same time. In the first case, we can get from y to m along the path obtained by removing the path from x to y from the path from x to m; in the second case we can similarly get from m to y, and in the third case, $m = y$. Thus $yf^*m \lor mf^*_=y$. We are about to turn to Frege's derivation of 133; before we do so, the reader might like to try his hand at giving a proper proof of 133, in the style of the proof of 98 given two paragraphs above. (One such proof is given in the appendix.)

One significant landmark in Frege's derivation of 133 is proposition 110: $\forall a(yfa \to xf_{\equiv}^* a)$ & $yf^* m \to xf_{\equiv}^* m$. 110 is itself got from 108: $zf_{\equiv}^* y$ & $yfv \to zf_{\equiv}^* v$, which has a straightforward proof.[5] 108 is fairly obvious; 110 is not at all obvious. (We cannot get yfm from $yf^* m$.) How does Frege get 110 from 108?

First of all, he reletters the variables in 108, replacing, z, y, and v by x, (German) d, and (German) a, and then universally quantifies upon a and d to get: $\forall d \forall a(xf_{\equiv}^* d$ & $dfa \to xf_{\equiv}^* a)$. He then takes 75: $\forall d \forall a(Fd$ & $dfa \to Fa) \to \text{Her}(F)$, which is one-half of the definition of $\text{Her}(F)$, substitutes $\{\, a : xf_{\equiv}^* a \,\}$ for F (as we would put it), and uses 108 to cut the antecedent of the result, thereby getting 109: $\text{Her}(\{\, a : xf_{\equiv}^* a \,\})$. Next he takes 78: $\text{Her}(F)$ & $\forall a(xfa \to Fa)$ & $xf^* y \to Fy$, which is a trivial consequence of the definition of the ancestral, respectively replaces x and y by y and m, again substitutes $\{\, a : xf_{\equiv}^* a \,\}$ for F, and drops $\text{Her}(\{\, a : xf_{\equiv}^* a \,\})$ from the result by 109, to get $\forall a(yfa \to xf_{\equiv}^* a)$ & $yf^* m \to xf_{\equiv}^* m$, as desired.

The complexity of the definition of the substituend $\{\, a : xf_{\equiv}^* a \,\}$ is noteworthy. '$xf_{\equiv}^* a$' abbreviates a disjunction one of whose disjuncts is a second-order universal quantification of a first-order formula. Were Frege merely substituting $\{\, a : Ga \,\}$ (G a one-place relation letter) for F, i.e. reletttering F as G, we should have no qualms about his procedure. But the substitution of so complicated a formula as $xf_{\equiv}^* a$ for a relation letter is a matter considerably more problematical.

Having obtained 110, Frege straightforwardly gets 129: FN & $(yf^* m \vee mf_{\equiv}^* y)$ & $yfx \to (xf^* m \vee mf_{\equiv}^* x)$.[6] 131: FN & $\text{Her}(\{\, a : af^* m \vee mf_{\equiv}^* a \,\})$ follows, again by a substitution, this time of $\{\, a : af^* m \vee mf_{\equiv}^* a \,\}$ for F in the quasi-definitional 75.

Frege then performs the same substitution to conclude the derivation. From 131, he uses propositional logic to infer 132: $[\text{Her}(\{\, a : af^* m \vee mf_{\equiv}^* a \,\})$ & $xf^* m$ & $xf^* y \to (yf^* m \vee mf_{\equiv}^* y)] \to [FN$ & $xf^* m$ & $xf^* y \to (yf^* m \vee mf_{\equiv}^* y)]$. To get 133, the consequent of 132, he must obtain the antecedent. This is how he does it. He has earlier established 81: Fx & $\text{Her}(F)$ & $xf^* y \to Fy$ (an easy consequence of the definition of the ancestral). By propositional logic

---

[5] A proof is given in the appendix.

[6] A proof is given in the appendix.

there follows 82: $(p \rightarrow Fx)$ & $\mathrm{Her}(F)$ & p & $xf^*y \rightarrow Fy$. (Frege uses 'a' instead of 'p.') He then substitutes hx for p and $\{\,a : ha \vee ga\,\}$ for F in 82 ('h' and 'g' are one-place relation letters, like 'F') and drops a tautologous conjunct of the antecedent to obtain 83: $\mathrm{Her}(\{\,a : ha \vee ga\,\})$ & hx & $xf^*y \rightarrow hy \vee gy$. The final logical move of the *Begriffsschrift* is the substitution in 83 of $\{\,a : af^*m\,\}$ and $\{\,a : mf^*_=a\,\}$ for h and g, which yields the antecedent of 132.

Of course, Frege could have condensed these two substitutions for F into one, by substituting $\{\,a : af^*m \vee mf^*_=a\,\}$ for F in 81 and using propositional logic to obtain the antecedent of 132. But to prove 133, Frege has had to make two essential uses of substitution, the first being the earlier substitution of $\{\,a : xf^*_=a\,\}$ for F, the second, that of $\{\,a : af^*m \vee mf^*_=a\,\}$. It is noteworthy that the—or at any rate, one—obvious attempt to prove 133 will require the same two substitutions, in the order in which they are found in Frege's derivation.

The fact that the *Begriffsschrift* contains a subtle and ingenious double induction—for that is what Frege's pair of substitutions amounts to—used to prove a significant result in the general theory of relations is not, I think, well-known, and the distinctively mathematical talent he displayed in discovering and proving the result is certainly not adequately appreciated. Frege's accomplishment may be likened to a feat the Wright brothers did not perform: inventing the airplane *and* ending its first flight with one loop-the-loop inside another.

Our Kantian has patiently had his hand up during this discussion of Frege's method in part 3, and it is time to give him his say.

The Kantian: 'I could not agree with you more about the excellences of proposition 133 and Frege's proof of it, but it is not a counterexample to any thesis that I hold or that a reasonable Kantian ought to hold. Indeed, if anything, it is confirming evidence for my view. I agree that 133 is precisely the sort of proposition that is possible only on the basis of an intuition. But I disagree that Frege has been able to prove it without the aid of any intuition at all. In fact, the feature of Frege's method that you have been at pains to emphasize, the substitution of formulae for relation letters, is precisely the point at which, I wish to claim, Frege appeals to intuition. I'd be prepared to concede, for the sake of avoiding an argument, that nowhere in the rest of the *Begriffsschrift* is an appeal to intuition made. But I do wish to claim that his use of the rule of substitution does involve him in just such an appeal.

'The difficulty that the rule of substitution presents can best be seen if we consider the axiom schema of comprehension: $\exists X \forall x (Xx \leftrightarrow A(x))$. It is well known that in the presence of the other standard rules of logic, the substitution rule and the comprehension schema are deductively equivalent; given either, one can derive the other. In outline, the proof of this equivalence runs as follows. From the provable $\forall x (Fx \leftrightarrow Fx)$, we obtain $\exists X \forall x (Xx \leftrightarrow Fx)$ by second-order existential generalization, whence by the substitution of $\{ a : A(a) \}$ for F, we have $\exists X \forall x (Xx \leftrightarrow A(x))$. Conversely, we observe that for any formulae $P[F]$ and $A(x)$, we can prove $\forall x (Fx \leftrightarrow A(x)) \rightarrow (P[F] \leftrightarrow P[\{ a : A(a) \}])$; the demonstration of this is an induction on subformulae of the formula $P[F]$. Now suppose that $P[F]$ is provable. Then so is $\forall x (Fx \leftrightarrow A(x)) \rightarrow P[\{ a : A(a) \}]$; and since the consequent $P[\{ a : A(a) \}]$ does not contain F, $\exists X \forall x (Xx \leftrightarrow A(x)) \rightarrow P[\{ a : A(a) \}]$ is also provable. Thus if we have as an axiom $\exists X \forall x (Xx \leftrightarrow A(x))$, as is guaranteed by comprehension, $P[\{ a : A(a) \}]$ is provable too, QED. Thus we cannot admit substitution as a logical rule unless we are prepared to admit that all instances of the comprehension schema $\exists X \forall x (Xx \leftrightarrow A(x))$ are logical truths, and that is precisely what I wish to deny.

'For what does $\exists X \forall x (Xx \leftrightarrow A(x))$ say? If we look at the *Begriffsschrift*, we find that when Frege wishes to decipher his relation letters and second-order quantifiers, he uses the terms "property," "procedure," "sequence"; he uses the terms "result of an application of a procedure" and "object" to tell us what sorts of things free variables like "x" and "y" denote. My point can be put as follows. Suppose that $A(x)$ is the formula: mf^*x. Then Frege would read the corresponding instance of the comprehension schema as "There is a property whose instances are exactly the objects that follow m in the f-sequence." This comprehension axiom is demonstrable in the *Begriffsschrift*. My question is: why should we believe that there is any such property? Now, *I* don't want to deny that there is such a property. I might well want to say that it's *obvious* or *evident* that there is one. And I would want to say to anyone who professed uncertainty concerning the existence of the property, "But don't you *see* that there has to be one?" In short, it is an intuition of precisely the kind Frege thinks he has shown unnecessary that licenses the rule of substitution. Thus Frege has not dispensed with intuition; he is up to his ears in it. (I may add that the inference from $\forall x(Fx \leftrightarrow Fx)$ to $\exists X \forall x(Xx \leftrightarrow Fx)$ also strikes me as problematical, but as it is legitimated by (the second-order analogue of) the standard logical rule of existential generalization, I have agreed not to object to it.)

'Moreover there is an important difficulty connected with the interpretation of the *Begriffsschrift*.[7] Frege does not discuss the question whether properties are objects, as one might put it. It is uncertain whether Frege thinks there can, for example, be sequences of properties, whether xfy might hold when x and y are themselves properties. One would have supposed so; but then, of course, taking "f" to mean "Is a property that is an instance of the property" produces a Russellian problem: $\exists X \forall x(Xx \leftrightarrow -xfx)$ is derivable in the *Begriffsschrift*, but would be read by Frege "There is a property whose instances are all and only those properties that are not instances of themselves," which is false, of course. Thus the system, although perhaps formally consistent, cannot be interpreted as Frege interprets it in the absence of some—I think the right word is "metaphysical"—doctrine of properties, which Frege does not supply. And what, pray, is the source of any such doctrine to be—pure logic? How then are we to interpret the *Begriffsschrift* so that its theorems all turn out to be truths that it does not require the aid of intuition to accept?

'I'm almost finished. Matters are no better and probably worse if Frege reads a second-order quantifier $\exists F$ as "There is a set F . . . ." For sets clearly are "objects"; thus the difficulty presented by Russell's paradox immediately arises if we take the range of "F" to be all *sets*. The only escape that I can see for Frege is for him to stipulate that the *Begriffsschrift* is to be employed in formalizing a certain theory only if the theory does not speak about *all* objects. The rule of substitution would then be licensed by the

---

[7]For an illuminating discussion of this difficulty, see I. S. Russinoff, "Frege's problem about concepts," MIT Ph.D. thesis, 1983.

Aussonderungsschema of set theory. But besides noting that this way out appears to be strongly at odds with his intentions in setting forth the *Begriffsschrift*, we may well wonder what justifies this appeal to the *Aussonderungsschema* if not *intuition* of some sort, for example the *picture* of the set-theoretic universe that yields the so-called "iterative conception of set." And now, I *am* finished.'

In reply: Russell's paradox does indeed show the difficulty of taking the second-order quantifiers of the *Begriffsschrift* as ranging over all sets or all properties and reading atomic formulae like Xx as meaning 'x is a member (or instance) of X.' We must find another way to interpret the formalism of the *Begriffsschrift*, on which we are not committed to the existence of such entities as sets or properties, and on which the comprehension schema $\exists X \forall x (Xx \leftrightarrow A(x))$ can plausibly be claimed to be a logical law.

Interpretation of a logical formalism standardly consists in a description of the objects over which the variables of the formalism are supposed to range and a specification that states to which of those objects the various relation letters of the formalism apply. Since Frege nowhere specifies what his relation letters 'f,' 'F,' etc. apply to, it is clear, I think, that he had no one 'intended' interpretation of the *Begriffsschrift* in mind: 'f,' for example, will have to be interpreted on each particular occasion by mentioning the pairs of objects that it is then intended to apply to. But it appears that Frege did intend the first-order variables of the *Begriffsschrift* to range over absolutely all of the 'objects,' or things, that there are. In any event even if Frege did envisage applications in which the first-order variables were to range over some but not all objects, it seems perfectly clear that he did allow for some applications in which they do range over absolutely all objects. And because a use of the *Begriffsschrift* in which the variables do not range over all objects that there are can, by introducing new relation letters to relativize quantifiers, be treated as one in which they do range over all objects, we shall henceforth assume that the *Begriffsschrift*'s first-order variables do range over all objects, whatever an object might happen to be.

But what do the second-order variables range over , if not all sets or all properties? I think that a quite satisfactory response to this question is to reject it, to say that no separate specification of items over which the second-order variables range is needed once it has

been specified what the first-order variables range over.[8] Instead
we must show how to give an intelligible interpretation of all the
formulae of the *Begriffsschrift* that does not mention special items
over which the second-order variables are supposed to range and on
which Frege's rule of substitution appears as a rule of logic and the
comprehension axioms appear as logical truths.

The key to such an interpretation can be found in the behaviour
of the logical particle 'the.'

If the rocks rained down, then there are some things that rained
down; if each of *them* [pointing] is a K and each K is one of them,
then there are some things such that each of them is a K and each
K is one of them; if Stiva, Dolly, Grisha, and Tanya are unhappy
with one another, then there are some people who are unhappy with
one another. Existential generalization can take place on plural pronouns and definite descriptions as well as on singular, and existential
generalization on plural definite descriptions is the analogue in natural language of Frege's rule of substitution. This type of inference is
not adequately represented by the apparatus of standard first-order
logic. However, a formalism like that of the *Begriffsschrift* can be
used to schematize plural existential generalization, and our understanding of the plural forms involved in this type of inference can
be appealed to in support of the claim that Frege's rule is properly
regarded as a rule of logic.

By a 'definite plural description' I mean either the plural form
of a definite singular description, for example 'the present kings of
France,' 'the golden mountains,' or a conjunction of two or more
proper names, definite singular descriptions, and (shorter) definite
plural descriptions, for example 'Russell and Whitehead,' 'Russell
and Whitehead and the present kings of France.'

Like the familiar condition: $\exists x \forall y (Ky \leftrightarrow y = x)$ which must be
satisfied by a definite singular description 'The K' for its use to be
legitimate, there is an analogous condition that must be satisfied by
definite plural descriptions. In the simplest case, in which a definite
plural description such as 'the present kings of France' is the plural
form of a definite singular description, the condition amounts only

---

[8]For more on this topic, see my "Nominalist Platonism," *The philosophical
review*, **94** (1985) 327–44.

to there being one object or more to which the corresponding count noun in the singular description applies. (Two or more, technically, if Moore and the Eleatic Stranger were right.) Thus like the definite singular description 'The K,' which has a legitimate use iff the K exists, i.e. iff there is such a thing as the K, 'The Ks' has a legitimate use iff the Ks exist, i.e. iff there are such things as the Ks, iff there is at least one K.

The obvious conjecture—I do not know whether or not it is correct—is that the general condition for the legitimate use of a conjunction of proper names, definite singular descriptions, and (shorter) definite plural descriptions is simply the conjunction of the conditions for the conjoined names and descriptions. We need not worry here whether the conjecture is true; for our purposes it will suffice to consider only definite plural descriptions of the simplest sort, plural forms of definite singular descriptions.

The connection between definite plural descriptions and the comprehension principle is that the condition under which the use of 'The Ks' is legitimate, viz. that there are some such things as the Ks, can also be expressed: there are some things such that each K is one of them and each one of them is a K. Thus 'if there is at least one K, then there are some things such that each K is one of them and each of them is a K' expresses a logical truth. Moreover, it is a logical truth that it is quite natural to symbolize as

$$\exists x K x \rightarrow \exists X (\exists x X x \ \& \ \forall x (X x \leftrightarrow K x)),$$

which is equivalent to the instance $\exists X \forall x (X x \leftrightarrow K x)$ of the comprehension scheme. Thus the idea suggests itself of using the construction 'there are some things such that ... them ... ' to translate the second-order existential quantifier $\exists X$ so that comprehension axioms turn out to have readings of the form 'if there is something ... , then there are some things such that each ... thing is one of them and each of them is something ... .' Let us see how this may be done.

We begin by supposing English to be augmented by the addition of pronouns 'it_x,' 'it_y,' 'it_z,' ... ; 'that_x,' 'that_y,' 'that_z,' ... ; '$\text{they}_X/\text{them}_X$,' '$\text{they}_Y/\text{them}_Y$,' '$\text{they}_Z/\text{them}_Z$,' ... ; '$\text{that}_X$,' '$\text{that}_Y$,' '$\text{that}_Z$,' .... (For each first-order variable v of the formalism, we introduce 'it'$\widehat{\ }_v$ and 'that'$\widehat{\ }_v$; and for each second-order variable V, 'they' $\widehat{\ }_V$, which is sometimes written 'them' $\widehat{\ }_V$, and 'that' $\widehat{\ }_V$.) The purpose of the subscripts is simply to disambiguate cross-reference and has nothing to do with the distinction between first- and second-order formulae or between singular and plural number. A similar augmentation would be required for translation into English of first-order formulae of the language of set theory containing multiple nested alternating quantifiers, for example formulae of the form $\forall w \exists x \forall y \exists z R(w, x, y, z)$. The extension of English we are contemplating is a conceptually minor one rather like lawyerese ('the former,' 'the latter,' 'the party of the seventeenth part'); our subscripts are taken for convenience to be the variables of the *Begriffsschrift* (instead of, say, numerals), but they no more *range over* any items than does 'seventeen' in 'the party of the seventeenth part.'

We now set out a scheme of translation from the language of the *Begriffsschrift* into English augmented with these subscripted pronouns.[9] Thus we specify the conditions under which sentences of the *Begriffsschrift* are true by showing how to translate them into a language we understand.

The translation of the atomic formula Xx is ⌜it$_x$ is one of them$_X$⌝. (The corner-quotes are Quinean quasi-quotes.)

The translation of the atomic formula $x = y$ is ⌜it$_x$ is identical with it$_y$⌝. The translation of any other atomic formula, for example Fx or xfy, is determined in an analogous fashion by the intended reading of the predicate letter it contains.

Let F_* and G_* be the translations of F and G. Then the translation of $-F$ is ⌜Not: F_*⌝ and that of $(F \ \& \ G)$ is ⌜Both F_* and G_*⌝. Similarly for the other connectives of the propositional calculus.

The translation of $\exists x F$ is ⌜There is an object that$_x$ is such that F_*⌝.

To obtain the translation of $\exists X F$: Let H be the result of substituting an occurrence of $-x = x$ for each occurrence of Xx in F and let H_* be the translation of H. (H has the same number of quantifiers as F.) Then the translation of $\exists X F$ is ⌜Either H_* or there are some objects that$_X$ are such that F_*⌝.

(Since ⌜There are some objects that$_X$ are such that F_*⌝ properly

[9]This scheme was given in my "To be is to be a value of a variable (or to be some values of some variables)," *The journal of philosophy*, **76** (1984) 430–49.

translates not $\exists X F$, but $\exists X(\exists x X x \quad \& \quad F)$, we need to disjoin a translation of H, which is equivalent to $\exists X(-\exists x X x \quad \& \quad F)$, with ⌜There are some objects that$_X$ are such that F_*⌝ to obtain a translation of $\exists X F$.)

When we apply this translation scheme to the notorious $\exists X - \exists x - (X x \leftrightarrow -x f x)$, with the predicate letter f given the reading: 'is a member of,' we obtain a long sentence that simplifies to 'if some object is not a member of itself, then there are some objects (that are) such that each object is one of them iff it is not a member of itself,' a trivial truth.

More generally, the translation of $\exists X \forall x(X x \leftrightarrow A(x))$ will, as desired, be a sentence that can be simplified to one that is of the form: either there is no object such that ... it ... or there are some objects such that an arbitrary object is one of them iff ... it .... And of course, our translation scheme respects the other rules of logic in the sense that if H follows from F and G by one of these rules, and the translations F_* and G_* of (the universal closures of) F and G are true, then the translation H_* of (the universal closure of) H is also true. Our scheme, therefore, respects Frege's rule of substitution of formulae for relation letters as well.

Thus there is a way of interpreting the formulae of the *Begriffsschrift* that is faithful to the usual meanings of the logical operators and on which each comprehension axiom turns out to say something that can also be expressed by a sentence of the form 'if there is something ... , then there are some things such that anything ... is one of them and any one of them is something ... .' Each sentence of this form, it seems fair to say, expresses a *logical* truth if any sentence of English does. It would, of course, be folly to offer a definition of logical truth—as Jerry Fodor once said, failing to take his own advice, 'Never give necessary and sufficient conditions for *anything*'—but I think one would be hard pressed to differentiate 'if there is a rock, then there are some things such that any rock is one of them and any one of them is a rock' from 'if there is a rock, then there is something such that if it is not a rock, then it is a rock' on the ground that the former but not the latter expresses a logical truth or on the ground that an intuition is required to see the truth of the former but not the latter.

Three final remarks about definite plural descriptions:

Valid inferences using the construction 'there are some things such that ... they ... ' that cannot be represented in first-order logic are not hard to come by. The interplay between this construction and definite plural descriptions is well illustrated by the inference

Every parent of someone blue is red.
Every parent of someone red is blue.
Yolanda is red.
Xavier is not red.
It is not the case that there are some persons such that
Yolanda is one of them,
Xavier is not one of them, and
every parent of any one of them is also one of them.
Therefore, Xavier is a parent of someone red.

To see that this is valid, note that it follows from the premisses and denial of the conclusion that Yolanda is either red or a parent of someone red, that Xavier is not, and that every parent of anyone who is red or a parent of someone red is also red or a parent of someone red. Thus there are some people, viz. the persons who are either red or a parent of someone red, such that Yolanda is one of them, Xavier is not one of them, and every parent of any one of them is also one of them, which contradicts the last premiss. This inference may be represented in second-order logic:

$\forall w \forall z (Bz \ \& \ wPz \rightarrow Rw)$

$\forall w \forall z (Rz \ \& \ wPz \rightarrow Bw)$

Ry

$- Rx$

$- \exists X (\exists z Xz \ \& \ Xy \ \& \ -Xx \ \& \ \forall w \forall z (Xz \ \& \ wPz \rightarrow Xw))$

Therefore, $\exists z (xPz \ \& \ Rz)$.

In deducing the conclusion from the premisses in the *Begriffsschrift*, one would, of course, substitute $\{ a : Ra \lor \exists z (aPz \ \& \ Rz) \}$ for the second-order variable X, thus making a move similar to those we have seen Frege make.

It appears that not much in general can be said about 'atomic' sentences that contain definite plural descriptions but do not express statements of identity. 'The rocks rained down,' for example, does not mean 'Each of the rocks rained down.' However, if the rocks rained down and the rocks under discussion are the items in pile x, then the items in pile x certainly rained down. If we have learned anything at all in philosophy, it is that it is almost certainly a waste of time to seek an analysis of 'The rocks rained down' that reduces it to a first-order quantification over the rocks in question. It is highly probable that an adequate semantics for sentences like 'They rained down' or 'the sets possessing a rank exhaust the universe' would have to take as primitive a new sort of predication in which, for example 'rained down' would be predicated not of particular objects such as this rock or that one, but rather of these rocks or those. Thus it would appear hopeless to try to say anything more about the meaning of a sentence of the form 'The Ks M' other than that it means that there are some things that are such that they are the Ks and they M. The predication 'they M' is probably completely intractable.

About statements of identity, though, something useful if somewhat obvious can be said: 'The Ks are the Ls' is true if and only if there is at least one K, there is at least one L, and every K is an L and vice versa: $\exists x K x$ & $\exists x L x$ & $\forall x (K x \leftrightarrow L x)$. 'They are the Ks' can also be naturally rendered with the aid of a free second-order variable X: $\exists x X x$ & $\forall x (X x \leftrightarrow K x)$. And of course if some things are the Ks and are also the Ls, then the Ks are the Ls. Frege was not far wrong when he laid down Basic Law V. Of course, from time to time, there will be no set of (all) the Ks, as the sad history of Basic Law V makes plain. We cannot always pass from a predicate to an extension of the predicate, a set of things satisfying the predicate. We can, however, always pass to the things satisfying the predicate (if there is at least one), and therefore we cannot always pass from the things to a set of them.

APPENDIX: PROOFS OF 108, 129 AND 133

Definitions:

$\text{Her}(F)$ $\forall d \forall a(Fd \;\&\; dfa \to Fa)$ (69 in *Begriffsschrift*)

$\text{In}(x, F)$ $\forall a(xfa \to Fa)$

xf^*y $\forall F(\text{Her}(F) \;\&\; In(x, F) \to Fy)$ (76)

$xf^*_=y$ $xf^*y \lor x = y$

FN $\forall d \forall e \forall a(dfe \;\&\; dfa \to a = e)$ (115)

108. $zf^*_= y \;\&\; yfv \to zf^*_= v.$

Proof. Assume $zf^*_= y$, yfv, $\text{Her}(F)$ and $\text{In}(x, F)$. If zf^*y, then Fy, and by yfv and $\text{Her}(F)$, Fv; but if $y = z$ then $\text{In}(y, F)$ and again Fv, as yfv. Thus zf^*v, whence $zf^*_= v$. $\square$

129. $FN \;\&\; (yf^*m \lor mf^*_= y) \;\&\; yfx \to (xf^*m \lor mf^*_= x).$

Proof. Assume FN, $(yf^*m \lor mf^*_= y)$, and yfx. We must show $xf^*m \lor mf^*_= x$. Suppose yf^*m. By 110 we need only show $\forall a(yfa \to xf^*_= a)$, for then $xf^*_= m$, whence xf^*m or $m = x$, and then $xf^*m \lor mf^*_= x$. So suppose yfa. Since yfx and FN, $x = a$, whence $xf^*_= a$. Now suppose $mf^*_= y$. We show mf^*x, whence $mf^*_= x$. Assume $\text{Her}(F)$ and $\text{In}(m, F)$. We are to show Fx. If mf^*y, then since $\text{Her}(F)$ and $\text{In}(m, F)$, Fy, and then, since yfx and $\text{Her}(F)$, Fx. But if $y = m$, then from yfx, mfx, whence again Fx, since $\text{In}(m, F)$. $\square$

The second main theorem of the *Begriffsschrift* (133). FN $\;\&\; xf^*m \;\&\; xf^*y \to [yf^*m \lor y = m \lor mf^*y].$

Proof after four lemmas.

Lemma 1. $bfa \to bf^*a.$ (91)

Proof. Suppose bfa. Assume $\text{Her}(F)$, $\text{In}(b, F)$; show Fa. Since bfa and $\text{In}(b, F)$. $\square$

Lemma 2. $cf^*d \;\&\; df^*a \to cf^*a.$ (98)

Proof. Suppose cf^*d and df^*a. Assume $\text{Her}(F)$ and $\text{In}(c, F)$; show Fa. Since cf^*d, $\text{Her}(F)$, and $\text{In}(c, F)$, Fd. If dfb, then since $\text{Her}(F)$, Fb; thus $\text{In}(d, F)$. Since $\text{Her}(F)$ and df^*a, Fa. $\square$

Lemma 3. $[c = d \vee cf^*d]$ & $dfa \rightarrow [c = a \vee cf^*a]$. (108)

Proof. Suppose $[c = d \vee cf^*d]$ and dfa. If $c = d$, then cfa, whence cf^*a by lemma 1; if cf^*d, then since dfa, df^*a by lemma 1, and by lemma 2, cf^*a again. In any event, $c = a \vee cf^*a$. $\quad \square$

Lemma 4. FN & cfb & $cf^*m \rightarrow [b = m \vee bf^*m]$. (124)

Proof. Suppose FN and cfb. Let $F = \{\, z : b = z \vee bf^*z \,\}$. Suppose $[b = d \vee bf^*d]$ and dfa. By lemma 3, $[b = a \vee bf^*a]$. Thus Her(F). If cfa, then by FN, $b = a$, whence $b = a \vee bf^*a$; thus In(c, F). Therefore in cf^*m, Fm, i.e. $b = m \vee bf^*m$. $\quad \square$

Proof of the theorem. Suppose FN. Let $F = \{\, z : zf^*m \vee z = m \vee mf^*z \,\}$. Suppose $[df^*m \vee d = m \vee mf^*d]$ and dfa. If df^*m, then by lemma 4, $[a = m \vee af^*m]$, whence $[af^*m \vee a = m \vee mf^*a]$; and if $d = m \vee mf^*d$, then $m = d \vee mf^*d$, and by lemma 3, $m = a \vee mf^*a$, whence again $[af^*m \vee a = m \vee mf^*a]$. Thus Her$(F)$. Now suppose xf^*m. Assume xfa. By lemma 4, $[a = m \vee af^*m]$, whence $[af^*m \vee a = m \vee mf^*a]$. Thus In$(x, F)$. At last, suppose xf^*y. Then Fy, i.e. $yf^*m \vee y = m \vee mf^*y$. $\quad \square$

7

Frege's Theory of Number

It is impossible to compare Frege's *Foundations of arithmetic* [1] with the writings on the philosophy of mathematics of Frege's predecessors—even with such great philosophers as Kant—without concluding that Frege's work represents an enormous advance in clarity and rigor. It is also hard to avoid the conclusion that Frege's analysis increases our understanding of the elementary ideas of arithmetic and that there are fundamental philosophical points that his predecessors grasped very dimly, if at all, which Frege is clear about.

I mention this impression which Frege's book makes because it is often forgotten in critical discussion of his ideas, and still more forgotten in discussion of "the Frege-Russell view," "the reduction of mathematics to logic," or "logicism." Frege's main thesis, that arithmetic is a part of logic, is not fashionable now. It seems to me

From *Philosophy in America*, Max Black, ed., Ithaca: Cornell University Press, 1965, 180–203. Reprinted by kind permission of George Allen and Unwin, Routledge & Kegan Paul and the author. The "Postscript" which follows the essay is from the author's *Mathematics in philosophy: selected essays*, Ithaca: Cornell University Press, 1983, 173–175. It is reprinted by kind permission of Cornell University Press and the author.

This paper benefited from the comments of hearers of earlier versions at Cornell and Columbia in 1962 and 1963 and from the comments of Professor Black and especially Burton Dreben on the penultimate version.

[1] *Die Grundlagen der Arithmetik*. Quotations are in Austin's translation, *The foundations of arithmetic*, Oxford: Blackwell, 1950, 2nd ed., 1953.

that this is justified, and the accumulated force of the criticisms of this thesis is overwhelming. But even though Frege is more studied now than at the time when his thesis was regarded by many as having been conclusively proved, I find that we still lack a clear view of what is true and what is false in his account of arithmetic. What follows is intended as a contribution toward such a view.

I

It will help with this task not to focus our attention too exclusively on the thesis that arithmetic is a part of logic. An examination of the argument of the *Foundations* shows that this thesis is introduced only after some of the confusions of his predecessors have been cleared up by other analyses. It seems to me that we might best divide Frege's view-into three theses, which I shall discuss in turn.

(1)   Having a certain cardinal number is a property of a *concept* in what we may take to be Frege's technical sense. It appears that the basic type of singular term referring to numbers is of the form 'the number of objects falling under the concept F,' or, more briefly, 'the number of Fs,' or in symbolic notation 'N_xFx.'

(2)   Numbers are *objects*—again in Frege's technical sense.

(3)   Arithmetic is a part of logic. This may be divided into two:
  (a) The concepts of arithmetic can be defined in terms of the concepts of logic.
  (b) The theorems of arithmetic can be proved by means of purely logical laws.

The first thesis does not require much discussion. The appeal to Frege's special sense of 'concept' is of course something which would give rise to difficulty and controversy, but it is not essential to the main point. Everyone will agree that we cannot get far in talking about cardinal numbers without introducing singular terms of the form mentioned, or others in which the general term is replaced by a term referring to a class or similar entity.

There is a further question whether, in elementary examples involving perceptual objects, we could attribute the number to something more concrete, or more in accord with the demands of nominalism, than a class or a concept. Frege himself was apparently not interested in this question, and it does not seem to me very important for the foundations of mathematics.

II

I shall now discuss the thesis that numbers are objects. It will prove to be closely connected with the third thesis, that arithmetic is a part of logic. The first thing to note is that for Frege the notion of an object is a *logical* one. He held that linguistic expressions satisfying certain syntactical conditions at least purport to refer to objects. I do not have a precise general account of what these conditions are. Being a possible subject of a proposition is the primary one, but it must be so in a *logical* sense; otherwise 'every man' in 'Every man is mortal' would refer to an object. The occurrence of the definite article is an important criterion. Such examples as 'The number 7 is a prime number' seem to show that numerical expressions satisfy the syntactical criteria.

Another criterion of great importance in the *Foundations* is that *identity* must have sense for every kind of object.

Frege took this in a very strong sense: if we think of '____' and ' ... ' as object-expressions, then '____ = ... ' must have sense even if the objects to which they purport to refer are of quite different categories, for example, if ____ is 'the Moon' and ' ... ' is 'the square root of 2.'[2] Moreover, the principle of substitutivity of identity must be satisfied.

> Now for every object there is one type of proposition which must have a sense, namely the recognition-statement, which in the case of numbers is called an identity .... When are we entitled to regard a content as that of a recognition-judgement? For this a certain condition has to be satisfied, namely that it must be possible in every judgement to substitute without loss

---

[2]This follows from Frege's general doctrine that a function must be defined for every object as argument, since identity is for Frege a function of two arguments. See *Funktion und Begriff*, pp. 19–20; *Grundgesetze* I, §§ 56–65. (See Peter Geach and Max Black, eds., *Translations from the philosophical writings of Gottlob Frege*, Oxford: Blackwell, 1952, 1st ed., pp. 159–170); *Foundations* § 66, 106.

of truth the right-hand side of our putative identity for its
left-hand side.[3]

This is a view which Quine expresses succinctly by the maxim, "No
entity without identity." One of the main efforts of the positive part
of the *Foundations* is to explain the sense of identities involving
numbers.

That terms satisfying these conditions, and perhaps others besides, occur in places accessible to quantification, and that we make
such inferences as existential generalization (e.g., from '2 is an even
prime number' to 'There is an even prime number') might be taken
to show that numerical terms purport to refer to objects. From
Frege's point of view, however, I should think that this shows only
that they purport to *refer*, for quantification could also occur over
functions, including concepts.

Shall we accept Frege's criteria for expressions to purport to refer to objects? We might, I think, separate those explicitly stated
by Frege from the criterion, added from Quine, of accessibility to
quantification. The latter has some complications having to do with
mathematical constructivity and predicativity. These do not make it
an unacceptable criterion, but might lead us to distinguish "grades
of referential involvement." With this reservation, I do not know
any better criteria than the ones I have mentioned. I am still not
very clear about their significance, what it is to *be* an object. In
particular, the central role of identity is something which I do not
know how to explain. Perhaps it has to do with the fact that the
cognitive activities of human beings are spread out over space and
time.

III

I have been very careful to speak of criteria for expressions to *purport* to refer to objects. Indeed, it would seem that if we explain
number-words in such a way that they will be shown in at least
some of their occurrences to satisfy these criteria, then we shall at
most have shown that they purport to refer to objects but not that
they actually *do* refer to them, i.e., that in these occurrences they

---

[3] *Foundations* § 106.

actually *have* reference. I shall now consider how this matter stands in Frege's analysis of number.

The simplest account of it is as follows: Frege finds[4] that a necessary and sufficient condition for the number of Fs to be *the same* as the number of Gs is that the concept F and the concept G stand in a relation he called *Gleichzahligkeit*, which may be translated as 'numerical equivalence.' The concept F is numerically equivalent to the concept G if there is a one-to-one correspondence of the objects satisfying F and the objects satisfying G. If we express this by '$Glz_x(Fx, Gx)$,' we may express the result of this stage of Frege's analysis as the principle

$$\text{(A)} \qquad N_x Fx = N_x Gx. \equiv Glz_x(Fx, Gx).$$

He then gives an *explicit definition* whose initial justification is that (A) follows from it. The number of Fs is defined as the extension of the concept *numerically equivalent to the concept F*, in other words the class of all concepts numerically equivalent to the concept F.

Then it seems that the problem of the existence of numbers is merely reduced to the problem of the existence of extensions. If this is so, Frege is in two difficulties.

The first is that the paradoxes make it not very clear what assumptions as to the existence of extensions of concepts are permissible. Frege sought a general logical law by which one could pass from a concept F to its extension $\hat{x}Fx$, but his Axiom V:

$$\hat{x}Fx = \hat{x}Gx. \equiv (x)(Fx \equiv Gx)^5$$

led directly to Russell's paradox. What stands in its place in later set theory is a variety of possible existence assumptions of varying degrees of strength.

---

[4] *Foundations* §§ 61–69.

[5] My symbolism, essentially that of Quine's *Methods of logic*, New York: Henry Holt, 1959, Rev. (2d) ed., follows the *Foundations* rather than the ideas of the *Grundgesetze*. Throughout I largely neglect the fact that in Frege's later writings concepts are a special kind of function; the axiom stated is more special than the actual Axiom V of the *Grundgesetze*, which relates arbitrary functions and what Frege calls their *Werthverläufe*. This expository convenience is not meant to exclude the *Grundgesetze* from the scope of my discussion.

Frege may have seen no alternative to the recourse to extensions in order to secure a reference for numerical terms. It may be for that reason that he saw Russell's paradox as a blow not just to his attempt to prove that arithmetic is a part of logic, but also to his thesis that numbers are objects:

> And even now I do not see how arithmetic can be scientifically established; how numbers can be apprehended as logical objects, and brought under review, unless we are permitted—at least conditionally—to pass from a concept to its extension.[6]

He never mentions the possibility of apprehending numbers as objects of a kind other than logical.

It is possible to identify at least finite numbers with quite unproblematic extensions. A natural way of doing this is to identify each number with some particular class having that number of members, as is done in von Neumann's construction of ordinal numbers.[7]

The second difficulty is that if we admit enough extensions, on some grounds or other, then there are too many possible ways of identifying them with numbers. In fact, *any* reasonably well-behaved sequence of classes can be chosen to represent the natural numbers.

Frege must have thought that his own choice was more natural than any alternative. The relation of numerical equivalence is reflexive, symmetric, and transitive; and he thought of a number as an equivalence class of this relation. He motivates it by discussing the notion of the direction of a line as an equivalence class of the relation of parallelism.[8] But this will not do. Ordinary equivalence classes are subclasses of some given class. But the application of numbers must be so wide that, if *all* concepts (or extensions of concepts) numerically equivalent to a concept F are members of $N_x Fx$ then it is by no means certain that $N_x Fx$ is not the sort of "unconditioned totality" that leads to the paradoxes. The difference is

---

[6] *Grundgesetze* II, Appendix p. 253; translation from Geach and Black, 1st ed., p. 234. (The Appendix in which this passage occurs is translated in full by Montgomery Furth in *The basic laws of arithmetic: exposition of the system*, Berkeley and Los Angeles: University of California Press, 1964.)

[7] See for example Quine, *Set theory and its logic*, Cambridge: Harvard University Press, 1969, chap. 7.

[8] Cf. Dummett, "Nominalism," *The philosophical review*, **65** (1956) 491–505, esp. p. 500.

reflected in the fact that in the most natural systems of set theory, such as those based on Zermelo's axioms, the existence of ordinary equivalence classes is easily proved, while, if anything at all falls under F, the nonexistence of Frege's $N_x F x$ follows.

It is odd that we should have to identify numbers with extensions in order to insure that number terms have a reference, but that we should then be able to choose this reference in almost any way we like. We might entertain the fantasy of a tribe of mathematicians who use the ordinary language of number theory and who also all accept the same set theory. In their public life, the question whether the numbers are to be identified with classes never arises. However, each one *for himself* identifies the natural numbers with a certain sequence of classes but does not tell the others which it is. If one says that two terms of number theory refer to the same number, whether another assents or dissents in no way depends on whether *his* natural numbers are the same as the speaker's.

The reader will be reminded of a well-known passage in the *Philosophical investigations* (I, 293). If it makes no difference in mathematics to which class a term of number theory refers, what is the relevance to the thesis that numbers are objects of the *possibility* of an identification of numbers and classes?

IV

Dummett, in the paper cited in note 8, suggests another way of looking at this problem. He appeals to Frege's principle that "only in a proposition have words really a meaning,"[9] that "we must never try to define the meaning of a word in isolation, but only as it is used in the context of a proposition."[10] Dummett takes this to mean that in order to determine the sense of a word, it is sufficient to determine the sense of the sentences in which the word is used. This is a matter of determining their truth-conditions. If a word functions syntactically as a proper name, then the sense of the sentences in which it occurs will determine its sense; and which sentences, as a matter of extralinguistic fact, express true propositions will determine its reference.

[9] *Foundations* § 60.
[10] *Foundations* § 106.

Although Dummett says that when this position is stated it is a "banality," it seems to me to have a serious ambiguity which makes it doubtful that it can get around the difficulties we have mentioned. Dummett apparently takes the principle to mean that if the specification of truth-conditions of the contexts of a name can be done at all, then the name *has* a reference. But this comes into apparent conflict with Frege's principles. For a sufficient explanation of the sense of a name by no means guarantees that it has a reference, and therefore that the sentences in which it occurs express propositions that are true or false. So it seems that there will be a further question, once the sense has been specified, whether the reference actually exists, which must be answered before one can begin to answer questions about the truth-values of the propositions. Or perhaps it will prove impossible to specify the sense of all the expressions required without in the process specifying the reference or presupposing that it exists. In either case, the usefulness of Frege's principle for repudiating philosophical questions about existence would be less than Dummett thinks.

However, I do not think that these considerations show that Dummett's interpretation of Frege is wrong. Another possible way of taking the principle is that if we can show that sentences containing a certain name have well-determined truth-values, then we are sure this name has a reference, and we do not have to discover a reference for the name antecedently to this. Frege indicates that the contextual principle is the guiding idea of his analysis of number in the *Foundations*; I shall try taking it this way. I shall also suppose that the same principle underlies Frege's attempted proof in the *Grundgesetze der Arithmetik*[11] that every well-formed name in the formal system there set forth has a reference.

There is a general difficulty in the application of the principle which will turn up in both these cases: it is hard to see how it can be applied unless it provides for the elimination of the names by contextual or explicit definitions. We shall see that, for a quite simple reason, the kind of contextual definition Frege's procedure

[11]I, §§ 28–32. I am indebted to M. D. Resnik for pointing out to me the relevance of this argument to my discussion. A detailed analysis of the argument occurs in his Ph.D. thesis "Frege's Methodology," Philosophy Dept., Harvard University, 1963.

might suggest is impossible. And in the case of the *Grundgesetze* argument, where the crucial case is that of names of extensions,[12] it seems obvious that explicit definition is impossible.

Frege regarded identity contexts of an object-name as those whose sense it is most important to specify. His procedure in the *Foundations* can be regarded as an effort to do this. These identities are of three forms:

(1) the number of Fs = the number of Gs,
(2) the number of Fs = 7,

where '7' could be replaced by any other such expression which we take to refer to a number.

(3) the number of Fs = ...

where ' ... ' represents a name of a quite different type, such as 'the moon,' 'Socrates,' or 'the extension of the concept *prime number*.'

The truth-conditions of identities of form (1) are determined by the above-mentioned principle (A): (1) is to be true if and only if there is a one-to-one correspondence of the Fs and the Gs. This in turn can be defined without appealing to the concept of number, as we shall see.

Frege analyzed identities of form (2) by defining individual numbers as the numbers belonging to certain particular concepts, so that they are in effect assimilated to identities of form (1). Thus 0 is the number of the concept *not identical with itself* and turns out to be the number of any concept under which nothing falls. $n + 1$ is the number of the concept *member of the series of natural numbers ending with n*. I shall not raise now any problems concerning such identities or Frege's way of handling them.

Thus we can regard the explicit definition mentioned above at the beginning of section III as necessary only to handle identities of form (3), as Frege intimates.[13] By supposing that the notion of the extension of a concept is already understood, Frege provides that the sense of (3) is to be the sense of:

the extension of the concept *numerically equivalent to the concept F* = ... ,

---

[12]More generally, *Werthverläufe* of functions; see note 5 above.
[13]*Foundations* § 106.

and the sense of the latter is already determined. Then, as I said, (A) follows from the explicit definition, so that this definition also serves for other contexts.

V

That Frege did not regard the introduction of extensions as essential to his argument is insisted upon by Peter Geach.[14] He seems to think that the analysis of identities of form (1) is by itself sufficient to establish the thesis that numbers are objects:

> Having analysed 'there are just as many As as Bs' in a way that involved no mention of numbers or of the concept number, Frege can now offer this analysis as a criterion for numerical identity—for its being the case that the number of As is the same number as the number of Bs. Given this sharp criterion for identifying numbers, Frege thought that only prejudices stood in the way of our regarding numbers as objects. I am strongly inclined to think he is right.[15]

It would seem that we could deal with identities of type (3) very simply, by specifying that they are all either nonsense or false. The first alternative is plausible enough in such examples as 'the number of planets = the moon' or '(the number of roots of the equation $x^2 + 3x + 2 = 0) =$ the class of prime numbers.' It would, however, be an abandonment of Frege's position that objects constitute a single domain, so that functions and concepts must be defined for all objects.[16] The second is plausible enough when ' ... ' is a closed term; it means rejecting the demand that we identify numbers with objects given in any other way.

However, both these solutions have a fatal defect: apparently we must explain the sense of (3) when ' ... ' is a *free variable*. We cannot declare that '$N_x F x = y$' is to be true of nothing, for that will contradict our stipulation concerning (1) and lead to the consequence that our universe contains no numbers. It seems that the

---

[14] "Class and Concept," *The philosophical review*, **64** (1955) 561–70; G. E. M. Anscombe and Geach, *Three philosophers*, Oxford: Blackwell, 1961, p. 158.

[15] Anscombe and Geach, *Three philosophers*, p. 161. It is not clear what Geach intends to be the final relation between numbers and extensions.

[16] See the references given in note 2 above.

only ways to take care of this fact are either to give an explicit definition of '$N_x F x = y$' or to assume we understand what the number of
Fs is and say that '$N_x F x = y$' is to be true of just that object. But
this in effect begs the question of reference. This is a quite general
difficulty which does not depend on the fact that numbers are to be
in the same universe of discourse with other objects. For it is essential to the use of *quantification* over a universe of discourse which
is to include numbers that '$N_x F x$' should occur in places where it
can be replaced by a variable. And not to quantify over numbers is
surely to renounce the thesis that numbers are objects.

However, it seems that explanation of '$N_x F x = \ldots$' where
'$\ldots$' is a *closed* term (in Frege's language, a proper name) might be
regarded as sufficient if it is presupposed that the quantifiers range
only over such objects as have names in one's formalism. Indeed,
if this is so then it seems that we have explained for what objects
'$N_x F x = y$' is to be true, so that the above objection has no force.
The condition, however, cannot be met if the formalism is to express (in the standard way) enough classical mathematics to include
a certain amount of set theory or the theory of real numbers, since
in that case the universe will contain indenumerably many objects,
while the formalism contains only denumerably many names.

In the *Grundgesetze* (I, §§ 28–32) Frege tries to show on the basis
of what seems to be a generalization of the contextual principle that
each well-formed name of his formal system has a reference. This
is, in particular, the only attempt he makes at a direct justification of his introduction of *Werthverläufe* and his axiom (V) about
them. The principle on which he operates is that an object name
has reference if every name which results from putting it into the
argument place of a referential first-level function name has reference. Similarly, a function name (of whatever level) has reference if
every result of putting in a referential name of the appropriate type
has reference. Since any sentence (i.e., name purporting to refer
to a truth-value) containing a given object name can be viewed as
the result of applying a function name to the given object name,
the first principle is a generalization of Dummett's principle that a
proper name has reference if every sentence in which it occurs has
a truth-value.

In applications it is sufficient to show that simple names have reference. The difficult case, and that which is of interest to this discussion, is that of abstracts. The problem of showing that '$\hat{x}(\ldots x \ldots)$' has a reference reduces to that of showing that '$\hat{x}(\ldots x \ldots) = \underline{}$' has a well-determined truth-value, whatever object the name '$\underline{}$' represents.

However, the argument fails for two reasons. First, the principles do not require him to determine the truth-value in the case where '$\underline{}$' is a free variable, so that the same problem arises in interpreting quantification. The other difficulty arises from impredicative constructions. Frege argues that if '$\ldots x \ldots$' is referential, then so is '$\hat{x}(\ldots x \ldots)$.' But to show that '$\ldots x \ldots$' is referential we need to show that '$\ldots \underline{} \ldots$' is referential if '$\underline{}$' is *any* referential object name. But one of these is '$\hat{x}(\ldots x \ldots)$,' if everything turns out right. So that it is not at all clear that the rules exclude circularities and contradiction, as indeed they do not if we let '$\ldots x \ldots$' be '$x \in x$' or '$-(x \in x)$.'

Frege could meet both difficulties, at the sacrifice of some of classical mathematics, by restricting himself to predicative set theory. This would, however, also not be in the spirit of his philosophy. It seems incompatible with Frege's realism about abstract objects to admit only such as have names and not to allow quantification over all of them (as would happen if predicativity were realized in the traditional way, by arranging the variables in a ramified hierarchy). Since the same considerations would dictate a predicative interpretation of quantification over concepts and relations, it seems there will no longer be a single relation of numerical equivalence, so that the notion of cardinal number will diverge from the (Cantorian) one which Frege intended. However, the elementary arithmetic of natural numbers would not be affected by this. So the thesis that numbers are objects might be sustained on this basis, if its application were restricted to natural numbers. But the divergence from Frege's intentions which this possibility involves justifies us in not pursuing it further.

From his realistic point of view, Frege cannot complete the specification of the senses of numerical terms and class abstracts. He cannot avoid making some assumptions as to their reference which could in principle be denied. The situation can, however, be viewed as follows: The information given about the *sense* of such terms by the principle (A) and the axiom (V) is not useless, since it enables one to eliminate the reference to numbers and classes from some contexts at least, and to decide the truth-values of some propositions referring to them. We can call (A) or (V) or similar principles *partial contextual definitions*. They give some justification for the assumption of entities of a certain kind. But they are no guarantee even against contradiction, as is shown by the fact that an instance of (V) gives rise to Russell's paradox.

I shall make some further remarks about explicitly defining 'the number of Fs' in terms of classes. The objection mentioned above loses much of its force if one drops the idea that it is the possibility of definition of this kind which guarantees that numbers exist. This does not mean that such an explicit definition cannot be used, even in a theory which is to serve as a philosophically motivated "rational reconstruction." The point is that the kind of general explanation and justification of the introduction of classes, to which that of numbers can be reduced by a definition, could be done directly for numbers. The identification of numbers with classes could still serve the philosophical purpose of showing that numbers are not more problematic entities than classes. There is still, however, a general difficulty about abstract entities which is illustrated by the fact that infinitely many different definitions, or perhaps none at all, are possible. This is that the concepts of different categories of abstract entities, even in highly developed mathematical theories, do not determine the truth-value of identities of entities of one category with those of another. There is perhaps some presumption in favor of regarding them as false or nonsense, but this is weakened by the fact that if it is disregarded for the sake of simplicity, no harm ensues to the logical coherence of the theories, and in practical application the worst that occurs is misunderstanding which can be dissipated easily. There is a similar difficulty, even within categories, which is illustrated by Frege's observation that if X is a one-one mapping of the universe of discourse onto itself, then if (V) holds and we set

$$\tilde{x}Fx = X(\hat{x}Fx)$$

then

$$\tilde{x}Fx = \tilde{x}Gx. \equiv (x)(Fx \equiv Gx)$$

also holds. So that the only condition he assumes about extensions does not suffice to determine which individual objects they are.[17]

Thus we still have a weaker form of our earlier difficulty: numbers and classes are regarded as definite objects while we seem able to choose freely between infinitely many incompatible assumptions about their identity and difference relations. One may regard this as just a failure of analogy between abstract and concrete entities[18] or regard it as resolvable by some kind of appeal to intuition.

VI

I shall now discuss the thesis that arithmetic is a part of logic. We are led into this by reminding ourselves that Frege's identification of numbers and extensions was part of the argument for this thesis. He regarded *Werthverläufe* as the most general kind of "logical objects"; the passage from a concept to its extension was the only way of inferring the existence of an object on logical grounds.[19] I shall give reasons for denying that set theory is "logic" once commitment to the existence of classes is introduced, but it is certainly a significant fact that in formal mathematical theories Frege's program of replacing postulation of objects by explicit definition in terms of classes can actually be carried out.

I do not know quite how to assess it. Dummett suggests that taking an equivalence relation of entities of one kind as a criterion of identity for entities of a new kind is the most general way of introducing reference to abstract entities, and if this is so then all abstract entities can be constructed as classes. This may not be enough to make axioms of class-existence logical principles, but if true it is striking and important.

The first part of Frege's argument for the thesis that arithmetic is a part of logic consists in his analysis of the general notion of cardinal number: he argues that the cardinal numbers belonging

---

[17] *Grundgesetze* I, § 10. Frege lessens this indeterminacy by identifying certain *Werthverläufe* with the True and the False.

[18] Similar difficulties may arise in the concrete case, e.g., between physical and psychical events.

[19] See in addition to the quotation above (p. 154) *Grundgesetze* II, § 147, and Frege's letter to Russell of July 28, 1902, in Gottfried Gabriel et al., eds., *Wissenschaftlicher Briefwechsel*, Hamburg: Meiner, 1976, p. 223.

to concepts must be objects satisfying the principle (A) and then gives the above-mentioned explicit definition of '$N_x Fx$' so that (A) becomes provable. In order to give an analysis of the notion of natural (finite) number, he must pick the natural numbers out from the class of cardinals.

We might ask whether the analysis so far is sufficient for the general notion of cardinal number. So far, I have talked as if what is essential to the notion of cardinal numbers is that they should be objects with the identity condition given by (A). Beyond that, it does not matter what objects they are, for example, whether a cardinal number is even identical with an object given in some other way. The emphasis of the *Foundations* suggests that Frege thinks this is all that is essential, but I doubt that he really does think so. To be clear about this, I shall consider a possible justification of (A) arising from the question, What is it to *know* the number of Fs for a given F? It could hardly be simply to know the name of an object with the given identity conditions, for 'the number of Fs' is already such a name.

The basis of indisputable fact on which Frege's analysis rests can be brought out by the following considerations. In the finite case, we know the number of Fs if we can name, in some standard fashion, a *natural number* n such that the number of Fs is n. The primary way of obtaining such a number is by *counting*. To determine by counting that there are n Fs involves correlating these objects, one by one, with the numbers from 1 to n. Thus we can take as a necessary condition for there to be n Fs that it should be possible to correlate the Fs one-to-one with $1 \ldots n$, i.e., that F and the concept *number from 1 to n* are numerically equivalent. The role of one-to-one correspondence in explaining the notion of number can be developed from this condition and the following mathematical considerations:

(1) If the Fs can be correlated one-by-one with the numbers from 1 to n, then the objects falling under the concept G can be correlated with the numbers from 1 to n if and only if they can be correlated with the Fs.

From (1) and our condition for there to be n Fs, it follows:

(2) If there are n Fs, then there are the same number of Fs as Gs if and only if there is a one-to-one correspondence of the Fs and

the Gs; that is, the principle (A) holds in the finite case.

(3) That a relation H establishes a one-to-one correspondence of the Fs and the Gs can be expressed by a formula of the first-order predicate calculus with identity. The condition is

$$(x)[Fx \supset (\exists! y)(Gy \cdot Hxy)] \cdot (y)[Gy \supset (\exists! x)(Fx \cdot Hxy)]$$

where '$(\exists! z)Jz$,' which can be read as 'there is one and only one z such that Jz,' is an abbreviation for

$$(\exists z)[Jz \cdot (x)(Jx \supset x = z)].$$

Thus, since F and G are numerically equivalent if and only if there is a relation H establishing a one-to-one correspondence of the F's and the G's, '$Glz_x(Fx, Gx)$' can be explicitly defined in the second-order predicate calculus with identity, in particular without appealing to the concept of number.

(4) For each natural n, a necessary and sufficient condition for there to be n Fs can also be expressed by a formula of the first-order predicate calculus with identity. If, following Quine,[20] we write 'there are exactly n objects x such that Fx' as '$(\exists x)_n Fx$,' then we have

(a) $(\exists x)_0 Fx \equiv -(\exists x)Fx$
(b) $(\exists x)_{n+1} Fx \equiv (\exists x)[Fx \cdot (\exists y)_n (Fy \cdot y \neq x)],$

so that for a particular n, the numerical quantifier can be eliminated step by step. Thus '$(\exists x)_2 Fx$' is equivalent to

$$(\exists x)\{Fx \cdot (\exists y)[Fy \cdot y \neq x \cdot -(\exists z)(Fz \cdot z \neq x \cdot z \neq y)]\}, \text{ i.e.}$$
$$(\exists x)(\exists y)[Fx \cdot Fy \cdot x \neq y \cdot (z)(Fz \supset .z = x \lor z = y)].$$

Frege followed Cantor in taking (A) as the basic condition cardinal numbers had to satisfy in the infinite case as well as the finite. He went beyond Cantor in using it as a basis for the ordinary arithmetic of natural numbers, and in observing that numerical equivalence could be expressed in terms merely of second-order logic. Without

---

[20] *Methods of logic*, § 39 (2d. ed.), § 44 (3d ed.), New York: Holt, Rinehart and Winston, 1972.

these two further steps, the procedure has, of course, no claim at all to be a reduction of arithmetic to logic.

As I have said, the remainder of Frege's argument, consists in picking out the natural numbers from the cardinals. Although this was not Frege's actual procedure, we can put it in the form of defining Peano's three primitives, '0,' 'natural number,' and 'successor,' and proving Peano's axioms. '0' and the successor relation are defined in terms of '$N_x Fx$':

$$0 = N_x(x \neq x)$$
$$S(x,y) \equiv (\exists F)[N_w Fw = y \cdot (\exists z)(Fz \cdot N_w(Fw \cdot w \neq z) = x)].$$

Then the natural numbers can be defined as those objects to which 0 bears the ancestral of the successor-relation, i.e.

$$NN(x) \equiv (F)\{F0 \cdot (x)(y)[Fx \cdot S(x,y). \supset Fy]. \supset Fx\}.$$

From these, Peano's axioms can be proved; it is not necessary to use any axioms of set existence except in introducing terms of the form '$N_x Fx$' and in proving (A), so that the argument could be carried out by taking (A) as an axiom. Lest this statement mislead, I should point out that I am provisionally counting the second-order predicate calculus as logic rather than as set theory; for from Frege's point of view the range of the higher-type variables will be *concepts*, while the extensions with which he identifies the numbers must be *objects* in the same domain with the objects numbered.

The definition of Peano's primitives and the proof of Peano's axioms can be carried out in one way or another not only in Frege's own formal system but also in Russell's theory of types and in the other systems of set theory constructed in order to remedy the situation produced by the paradoxes, which of course showed Frege's system inconsistent. We have seen that it is possible to give these definitions in infinitely many ways. It is sometimes said that what the logicists achieved in trying to prove that arithmetic is a part of logic is the proof that arithmetic can be modeled in set theory. But it should be pointed out that the modeling of the Peano arithmetic in set theory does not need to make use of the facts (1)–(4) above cited or the general analysis of cardinal number in terms of numerical equivalence, if Frege's choice of the sets to represent the numbers is abandoned.

VII

I intend to discuss two main lines of criticism of the thesis that arithmetic is a part of logic. The first points to the fact that the formal definitions and proofs by which the thesis is justified make use of the notion of extension, class, or set, and assume the existence of such entities. It is denied that set theory is logic. It is also denied that a reduction merely to set theory will suffice for a philosophical foundation of arithmetic or for a refutation of the epistemological theses about arithmetic (e.g., Kant's and Mill's views) against which the reduction is directed.

In discussing the thesis that numbers are objects, we found a difficulty for Frege in justifying assumptions of the existence of classes. Indeed, set theory will be logic only if propositions which assert the existence of classes are logical laws. Paul Benacerraf points out[21] that this would not be in accord with the usual definition of logical validity, according to which a formula is logically valid if and only if it is true under all interpretations in any nonempty universe, i.e., regardless of what objects, and how many, there are in the universe.[22] This definition applies to higher-order logic such as the one we have used in formulating Frege's views. But if numbers and classes are to be objects, a law which provides for the existence of any at all will require the universe over which the quantifiers range to contain *specific* objects, and if it provides for the existence of enough for even elementary number theory, it will require the universe to be infinite.

Thus there is a clear sense in which the predicate calculus is a more general theory than any set theory, and therefore more entitled to be called "logic." This observation is confirmed by the fact that there are infinitely many possible assumptions of the existence

[21]In his Ph.D. thesis, "Logicism: some considerations," Philosophy Dept., Princeton University, 1960, p. 196n.

[22]The qualification "nonempty" may be regarded merely as excluding a special case for the sake of simplicity.

of classes which can be ordered by logical strength, if the notion of strength is construed as what one may prove from them by the predicate calculus. Moreover, the stronger assumptions are in many ways more complex, obscure, and doubtful. Each well-defined system of such assumptions is incomplete in a strong sense and extendable in a natural way, in contrast to the completeness of the predicate calculus. But it is hard to see that the principles involved in such extensions are self-evident or logically compelling. This is particularly true when they allow impredicatively defined classes. Thus the existential commitments of set theory are connected with a number of important formal differences between it and the predicate calculus, and it seems that the predicate calculus is much closer to what was traditionally conceived as formal logic.

As a concession to Frege, I have accepted the claim of at least some higher-order predicate calculi to be purely logical systems. Our criticisms of Frege hold under this condition. The justification for not assimilating higher-order logic to set theory would have to be an ontological theory like Frege's theory of concepts as fundamentally different from objects, because "unsaturated." But even then there are distinctions among higher-order logics which are comparable to the differences in the strength of set theories. Higher-order logics have existential commitments. Consider the full second-order predicate calculus, in which we can define concepts by quantification over *all* concepts. If a formula is interpreted so that the first-order variables range over a class D of objects, then in interpreting the second-order variables we must assume a well-defined domain of concepts applying to objects in D which, if it is not literally the domain of *all* concepts over D, is comprehensive enough to be closed under quantification. Both formally and epistemologically, this presupposition is comparable to the assumption which gives rise to both the power and the difficulty of set theory, that the class of all subclasses of a given class exists. Thus it seems that even if Frege's theory of concepts is accepted, higher-order logic is more comparable to set theory than to first-order logic.

It is also sometimes claimed that the concept of class is intrinsically more problematic than that of numbers, so that the reduction of arithmetic to set theory is not a suitable philosophical foundation for arithmetic. There are several reasons which can be cited to support this contention—the multiplicity of possible existence assumptions, questions about impredicative definitions, the paradoxes, the possible indeterminacy of certain statements in set theory such as the continuum hypothesis. However, it should be pointed out that a quite weak set theory suffices for elementary number theory. Most of the difficulties arise only in the presence of impredicatively defined classes, and for the development of elementary number theory we do not have to suppose that any impredicative classes exist. If the modeling in set theory of a part of mathematics requires the existence of such classes, this can only be because the mathematics itself involves impredicativity, so that this is not a difficulty about the reduction to set theory.

However, it still seems that in order to understand even the weak theory, one must either have a general concept of set or assume it to be restricted in some way which involves the notion of number. The theory may be such that only finite sets can be proved to exist in it, but the reduction is not very helpful if the quantifiers of the theory are interpreted as ranging over finite sets. It seems that for a proposition involving quantification over *all* sets to have a definite truth-value, it must be objectively determined what sets exist. I think one might get around this, but it is certain that the assumption that any such proposition *has* a definite truth-value, which seems to be involved already in applying classical logic to such propositions, is stronger and more doubtful than any principle which needs to be assumed in elementary number theory.

VIII

The second criticism which I want to consider denies that arithmetic is reducible to set theory in the most important sense. This objection is as old as those concerning the foundations of set theory, but on the surface at least independent of them. It is to be found in Brouwer and Hilbert, but was probably argued in greatest detail by Poincaré.[23] Recently it has been taken up by Papert.[24] It is closely

[23] *Science et méthode*, Paris: Flameron, 1908, chap. 4.
[24] "Sur le réductionnisme logique," in P. Greco et al., eds., *Problèmes de la construction du nombre*, Paris: Presses Universitaires de France, 1960.

related to the criticisms of Wittgenstein[25] and Wang.[26]

This objection holds that the reduction is circular because it makes use of the notion of natural number. This obviously does not mean that the notion of natural number, or one defined in terms of it, appears as a primitive term of any set theory by which the reduction could be carried out, for this would be obviously false. Rather, the claim is that we must use the notion of natural number either to set up the set theory, to see the truth of the set-theoretical propositions to which number-theoretical propositions are reduced by explicit definitions, or to see the equivalence of the set-theoretical propositions and their number-theoretical correlates. This can in fact be seen by a quite simple argument. Inductive definitions, especially, play an essential role both in setting up a system of set theory and in establishing the correspondence between it and the system of number theory. For example, typically the definition of *theorem* for each system will be an inductive definition of the following form: Certain axioms and rules of inference are specified. Then to be a theorem is either to be an axiom or to be obtainable from theorems by a single application of one of the rules of inference. Then the model of number theory in set theory is established by defining for each formula A of number theory a set-theoretical translation $T(A)$. To prove that if A is a theorem of number theory then $T(A)$ is a theorem of set theory, we first prove this for the case where A is an axiom. Then suppose A follows from, say, B and C by a rule of inference in number theory, where B and C are theorems. We then show that $T(A)$ is deducible from $T(B)$ and $T(C)$ in the set theory. By hypothesis of induction, $T(B)$ and $T(C)$ are theorems of set theory. Therefore so is $T(A)$. By an induction corresponding to the definition of theorem of number theory, $T(A)$ is a theorem of set theory whenever A is a theorem of number theory.

Although the observation on which this objection is based is true, this is not sufficient to refute the reductionist. For the latter may maintain that the notion he defines is capable of *replacing* that of

---

[25] *Remarks on the foundations of mathematics*, Oxford: Blackwell, 1956, G. E. M. Anscombe, tr., part II of 1st ed.

[26] "Process and existence in mathematics," in Y. Bar-Hillel et al., eds., *Essays on the foundations of mathematics dedicated to A. A. Fraenkel*, Jerusalem: Magnes Press, 1961.

natural number (and equivalent notions involving induction) in all contexts, in particular those uses which are involved in describing the logical systems and in establishing the correspondence. So that we can imagine that in both the number-theoretical and the set-theoretical formal systems 'A is a theorem' is defined, say, as 'A belongs to every class of formulas which contains all axioms and is closed under the rules of inference,' as is done in some writings of Tarski. Whenever natural numbers are used ordinally for indexing purposes, they are to be replaced by their set-theoretical *definientia*.

We might consider in this connection a form of Poincaré's objection due to Papert (see note 24). Papert says in effect that the Frege-Russell procedure defines *two* classes of natural numbers, such that mathematical induction is needed to show them identical. For we give explicit definitions in the set theory of '0,' '$S0$', '$SS0$,' ... and also define the predicate '$NN(x)$.' How are we to be sure that '$NN(x)$' is true of what 0, $S0$, $SS0$, ... are defined to be and only these? Well, we can prove '$NN(0)$' and '$NN(x) \supset NN(Sx)$.'

If $NN(S^{(n)}0)^{27}$ is the last line of a proof, then by substitution and *modus ponens*, we have a proof of '$NN(S^{(n+1)}0)$.' By induction, we have a proof of '$NN(S^{(n)}0)$,' for every n.

We can likewise prove by induction that every x for which '$NN(x)$' is true is denoted by a numeral. For 0 clearly is. If $n = S^{(m)}0$, then $Sn = S^{(m+1)}0$. So by induction (the derived rule of the set theory), if $NN(x)$, then $x = S^{(m)}0$ for some $m > 0$.

What we have proved is a metalinguistic proposition. The reply to Papert would be to say that we can define the class of symbols $S^{(m)}0$ by the same device; i.e., if $Num(x)$ is to be true of just these symbols, we define:

$$Num(x) \equiv (F)\{F(`0') \cdot (x)[Fx \supset F(`S'^\frown x)]. \supset Fx\}.^{28}$$

Then the first of the above two inductive proofs will be an application of this definition, just as the second is an application of the definition of $NN(x)$.

---

[27]'$S^{(n)}0$' is an abbreviation for '0' preceded by n occurrences of 'S.'

[28]If x and y are expressions, $x^\frown y$ is their concatenation, that is, the result of writing x followed by y.

Papert can raise the same question again. We can prove formally that if $NN(x)$ holds, then x is denoted by an object y such that $Num(y)$. Moreover, we can prove formally that if $Num(y)$ then the result of substituting y for the variable 'x' in '$NN(x)$' is provable. But how do we know that the extension of '$Num(x)$' consists just of '0,' '$S0$,' '$SS0$,' etc.?

This is clearly the beginning of a potentially infinite regress. What is happening is this. If we replace an inductive definition by an explicit one, it takes an inductive *proof* to show that they define coextensive concepts. *This* difficulty can be avoided by denying that we have an independent understanding of inductive definitions. That is, if we ask Papert what he means by ' "0," "$S0$," "$SS0$," etc.,' he would have to reply by an explicit definition which would turn out to be equivalent to '$Num(x)$.' But this last reply to Papert depends on the claim that the apparatus in terms of which such explicit definitions as that of '$NN(x)$' are given can be understood independently of even the most elementary inductive definitions. This is implausible since the explicit definitions involve quantification over all concepts. It is hard to see what a concept is, or what the totality of concepts might be, without something like the inductive generation of linguistic expressions which (on Frege's view) refer to them.

There is another difficulty which Papert's exposition shows he has in mind.[29] It is independent of the question whether inductive definitions are replaced by explicit ones. Consider some numeral, say '$SSS0$,' and a proposition of the form '$F(SSSS0)$.' If we have proved '$F0$' and '$(n)[Fn \supset F(Sn)]$,' we have two independent ways of proving a proposition '$F(SSSS0)$.' We might infer '$(n)Fn$' by induction and '$F(SSSS0)$' by universal instantiation. Or we might prove it without induction, by successive applications of *modus ponens*:

---

[29] Especially in the subsequent essay, "Problèmes épistémologiques et génétiques de la récurrence," in *Problèmes de la construction du nombre*.

$$\dfrac{\dfrac{\dfrac{\dfrac{F0 \quad F0 \supset F(S0)}{F(S0) \quad F(S0) \supset F(SS0)}}{F(SS0) \quad F(SS0) \supset F(SSS0)}}{F(SSS0) \quad F(SSS0) \supset F(SSSS0)}}{F(SSSS0).}$$

We say that the proof by way of induction gives us the assurance that we *can* construct such a proof by successive applications of *modus ponens*. And the complexity of such proofs is unbounded. (But how do we see that we *always* can? By induction!) The application of the procedure involves the iteration of certain steps n times for the proof of $F(S^{(n)}0)$. Neither the reduction of induction to an explicit definition nor the Wittgensteinian doctrine[30] that '$F0 \cdot (n)[Fn \supset F(Sn)]$' constitutes the *criterion* for the truth of 'for all natural numbers n, Fn' gives us an assurance that there will be no conflict between the two methods for proving individual cases. When Poincaré said that a step of induction contained an infinity of syllogisms, he was saying that it guaranteed the possibility of all the proofs of the instances by reiterated *modus ponens*. He was right at least to this extent: if we do have an a priori assurance that there will be no conflict between construction in individual cases and inference from general propositions proved by induction, then this assurance is not founded on logic or set theory.

IX

Let us now return to the question which we left in the air in our discussion of the thesis that numbers are objects: What *does* guarantee that the singular terms of arithmetic have reference?

If we consider two different systems of numerals,

0, $S0$, $SS0$, ...
0, $T0$, $TT0$, ...

then the order-preserving correspondence between them fixes when two numerals shall *denote the same number*. Indeed, I do not think

---

[30]Waismann, *Introduction to mathematical thinking*, New York: Ungar, 1951, Theodore J. Bena, tr., chap. 9.

there is a more fundamental criterion of this. For this reason I should say that if the critics of logicism intend to deny Frege's assertion that the notion of one-to-one correspondence plays a constitutive role in our notion of number, they are wrong. It also follows that the ordinal and cardinal notions of number are interdependent. One might hold that cardinals are "applied ordinals"; i.e., if we are given the numbers with their order, then the relation to one-to-one correspondence and therefore to numerical equivalence is something that arises only in the application of numbers and does not belong to their nature as pure mathematical objects. However, I do not see how we could be said to be *given* the numbers except through a sequence of numerals or some other representatives of the same type of order. And if it is not part of the concept of number that the sequence in question should be a paradigm which could be replaced by some other equivalent sequence, the unnatural conclusion follows that features of the representative sequence which one would expect to be quite accidental, such as the particular design of the numerals, *are* relevant. But Frege's definitions *do* bring some avoidable complexity into the notion of natural number, because it defines them in terms of *explicit assertions* of the existence of one-to-one correspondences, and this need *not* be part of the sense of statements about natural numbers.

Suppose we regard it as essential to a system of numerals that it be formed by starting with some initial numeral and iterating some basic operation (representing the successor function). Let us try the following model of how the numerals come to have reference: At first they function like indexical expressions, in that in counting each refers to the object with which it is correlated, as is actually true of the expressions 'the first,' 'the second,' etc. Then by a kind of abstraction, the reference of the numeral is taken to be the *same* on all occasions. What guarantees the *existence* of the number n is the existence of an ordered set in which some object is the nth. For any numeral, the numerals up to that one will be a set. Then no ulterior fact beyond the generation of the numerals is needed to guarantee that they have reference.[31]

---

[31] Cf. Benacerraf, "Logicism: some considerations," pp. 162–174, and "What numbers could not be," *The philosophical review*, **74** (1965) 47–73.

In the case of small numbers, the guarantee of their existence is somewhat analogous to that which sense-experience gives of the existence of physical objects. But because any instance of a certain type of order will do, and because such instances can be found in sense-experience, and we cannot imagine what it would be like for us to be *unable* to find such instances, we are inclined to regard the evidence of the existence of small numbers as a priori.

We speak of its being *possible* to continue the construction of numerals to infinity. On the basis of this possibility we can say that numbers exist for which we have no numerals. However, this is an extrapolation from the concrete possibilities, and the only reason for favoring *it* as the guarantee for the existence of large numbers over the hypothesis of an actual infinity of sets, from which representatives of the numbers can be chosen, is that it is weaker. It is comparable to Kant's assumption that the "possibility of experience" extends to regions of space too small or too distant for us to perceive.

With respect to the thesis that arithmetic is a part of logic, our conclusion is that although the criticisms which have been made of it over the years *do* suffice to show that it is false, it ought not to be rejected in the unqualified way in which it appears to have been rejected by Poincaré, Brouwer, and Hilbert. It seems to me that Frege *does* show that the logical notion of one-to-one correspondence is an essential constituent of the notion of number, ordinal as well as cardinal. The content of arithmetic that is clearly additional to that of logic, which Frege failed to acknowledge, is in its existence assumptions, which involve an appeal to intuition and extrapolation. (The same kind of intuitive construction as is involved in arithmetic is also involved in *perceiving the truth* of logical truths.)

The same appeal is involved in set theory, although if we go beyond general set theory and assume the existence of infinite sets, the extrapolation is of course greater. Concerning the relation of the notion of natural number to the notion of finite set, it seems clear to me that they go together, and that neither can be understood without the other. I should like to argue, although I do not have the space to do so now, that there is a reciprocal relation between such arithmetical identities as '2 + 2 = 4' and the logical truths closely associated with them under the Frege-Russell analysis, so that it will not do to regard one as a mere abbreviation of the other.

POSTSCRIPT

The treatment in section VIII of this essay of "Poincaré-type" objections to logicism was subjected to extended critical scrutiny by Mark Steiner.[32] Although this criticism brings out unclarities in my exposition, Steiner is not very explicit about when I am presenting Papert's views and when my own, so that his reader could easily get the impression (probably unintended) that I endorse Papert's claims unequivocally.[33] Some of his criticisms are elaborations of moves in the dialectic I engaged in beginning with Papert's points.

However, in the remarks about quantifications over all concepts (p. 204 above) which Steiner quotes on pp. 33–34, I am speaking for myself. There is no doubt that what I say is too cryptic. It should be clear how complex I subsequently found such issues.[34] But it is Steiner's "more modest interpretation" (p. 35) that I had in mind. However, I was thinking rather specifically of Frege's notion of a concept, which is tied to language in that his basic explanation of what a concept is is that it is the reference of a predicate. Steiner is right that it is another question whether definitions in the spirit of Frege serve to reduce induction to *set theory*. And that "the notion of a set can precede the realization that language unfolds inductively" (p. 35) is certainly not refuted by any considerations I give, and indeed it is suggested by the idea of a set as constituted by its elements, to which I am generally sympathetic. But inductive generation enters into our understanding of set theory in other respects, for example in the motivation of the axioms by the "iterative" conception of sets as arising in a well-ordered sequence of stages.[35] Moreover, the linguistic considerations that arise in connection with Frege's notion of a concept also arise with respect to

[32] *Mathematical knowledge*, Ithaca: Cornell University Press, 1975, pp. 28–41.

[33] Thus on p. 14 he represents as mine a view I explicitly present as Papert's; in the citation for the same view (p. 28), he leaves off the opening part of the first sentence, which is "Papert says in effect that."

[34] See, for example, "Sets and classes," *Noûs*, **8** (1974) 1–12, and "The liar paradox," *Journal of philosophical logic*, **3** (1974) 381–412.

[35] See my paper, "What is the iterative conception of set," in R. E. Butts and Jaakko Hintikka, eds., *Logic, foundations of mathematics, and computability theory*, Dordrecht: Reidel, 1977, 335–67.

the second-order principles of set theory, the axiom of separation, and the axiom of replacement.

Much of Steiner's discussion proceeds on the assumption that the logicist *knows* the truth of his axioms. Papert's and my arguments could be taken as questioning how this assumption could be true given what the logicist admits as logic. Moreover, the assumption does not make the issue about consistency that Steiner discusses on p. 40 (with reference to the last paragraph of section VIII) quite so unproblematic as he makes it appear. For to conclude from it that the logicist knows that his theory is consistent is to assume his knowledge closed under some sort of semantic reflection. Suppose for simplicity that our logicist L has a theory with finitely many axioms $A_1 \ldots A_n$. Then for each $i \leqslant n$, L knows that A_i. But by Gödel's second incompleteness theorem, even the assumption that L knows any logical consequence of what he knows is not sufficient to give him the knowledge that the theory with these axioms is consistent; he would need something like the inference to 'A_i is true' and then some reasoning with the concept of truth.

However, this is not quite the nub of the difficulty, since he can do all this with the help of more set theory, though the assumption that he knows the stronger axioms might get more and more tenuous. I think what Papert had in mind was something like a skeptical doubt: if one proves a generalization with infinitely many instances which can be decided independently of the generalization, how does one know that the independent procedure will not turn up a counterinstance? Any *proof* that it will not will give rise to the same questions. Papert's view seems to be that there is something irreducibly empirical in any assurance one has about such matters. The doubt is of the sort that arises in Wittgenstein's discussion of following a rule.[36] On this subject, I do not have much to add now

[36]Indeed, Steiner has pointed out to me that just this doubt, in the case of induction, is discussed by Wittgenstein in *Lectures on the foundations of mathematics*, Cora Diamond, ed., Ithaca: Cornell University Press, 1976, lecture 31. Wittgenstein says that we take the result of the general proof (by induction) as a criterion for the correctness of a result obtained by figuring out a particular case, e.g., deducing in 3,000 steps that a proposition holds for 3,000. What is irreducibly empirical is the general agreement of results that such legislation rests on (see esp. ibid., pp. 291–292).

to the cautious comment in the paper, that doubts of this kind are
not resolved by appeals to logic and set theory.

Nonetheless, my view of the force of "Poincaré-type" objections
has changed since writing this paper. The analogy is very close between the kind of circle that arises in the logicist's treatment of induction and the circle that arises in justifying logical laws by means
of semantical definitions of the logical operators, such as the characterizations of sentential connectives by truth-tables: to show the
most elementary inferences valid, one has to use the same inferences,
or others posing the same questions, in the metalanguage. This is
just what happens with the possible justifications of mathematical
induction that arise (not only the Fregean one). So although the
considerations adduced by Poincaré do set a limit on the reducibil-
ity of induction to other principles, they do not show that it is not
importantly like a logical principle. In a natural deduction formalization of arithmetic, induction can be treated as an elimination
rule for the predicate "is a natural number," in analogy with the
well-known elimination rules for logical connectives.[37] The matter
deserves further exploration.

[37]See, for example, P. Martin-Löf, "Hauptsatz for iterated inductive definitions," in J. E. Fenstad, ed., *Proceedings of the second Scandinavian logic
symposium*, Amsterdam: North-Holland, 1971, p. 190 (in the context of a general treatment of inductive definitions).

Formulations of constructive mathematics by what are called theories of constructions treat induction and recursion together, but the analogy between the
natural number predicate and logical connectives can still be made. See for
example W. Tait, "Finitism," *The journal of philosophy*, **78** (1981) 524–46,
esp. pp. 531–532, 537, and P. Martin-Löf, "An intuitionistic theory of types,"
in H. E. Rose and J. C. Shepardson, eds., *Logic colloquium '73*, Amsterdam:
North-Holland, 1975, esp. pp. 95-96.

8

The Consistency of Frege's
Foundations of Arithmetic

Is Frege's *Foundations of arithmetic* inconsistent? The question may seem to be badly posed. The *Foundations*, which appeared in 1884, contains no formal system like those found in Frege's *Begriffsschrift* (1879) and *Basic laws of arithmetic* (Vol. 1, 1893, Vol. 2, 1903). As is well known, Russell showed the inconsistency of the system of the *Basic laws* by deriving therein what we now call Russell's paradox. The system of the *Begriffsschrift*, on the other hand, can plausibly be reconstructed as an axiomatic presentation of second-order logic, which is therefore happily subject to the usual consistency proof, consisting in the observation that the universal closures of the axioms and anything derivable from them by the rules of inference are true in any one-element model.[1] Since the *Foundations*

From *On being and saying: essays in honor of Richard Cartwright*, Judith Jarvis Thomson, ed., Cambridge: MIT Press, 1987, 3–20. Reprinted by kind permission of MIT Press and the author.

The papers by Paul Benacerraf, Harold Hodes, and Charles Parsons cited below have been major influences on this one. I would like to thank Paul Benacerraf, Sylvain Bromberger, John Burgess, W. D. Hart, James Higginbotham, Harold Hodes, Paul Horwich, Hilary Putnam, Elisha Sacks, Thomas Scanlon, and Judith Jarvis Thomson for helpful comments. Research for this paper was carried out under grant SES-8607415 from the National Science Foundation.

[1]I. S. Russinoff, "Frege's problem about concepts," Ph.D. dissertation, MIT, Department of Linguistics and Philosophy, 1983.

contains no formal system at all, our question may be thought to need rewording before an answer to it can be given.

One might nevertheless think that, however reworded and badly posed or not, it must be answered yes. The *Basic laws*, that is, the system thereof, *is* inconsistent and is widely held to be a formal elaboration of the mathematical program outlined in the earlier *Foundations*, which contains a more thorough development of its program than one is accustomed to find in programmatic works. Thus the inconsistency which Russell found in the later book must have been latent in the earlier one.

Moreover, the characteristic signs of inconsistency can be found in the use Frege makes in the *Foundations* of the central notions of 'object,' 'concept,' and 'extension.' Objects fall under concepts, but some extensions—numbers, in particular and crucially—contain concepts, and these extensions themselves are objects, according to Frege. Thus, although a division into two types of entity, concepts and objects, can be found in the *Foundations*, it is plain that Frege uses not one but two instantiation relations, 'falling under' (relating some objects to some concepts) and 'being in' (relating some concepts to some objects), and that both relations sometimes obtain reciprocally: The number 1 is an object that falls under 'identical with 1,' a concept that is in the number 1. Even more ominously (because of the single negation sign), the number 2 does not fall under 'identical with 0 or 1,' which is in 2. Thus the division of the *Foundations*'s entities into two types would appear to offer little protection against Russell's paradox.

It is not only Russell's paradox that threatens. Recall that Frege defines 0 as the number belonging to the concept 'not identical with itself.'[2] If there is such a number, would there not also have to be a number belonging to the concept 'identical with itself,' a *greatest* number? Cantor's paradox also threatens.

It is therefore quite plausible to suppose that it is merely through its lack of formality that the *Foundations* escapes outright inconsistency and that, when suitably formalized, the principles employed

---

[2]Plurals find happy employment here, as elsewhere in the discussion of concepts: For example, instead of "the number belonging to the concept 'horse,'" one can say "the number of horses." 0 is thus *defined* by Frege to be the number of things that are not self-identical. And Frege was right!

by Frege in the *Foundations* must be inconsistent.

This plausible and, I suspect, quite common supposition is mistaken, as we shall see. Although Frege freely assumes the existence of needed concepts at every turn, he by no means avails himself of extensions with equal freedom. With one or two insignificant but possibly revealing exceptions, which I discuss later, the *only* extensions whose existence Frege claims in the central sections of the *Foundations* are the extensions of higher level concepts of the form 'equinumerous with concept F.' (I use the term "equinumerous" as the translation of Frege's *gleichzahlig*.) It turns out that the claim that such extensions exist can be consistently integrated with existence claims for a wide variety of first level concepts in a way that makes possible the execution of the mathematical program described in §§ 68–83 of the *Foundations*. Indeed I shall now present a formal theory, *FA* ("Frege Arithmetic") that captures the whole content of these central sections and for which a simple consistency proof can be given, one that shows *why* FA is consistent.

FA is a theory whose underlying logic is standard axiomatic second-order logic written in the usual Peano-Russell logical notation. FA could have been presented as an extension of the system of Frege's *Begriffsschrift*. Indeed, there is some evidence that Frege thought of himself as translating *Begriffsschrift* notation into the vernacular when writing the *Foundations*. Not only does the later work abound with allusions and references to the earlier, along with repetitions of claims and arguments for its significance, when Frege defines the ancestral in § 79, he uses the variables x, y, d, and F in exactly the same logical roles they had played in the *Begriffsschrift*.

FA is a system with three sorts of variable: first-order (or object) variables a, b, c, d, m, n, x, y, z, ... ; unary second-order (or concept) variables F, G, H, ... ; and binary second-order (or relation) variables φ, ψ, ... The sole nonlogical symbol of the language of FA is η, a two-place predicate letter attaching to a concept variable and an object variable. (η is intended to be reminiscent of $\in$ and may be read "is in the extension." Frege's doctrine that extensions are objects receives expression in the fact that the second argument place of η is to be filled by an object variable.) Thus the atomic formulas of FA are of the forms Fx (F a concept variable), $x\varphi y$, and $F\eta x$. Formulas of FA are constructed from the atomic formulas by means of propositional connectives and quantifiers in the usual manner.

Identity can be taken to have its standard second-order definition: $x = y$ if and only if $\forall F(Fx \leftrightarrow Fy)$. Frege endorses Leibniz's definition (" ... *potest substitui ... salva veritate*") in § 65 of the *Foundations* but does not actually do what he might easily have done, viz. state that Leibniz's definition of the identity of x and y can be put: y falls under every concept under which x falls (and vice versa).

The logical axioms and rules of FA are the usual ones for such a second-order system. Among the axioms we may specially mention (i) the universal closures of all formulas of the form

$$\exists F \forall x (Fx \leftrightarrow A(x)),$$

where $A(x)$ is a formula of the language of FA not containing F free; and (ii) the universal closures of all formulas of the form

$$\exists \varphi \forall x \forall y (x\varphi y \leftrightarrow B(x,y)),$$

where $B(x,y)$ is a formula of the language not containing φ free. Throughout §§ 68-83 of the *Foundations* Frege assumes, and needs to assume, the existence of various particular concepts and relations. The axioms (i) and (ii) are called comprehension axioms; these will do the work in FA of Frege's concept and relation existence assumptions.

The sole (nonlogical) axiom of the system FA is the single sentence

Numbers: $\forall F \exists! x \forall G(G\eta x \leftrightarrow F \operatorname{eq} G),$

where $F \operatorname{eq} G$ is the obvious formula of the language of FA expressing the equinumerosity of the values of F and G, viz.

$$\exists \varphi [\forall y(Fy \to \exists! z(y\varphi z \wedge Gz)) \wedge \forall z(Gz \to \exists! y(y\varphi z \wedge Fy))].$$

Here the sign η is used for the relation that holds between a concept G and the extension of a (higher level) concept under which G falls; before we used the term "is in" for this relation and "contains" for its converse. In § 68 Frege first asserts that F is equinumerous with G if and only if the extension of 'equinumerous with F' is the same as that of 'equinumerous with G' and then defines the number belonging to the concept F as the extension of the concept 'equinumerous with the concept F.' Since Frege, like Russell, holds that existence and uniqueness are implicit in the use of the definite article, he supposes that for any concept F, there is a unique extension of the concept 'equinumerous with F.' Thus the sentence Numbers expresses this supposition in the language of FA; it is the sole nonlogical assumption[3] utilized by Frege in the course of the mathematical work done in §§ 68–83.

How confident may we be that FA is consistent? Recent observations by Harold Hodes and John Burgess bear directly on this question. To explain them, it will be helpful to consider a certain formal sentence, which we shall call Hume's principle:

$$\forall F \forall G(NF = NG \leftrightarrow F \operatorname{eq} G).$$

Hume's principle is so called because it can be thought of as explicating a remark that Hume makes in the *Treatise* (I, III, I, para. 5), which Frege quotes in the *Foundations*:

> We are possest of a precise standard by which we can judge of the equality and proportion of numbers .... When two numbers are so combined, as that the one has always an unite answering to every unite of the other, we pronounce them equal
>
> ....

The symbol N in Hume's principle is a function sign which when attached to a concept variable makes a term of the same type as object variables; thus $NF = NG$ and $x = NF$ are well-formed. Taking N ... as abbreviating "the number of ... s," we may read Hume's principle: The number of Fs is the number of Gs if and only if the Fs can be put into one-one correspondence with the Gs. (As Hume said, more or less).

In his article "Logicism and the ontological commitments of arithmetic,"[4] Hodes observes that a certain formula, which he calls "(D)"

---

[3] It is nonlogical by my lights, though not, of course, by Frege's.

[4] *The journal of philosophy*, **81** (1984) 123–49, see p. 138.

is satisfiable. He writes:

$$\begin{matrix} \forall X \exists x \\ \\ \forall Y \exists y \end{matrix} \quad (x = y \leftrightarrow X \operatorname{eq} Y)$$

(D)

is satisfiable. In fact if we accept standard set theory, it's true.

(I have replaced Hodes's "$(Q_E z)(Xz, Yz)$" by "$X \operatorname{eq} Y$." The label "(D)" is missing from the text of his article.) Branching quantifiers, which are notoriously hard to interpret, may always be eliminated in favor of ordinary function quantifiers. Eliminating them from (D) yields the formula $\exists N \exists M \forall X \forall Y (NX = MY \leftrightarrow X \operatorname{eq} Y)$. Now (D) is satisfiable if and only if Hume's principle is satisfiable. For if (the function quantifier equivalent of) (D) holds in a domain U, then for some functions N, M, $\forall X \forall Y (NX = MY \leftrightarrow X \operatorname{eq} Y)$ holds in U; $\forall Y (Y \operatorname{eq} Y)$ holds in U, so does $\forall Y (NY = MY)$, and therefore so does Hume's principle $\forall X \forall Y (NX = NY \leftrightarrow X \operatorname{eq} Y)$. Conversely, Hume's principle implies (D). Thus a bit of deciphering enables us to see that Hodes's claim is tantamount to the assertion that Hume's principle is satisfiable.

Hodes gives no proof that (D), or Hume's principle, is satisfiable. But Burgess, in a review of Crispin Wright's book *Frege's conception of numbers as objects*[5] shows that it is. He writes:

> Wright shows why the derivation of Russell's paradox cannot be carried out in $N^=$ [Wright's system, obtained by adjoining a version of Hume's principle to second-order logic], and ought to have pointed out that the system is pro[v]ably consistent. (It has a model whose domain of objects consists of just the cardinals zero, one, two, ... and aleph-zero.)[6]

It will not be amiss to elaborate this remark. To produce a model $\mathcal{M}$ for Hume's principle that also verifies all principles of axiomatic second-order logic, take the domain U of $\mathcal{M}$ to be the set $\{0, 1, 2, \ldots, \aleph_0\}$. To ensure that $\mathcal{M}$ is a model of axiomatic second-order logic, take the domain of the concept variables to be the set

[5] Aberdeen: Aberdeen University Press, 1983.

[6] *The philosophical review*, **93** (1984) 638–40. The text of the review has "probably consistent," which is an obvious misprint.

of all subsets of U, and similarly take the domain of the binary (or, more generally, n-ary) relation variables to be the set of all binary (or n-ary) relations of U, that is, the set of sets of ordered pairs (or n-tuples) of members of U.

To complete the definition of $\mathcal{M}$, we must define the function f by which the function sign N is to be interpreted in $\mathcal{M}$. The *cardinality* of a set is the number of members it contains. U has the following important property: *The cardinality of every subset of U is a member of U.* (Notice that the set of natural numbers *lacks* this property.) Thus we may define f as the function whose value for every subset V of U is the cardinality of V. We must now see that Hume's principle is true in $\mathcal{M}$.

Observe that an assignment s of appropriate items to variables satisfies $NF = NG$ in $\mathcal{M}$ if and only if the cardinality of $s(F)$ equals the cardinality of $s(G)$ and satisfies F eq G in $\mathcal{M}$ if any only if $s(F)$ can be put into one-one correspondence with $s(G)$. Since the cardinality of $s(F)$ is the same as that of $s(G)$ if and only if $s(F)$ can be put into one-one correspondence with $s(G)$, every assignment satisfies $(NF = NG \leftrightarrow F$ eq $G)$ in $\mathcal{M}$, and $\mathcal{M}$ is a model for Hume's principle.

A similar argument shows the satisfiability of Numbers: Let the domain of $\mathcal{M}$ again be U, and let $\mathcal{M}$ specify that η is to apply to a subset V of U and a member u of U if and only if the cardinality of V is u. Then Numbers is true in $\mathcal{M}$. (On receiving the letter from Russell, Frege should have immediately checked into Hilbert's Hotel.)

(It may be of interest to recall the usual proof that the comprehension axioms (i) are true in standard models (like $\mathcal{M}$) for second-order logic: Let $A(x)$ be a formula not containing free F, and let s be an assignment. Let C be the set of objects of which $A(x)$ is true, and let s' be just like s except that $s'(F) = C$. Since $A(x)$ does not contain free F, s' satisfies $\forall x (Fx \leftrightarrow A(x))$ and s satisfies $\exists F \forall x (Fx \leftrightarrow A(x))$. Similarly for the comprehension axioms (ii).)

There is a cluster of worries or objections that might be thought to arise at this point: Does not the appeal to the natural numbers in the consistency proof vitiate Frege's program? How can one invoke the existence of the numbers in order to justify FA? There is a quick answer to this objection: You mean we *shouldn't* give a consistency proof? More fully: We are simply trying to use what we know in order to allay all suspicion that a contradiction is formally derivable in FA, about whose consistency anyone knowing the history of logic might well be quite uncertain. We are not attempting to show that FA is true.

But there is perhaps a more serious worry. At a crucial step of the proof of the consistency (with second-order logic) of the formal sentence called Hume's principle, we made an appeal to an informal principle connecting cardinality and one-one correspondence which can be symbolized as—Hume's principle. (We made this appeal when we said that the cardinality of $s(F)$ is the same as that of $s(G)$ if and only if $s(F)$ can be put into one-one correspondence with $s(G)$.) Should this argument then count as a *proof* of the consistency of Hume's principle? What assurance can any argument give us that a certain sentence is consistent, if the argument appeals to a principle one of whose formalizations is the very sentence we are trying to prove consistent?

The worry is by no means idle. We have attempted to prove the consistency of Hume's principle by arguing that a certain structure $\mathcal{M}$ is a model for Hume's principle; in proving that $\mathcal{M}$ is a model for Hume's principle we have appealed to an informal version of Hume's principle. A similar service, however, can be performed for the notoriously inconsistent naive comprehension principle $\exists y \forall x (x \in y \leftrightarrow \ldots x \ldots)$ of set theory: By informally invoking the naive comprehension principle, we can argue that all of its instances are true under the interpretation I under which the variables range over all sets that there are and $\in$ applies to a, b if and only if b is a set and a is a member of b. Let $\ldots x \ldots$ be an arbitrary formula not containing free y. (By the naive comprehension principle) let b be the set of just those sets satisfying $\ldots x \ldots$ under I. Then for every a, a and b satisfy $x \in y$ under I if and only if a satisfies $\ldots x \ldots$ under I. Therefore b satisfies $\forall x (x \in y \leftrightarrow \ldots x \ldots)$ under I, and $\exists y \forall x (x \in y \leftrightarrow \ldots x \ldots)$ is true under I. Thus I is a model of all instances of the naive comprehension principle. (Doubtless Frege convinced himself of the truth of the fatal Rule (V) of *Basic laws* by running through some such argument.) Of course we can now see that, *pace* the principle, there is not always a set of just those sets satisfying $\ldots x \ldots$. But how certain can we be that the proof of the consistency of Hume's principle and FA does not contain some similar gross (or subtle) mistake, as does the "proof" just given of the consistency of the naive comprehension principle?

Let us first notice that the argument can be taken to show not merely that FA is consistent, but that *it is provable in standard set theory* that FA is consistent. (Standard set theory is of course ZF, Zermelo-Fraenkel set theory.) The argument can be "carried out" or "replicated" *in* ZF. Thus, if FA is inconsistent, ZF is in error. (Presumably the word "provably" in Burgess's observation refers to an informal, model-theoretic proof, which could be formalized in ZF, or to a formal ZF proof.) Thus anyone who is convinced that nothing false is provable in ZF must regard this argument as a proof that FA is consistent. Moreover, if ZF makes a false claim to the effect that FA, or any other formal theory, is consistent, then ZF is not merely in error but is itself inconsistent, for ZF will then certainly also make the correct claim that there exists a derivation of ⊥ in FA. (Indeed systems much weaker than ZF, for example, Robinson's arithmetic Q, will then make that correct claim.)

Something even stronger may be said. We shall show that any derivation of an inconsistency in FA can immediately be turned into a derivation of an inconsistency in a well-known theory called "second-order arithmetic" or "analysis," about whose consistency there has never been the slightest doubt. In the language of analysis there are two sorts of variables, one sort ranging over (natural) numbers, the other over sets of and relations on numbers. The axioms of analysis are the usual axioms of arithmetic, a sentence expressing the principle of mathematical induction ("Every set containing 0 and the successor of every member contains every natural number"), and, for each formula of the language, a comprehension axiom expressing the existence of the set or relation defined by the formula.[7] If ZF is consistent, so is analysis; but ZF is stronger than analysis, and the consistency of analysis can be proved in ZF. It is (barely) conceivable that ZF is inconsistent; but unlike ZF, analysis did not arise as a direct response to the set-theoretic antinomies, and the discovery of the inconsistency of analysis would be the most

---

[7] A standard reference concerning analysis is section 8.5 of J. R. Shoenfield's *Mathematical logic*, Reading Mass.: Addison-Wesley, 1967.

surprising mathematical result ever obtained, precipitating a crisis
in the foundations of mathematics compared with which previous
"crises" would seem utterly insignificant.

Let us sketch the construction by which proofs of $\bot$ in FA can be
turned into proofs of $\bot$ in analysis. The trick is to "code" $\aleph_0$ by 0
and each natural number z by $z+1$ so that the argument given may
be replicated in analysis. It is easy to construct a formula $A(z, F)$ of
the language of analysis that expresses the relation "exactly z natural numbers belong to the set F": Simply write down the obvious
symbolization of "there exists a one-one correspondence between the
natural numbers less than z and the members of F." Let $\mathrm{Eta}(F, x)$
be the formula

$$[\neg \exists z A(z, F) \wedge x = 0] \vee [\exists x (A(z, F) \wedge x = z + 1)].$$

Then, since $\exists! x\, \mathrm{Eta}(F, x)$ and

$$[\exists x (\mathrm{Eta}(F, x) \wedge \mathrm{Eta}(G, x)) \leftrightarrow F \operatorname{eq} G]$$

are theorems of analysis, so is the result

$$\forall F \exists! x \forall G (\mathrm{Eta}(G, x) \leftrightarrow F \operatorname{eq} G)$$

of substituting $\mathrm{Eta}(G, x)$ for $G\eta x$ in Numbers, as the following argument, which can be formalized in analysis, shows: Let F be any set
of numbers. Let x be such that $\mathrm{Eta}(F, x)$ holds. Let G be any set.
Then $\mathrm{Eta}(G, x)$ holds if and only if $F \operatorname{eq} G$ does. And since $F \operatorname{eq} F$
holds, x is unique. Of course each of the comprehension axioms of
FA is provable in analysis under these substitutions, since they turn
into comprehension axioms of analysis. Thus a proof of $\bot$ in FA
immediately yields a proof of $\bot$ in analysis.

It is therefore as certain as anything in mathematics that, if analysis is consistent, so is FA. Later we shall see that the converse holds.
(A sketch of a major part of the proof of the converse was given by
Frege, in the *Foundations*. Of course.) The connection between
FA and Russell's paradox is discussed later. Since the possibility
that analysis might be inconsistent at present strikes us as utterly
inconceivable, we may relax in the certainty that neither Russell's
nor any other contradiction is derivable in FA.

We now want to show that the definitions and theorems of §§ 68–83 of the _Foundations_ can be stated and proved in FA, _in the manner indicated by Frege_. I am not sure that it is possible to appreciate the magnitude and character of Frege's accomplishment without going through at least some of the hard details of the derivation of arithmetic from Numbers, in particular those of the proof that every natural number has a successor, but readers who wish take it on faith that the derivation can be carried out in FA along a path _very_ close to Frege's may skim over some of the next seventeen paragraphs. Do not forget that it is Frege himself who has made formalization of his work routine.

In the course of replicating in FA Frege's treatment of arithmetic, we shall of course make definitional extensions of FA. For example, as Frege defined the number belonging to the concept F as the extension of the concept 'equinumerous to F,' so we introduce a function symbol N, taking a concept variable and making a term of the type of object variables, and then define $NF = x$ to mean $\forall G(G\eta x \leftrightarrow F \operatorname{eq} G)$; the introduction of the symbol N together with this definition is of course licensed by Numbers. it will also prove convenient to introduce terms $[x : A(x)]$ for concepts: $[x : A(x)]t$ is to mean $A(t)$; $F = [x : A(x)]$ is to mean $\forall x(Fx \leftrightarrow A(x))$; $[x : A(x)]\eta y$ is to mean $\exists F(F = [x : A(x)] \wedge F\eta y)$; $[x : A(x)] = [x : B(x)]$ is to mean $\forall x(A(x) \leftrightarrow B(x))$, etc. The introduction of such terms is of course licensed by the comprehension axioms (i).

§§ 70-73 provide the familiar definition of equinumerosity. In § 73, Frege proves Hume's principle. Note that the comprehension axioms (ii) provide the facts concerning equinumerosity needed for this theorem to be provable. Once Hume's principle is proved, _Frege makes no further use of extensions_.[8,9]

---

[8]See section VI of Charles Parsons, "Frege's theory of number," Chapter 7, above.

[9]In his estimable _Frege's conception of numbers as object_, Wright sketches a derivation of the Peano axioms in a system of higher-order logic to which a version of Hume's principle is adjoined as an axiom. Wright discusses the question of whether such a system would be consistent, attempts to reproduce various well-known paradoxes in such a system, is unsuccessful, and concludes on page 156 that "there are grounds, if not for optimism, at least for a cautious confidence that a system of the requisite sort is capable of consistent formulation." Wright's instincts are correct, as Hodes and Burgess have seen. It may be of

In § 72 Frege defines "number": "n is a number" is to mean "there exists a concept such that n is the number which belongs to it." In parallel, we make the definition in FA: $Zx \leftrightarrow \exists F(NF = x)$. In § 74 Frege defines 0 as the number belonging to the concept 'not identical with itself'; we define in FA: $0 = N[x : x \neq x]$. The content of § 75 is given in the easy theorem of FA:

$$\forall F \forall G([\forall x \neg Fx \rightarrow ((\forall x \neg Gx \leftrightarrow F \text{ eq } G) \land NF = 0)]$$
$$\land [NF = 0 \rightarrow \forall x \neg Fx]).$$

In § 76 Frege defines "the relation in which every two adjacent members of the series of natural numbers stand to each other."[10] Correspondingly, we define nSm (read "n succeeds m"):

$$\exists F \exists x \exists G(Fx \land NF = n \land \forall y(Gy \leftrightarrow Fy \land y \neq x) \land NG = m).$$

$\neg 0Sa$ immediately follows in FA from this definition: Zero succeeds nothing. In § 77 Frege defines the number 1. We make the corresponding definition: $1 = N[x : x = 0]$. $1S0$ is easily derived in FA.

The theorems corresponding to those of § 78 are proved without difficulty:

(1) $aS0 \rightarrow a = 1$,
(2) $NF = 1 \rightarrow \exists x Fx$,
(3) $NF = 1 \rightarrow (Fx \land Fy \rightarrow x = y)$,
(4) $\exists x Fx \land \forall x \forall y(Fx \land Fy \rightarrow x = y) \rightarrow NF = 1$,
(5) $\forall a \forall b \forall c \forall d(aSc \land bSd \rightarrow (a = b \leftrightarrow c = d))$,
(6) $\forall n(Zn \land n \neq 0 \rightarrow \exists m(Zm \land nSm))$.

Although Frege and we have now defined "number," defined 0 and 1, proved that they are different numbers, proved that "succeeds" is one-one, and proved that every non-zero number is a successor,

---

interest to note that FA supplies the answer to a question raised by Wright on page 156 of his book. It is a theorem of FA that the number of numbers that fall under none of the concepts of which they are the numbers is *one*. (Zero is the only such number.)

[10]Note that, although Frege here introduces the expression "folgt in der naturlichen Zahlenreihe unmittelbar auf" for the *succeeds* relation, he will define "finite" number only at the end of § 83.

"finite number," that is, "natural number," has not yet been defined; nor has it been shown that every natural number has a successor.

In § 79 Frege defines the ancestral of φ, "y follows x in the φ-series," as in the *Begriffsschrift*. Thus in FA we define $x\varphi^*y$:

$$\forall F(\forall a(x\varphi a \rightarrow Fa) \wedge \forall d\forall a(Fd \wedge d\varphi a \rightarrow Fa) \rightarrow Fy).$$

§ 80 is a commentary on § 79. At the beginning of § 81 Frege introduces the terminology "y is a member of the φ-series beginning with x" and "x is a member of the φ-series ending with y" to mean: either y follows x in the φ-series or y is identical with x. Frege uses the phrase "in the series of natural numbers" instead of "in the φ-series" when φ is the converse of the succeeds relation. In FA we define mPn to mean nSm, $m < n$ to mean mP^*n, and $m \leqslant n$ to mean $m < n \vee m = n$. Frege defines "n is a finite number" only at the end of § 83. In FA we define Fin n to mean $0 \leqslant n$.

In 82 and 83 Frege outlines a proof that every finite number has a successor. He adds that, in proving that a successor of n always exists (if n is finite), it will have been proved that "there is no last member of this series." (He obviously means the sequence of finite numbers). This will certainly have been shown if it is also shown that no finite number follows itself in the series of natural numbers; in § 83 Frege indicates that this proposition is necessary and how to prove it.

Frege's ingenious idea is that we can prove that every finite number has a successor by proving that if n is finite, the number of numbers less than or equal to n—in Frege's terminology "the number which belongs to the concept 'member of the series of natural numbers ending with n'"—succeeds n. Frege's outline can be expanded into a proof in FA of: Fin $n \rightarrow N[x : x \leqslant n]Sn$. Since $ZN[x : x \leqslant n]$ is provable in FA, so is (Fin $n \rightarrow \exists x(Zx \wedge xSn)$).

In § 82 Frege claims that certain propositions are provable; the translations of these into FA are $aSd \wedge N[x : x \leqslant d]Sd \rightarrow N[x : x \leqslant a]Sa$ and $N[x : x \leqslant 0]S0$. Frege adds that the statement that for finite n the number of numbers less than or equal to n succeeds n then follows from these by applying the definition of "follows in the series of natural numbers."

$N[x : x \leqslant 0]S0$ is easily derived in FA: $xP^*y \rightarrow \exists a\, aPy$ follows from the definition of the ancestral; consider $[z : \exists a\, aPz]$. Since

$\neg 0Sa$ and $1S0$ are theorems, so are $\neg aP0$, $\neg aP^*0$, $x \leqslant 0 \leftrightarrow x = 0$, and $N[x : x \leqslant 0] = N[x : x = 0]$, from which, together with the definition of 1, $N[x : x \leqslant 0]S0$ follows.

But the derivation of $aSd \wedge N[x : x \leqslant d]Sd \rightarrow N[x : x \leqslant a]Sa$ is not so easy. Frege says that, to prove it, we must prove that $a = N[x : x \leqslant a \wedge x \neq a]$; for which we must prove that $x \leqslant a \wedge x \neq a$ if and only if $x \leqslant d$, for which in turn we need $\text{Fin}\, a \rightarrow \neg a < a$. This last proposition is again to be proved, says Frege, by appeal to the definition of the ancestral; it is the fact that we need the statement that no finite number follows itself, he writes, that obliges us to attach to $N[x : x \leqslant n]Sn$ the antecedent $\text{Fin}\, n$.

An interpretive difficulty now arises: It is uncertain whether or not Frege is assuming the finiteness of a and d in § 82. Although he does not say so, it would appear that he must be assuming that d, at least, is finite, for he wants to show $(aSd \wedge N[x : x \leqslant d]Sd \rightarrow N[x : x \leqslant a]Sa)$ by showing $aSd \rightarrow (x \leqslant a \wedge x \neq a \leftrightarrow x \leqslant d)$. Without assuming the finiteness of a and d, he can certainly show $aSd \rightarrow \forall x(x < a \leftrightarrow x \leqslant d)$. However, $\neg a < a$, or something like it, is needed to pass from $x < a$ to $(x \leqslant a \wedge x \neq a)$, and Frege would therefore appear to need $\text{Fin}\, a$. But since $\text{Fin}\, 0$ is trivially provable and $\forall d \forall e(dPa \wedge \text{Fin}\, d \rightarrow \text{Fin}\, a)$ easily follows from propositions 91 and 98, $(xPy \rightarrow xP^*y)$ and $(xP^*y \wedge yP^*z \rightarrow xP^*z)$, of the *Begriffsschrift* Frege's argument can be made to work in FA, provided that we take him as assuming that d (and therefore a) is finite. Let us see how.

From propositions 91 and 98, $dPa \rightarrow (xP^*d \vee x = d \rightarrow xP^*a)$ easily follows. We also want to prove

$$(*) \qquad\qquad dPa \rightarrow (xP^*a \rightarrow xP^*d \vee x = d),$$

for which it suffices to take $F = [z : \exists d\, dPz \wedge \forall d(dPz \rightarrow xP^*d \vee x = d))$, and show $(xP^*a \rightarrow Fa)$ by showing, as usual, $(xPb \rightarrow Fb)$ and $(Fa \wedge aPb \rightarrow Fb)$.

$(xPb \rightarrow Fb)$: Suppose xPb. Then the first half of Fb is trivial; and if dPb, then by § 78(5) of the *Foundations*, $x = d$, whence $xP^*d \vee x = d$. As for $(Fa \wedge aPb \rightarrow Fb)$, suppose Fa and aPb. The first half of Fb is again trivial; now suppose dPb. By § 78(5), $d = a$. Since Fa, for some c, cPa, and then $xP^*c \vee x = c$. Since cPa and

$d = a$, cPd. But then by §§ 91 and 98, xP^*d, whence $xP^*d \vee x = d$. Thus $(xP^*a \to Fa)$, whence $dPa \to (xP^*a \to xP^*d \vee x = d)$ and $dPa \to (xP^*a \leftrightarrow xP^*d \vee x = d)$ follow.

We must now prove

$$(**) \qquad\qquad\qquad \text{Fin}\, a \to \neg aP^*a.$$

Since $\neg 0P^*0$, it suffices to show $0P^*a \to \neg aP^*a$. We readily prove $(0Pb \to \neg bP^*b)$ and $(\neg aP^*a \wedge aPb \to \neg bP^*b)$: If $0Pb$ and bP^*b, then by $(*)$, $bP^*0 \vee b = 0$, whence by §§ 91 and 98, $0P^*0$, impossible; if $\neg aP^*a$, aPb, and bP^*b, then by $(*)$, $bP^*a \vee b = a$, whence by §§ 91 and 98, aP^*a, contradiction.

Combining $(*)$ and $(**)$ yields

$$dPa \wedge \text{Fin}\, a \to ((xP^*a \vee x = a) \wedge x \neq a \leftrightarrow xP^*d \vee x = d).$$

Abbreviating, we have

$$dPa \wedge \text{Fin}\, a \to (x \leqslant a \wedge x \neq a \leftrightarrow x \leqslant d])$$

and then by Hume's principle

$$dPa \wedge \text{Fin}\, a \to N[x : x \leqslant a \wedge x \neq a] = N[x : x \leqslant d].$$

Thus, if $\text{Fin}\, d$, $N[x : x \leqslant d]Sd$, and dPa, then $\text{Fin}\, a$ and aSd; since $a \leqslant a$,

$$N[x : x \leqslant a]SN[x : x \leqslant a \wedge x \neq a] = N[x : x \leqslant d];$$

since aSd, by § 78(5), $N[x : x \leqslant d] = a$, and therefore $N[x : x \leqslant a]Sa$. Since $\text{Fin}\, 0$ and $N[x : x \leqslant 0]S0$, we conclude

$$\text{Fin}\, n \to (\text{Fin}\, n \wedge N[x : x \leqslant n]Sn),$$

whence $\text{Fin}\, n \to N[x : x \leqslant n]Sn$.

O.K., stop skimming now. One noteworthy aspect of Frege's derivation of what are in effect the Peano postulates is that so much can be derived from what appears to be so little. Whether or not Numbers is a purely logical principle is a question that we shall consider at length in what follows. I now want to consider the status of the other principles employed by Frege, which, having argued the matter elsewhere, I shall assume are properly regarded as logical. Frege shows these principles capable of yielding conditionals whose antecedent is the apparently trivial and in any event trivially consistent Numbers and whose consequents are propositions like $\forall m(\text{Fin}\, m \rightarrow \exists n(Zn \wedge nSm))$. The consequents would "not in any wise appear to have been thought in" Numbers; thus these conditionals at least look synthetic, and Frege himself would appear to have shown the principles and rules of logic that generate such weighty conditionals to be synthetic. But if the principles of Frege's logic count as synthetic, then a reduction of arithmetic to logic gives us no reason to think arithmetic analytic. There is a criticism of Kant to which Frege is nevertheless entitled: Kant had no conception of this sort of analysis and no idea that content could be thus created by deduction.

The hard deductions found in the *Begriffsschrift* and the *Foundations* would make evident, if it were not already so, the utter vagueness of the notions of *containment* and of *analyticity*: Even though *containment* appears to be closed under obvious consequence, it is certainly not closed under consequence; there is often no saying just when conclusions stop being contained in their premises.

In particular, the argument Frege uses to prove the existence of successors—show by induction on finite numbers n that the number belonging to the concept $[x : x \leqslant n]$ succeeds n—is a fine example of the way in which content is created. "Through the present example" wrote Frege in § 23 of the *Begriffsschrift*

> we see how pure thought ... can, solely from the content that results from its own constitution, bring forth judgments that at first sight appear to be possible only on the basis of some intuition. This can be compared with condensation, through which it is possible to transform the air that to a child's consciousness appears as nothing into a visible fluid that forms drops.

That successors appear to have been condensed by Frege out of less than thin air may well have heightened some of its readers' suspicions that the principles employed in the *Foundations* are inconsistent.

On the other hand, Frege's construction of the natural numbers foreshadows von Neumann's well-known construction of them, the consistency of which was never in doubt. Frege defines 0 as the number of things that are non-self-identical; von Neumann defines 0 as the set of things that are non-self-identical. Frege shows that n is succeeded by the number of numbers less than or equal to n; von Neumann defines the successor of n as the set of numbers less than or equal to n. Peano arithmetic based on the von Neumann definition of the natural numbers can be carried out (interpreted) in a surprisingly weak theory of sets sometimes called General Set Theory, the axioms of which are:

Extensionality: $\forall x \forall y (\forall z (z \in x \leftrightarrow z \in y) \rightarrow x = y)$,

Adjunction: $\forall w \forall z \exists y \forall x (x \in y \leftrightarrow x \in z \lor x = w)$, and all

Separation axioms: $\forall z \exists y \forall x (x \in y \leftrightarrow x \in z \land A(x))$.

There is a familiar model for general set theory in the natural numbers: $x \in y$ if and only if starting at zero and counting from right to left, one finds a 1 at the xth place of the binary numeral for y. It is obvious that extensionality, adjunction, and separation hold in this model. Thus it has been clear all along that something *rather* like what Frege was doing in the *Foundations* could consistently be done.

The results of the *Foundations* that the series of finite numbers has no last member and that the "less than" relation on the finite numbers is irreflexive complement those of the *Begriffsschrift*, whose main theorems, when applied to the finite numbers, are that "less than" is transitive (98) and connected (133). Much more of mathematics can be developed in FA than Frege carried out in his three logic books. (It would be interesting to know how much of the *Basic laws* can be salvaged in FA.) Since addition and multiplication can be defined in any of several familiar ways and their basic properties proved from the definitions, the whole of analysis can be proved (more precisely, interpreted) in FA. (The equiconsistency of analysis and FA can be proved in Primitive Recursive Arithmetic.) Thus it is a vast amount of mathematics that can be carried out in FA.

Instead of discussing this rather familiar material, I want instead to take a look at certain strange features of FA, one of which was alluded to earlier. Frege defined 0 as the number belonging to the concept 'not identical with itself.' What is the number belonging to the concept 'identical with itself'? What is the number belonging to the concept 'finite number'? Frege introduces the symbol ∞_1 to denote the latter number, shows that ∞_1 succeeds itself, and concludes that it is not finite. But, although Frege does not consider the former number and hence does not deal with the question of whether the two are identical, it is clear that he must admit the existence of such a number. The statement that there is a number that is the number of all the things there are (among them itself) is antithetical to Zermelo-Fraenkelian doctrine, but as a view of infinity it is not altogether uncommonsensical. The thought that there is only one infinite number, *infinity*, which is the number of all the things there are (and at the same time the number of *all* the finite numbers), is not much more unreasonable than the view that there is no such thing as infinity or infinite numbers. In any event the view is certainly easier to believe than the claim that there are so many infinite numbers that there is no set or number, finite or infinite, of them all.

But can we decide the question of whether these numbers are the same? Not in FA. $N[x : x = x] = N[x : \mathrm{Fin}\, x]$ is true in some models of FA, for example, the one given, and false in others, as we can readily see. Let U' be the set of all ordinals $\leqslant \aleph_1$, and let η be true of V, u ($V \subseteq U'$, $u \in U'$) if and only if the (finite or infinite) cardinality of V is u. Numbers is then true in this structure, $\mathrm{Fin}\, x$ is satisfied by the natural numbers, $N[x : \mathrm{Fin}\, x]$ denotes $\aleph_0$, but $N[x : x = x]$ denotes $\aleph_1$. $N[x : x = x] = N[x : \mathrm{Fin}\, x]$ is thus an undecidable sentence of FA. Of course, so is $\exists x \neg Z x$, but $N[x : x = x] = N[x : \mathrm{Fin}\, x]$ is an undecidable sentence about numbers. From Frege's somewhat sketchy remarks on Cantor, one can conjecture that Frege would have probably regarded $N[x : x = x] = N[x : \mathrm{Fin}\, x]$ as false.

I now turn to the way Russell's paradox bears on the philosophical aims of the *Foundations*. My view is a more or less common one: As a result of the discovery of Russell's paradox our idea of logical truth has changed drastically, and we now see arithmetic's commitment to the existence of infinitely many objects as a greater difficulty for logicism than Russell's paradox itself.

But is not Frege committed to views that generate Russell's paradox? Does he not suppose that every predicate determines a concept and every concept has a unique extension? In § 83 he says:

And for this, again, it is necessary to prove that this concept has an extension identical with that of the concept 'member of the series of natural numbers ending with d.'

In § 68 he mentions the extension of the concept 'line parallel to line a.' And the number belonging to the concept F is defined as the extension of the concept 'equinumerous with the concept F.' How, in view of his avowed opinions on the existence of extensions, can he be thought to escape Russell's paradox?

The first quotation can be dealt with quickly, as a turn of phrase. Had Frege written " ... to prove that an object falls under this concept if and only if it is a member of the series of ... ," it would have made no difference to the argument. The extension of the concept 'line parallel to the line a' is used merely to enable the reader to understand the point of the definition of number. (These are the insignificant but possibly revealing exceptions mentioned to the claim that the only extensions to whose existence Frege explicitly commits himself in §§ 68–83 are those of concepts of the form 'equinumerous to the concept F.') Thus, if there is a serious objection to Frege's introduction of extensions of concepts, it must concern the definition of numbers as extensions of concepts of the form 'equinumerous with the concept F.'

And of course there is one. According to Frege, for every concept F there is a unique object x, an 'extension,' such that for every concept G, G bears a certain relation, 'being in,' designated by η, to x if and only if the objects that fall under F are correlated one-one with those that fall under G; that is, Numbers holds. And although the language of FA, in which Numbers is expressed, is not one in which the most familiar version of Russell's contradiction $\exists x \forall y(y\eta x \leftrightarrow \neg y\eta y)$ is a well-formed sentence, it is not true that Frege is now safe from all versions of Russell's paradox.

For consider Rule (V) of Frege's *Basic laws*:

$$\forall F \exists! x \forall G(G\eta x \leftrightarrow \forall y(Fy \leftrightarrow Gy)),$$

which yields an inconsistency in the familiar way.

Suppose Rule (V) true. By comprehension, let $F = [y : \exists G(G\eta y \wedge \neg Gy)]$. Then for some x,

(†) $\forall G(G\eta x \leftrightarrow \forall y(Fy \leftrightarrow Gy))$.

Since $\forall y(Fy \leftrightarrow Fy)$, by (†) $F\eta x$. If $\neg Fx$, then $\forall G(G\eta x \rightarrow Gx)$, whence Fx; but if Fx, then for some G, $G\eta x$ and $\neg Gx$, whence by (†), $\neg Fx$, contradiction.

Or consider the simpler

SuperRussell: $\exists x \forall G(G\eta x \leftrightarrow \exists y(Gy \wedge \neg G\eta y))$.

Suppose SuperRussell true. Let x be such that for every G, $G\eta x$ if and only if $\exists y(Gy \wedge \neg G\eta y)$. By comprehension, let $F = [y : y = x]$. Then, $F\eta x$ iff $\exists y(Fy \wedge \neg F\eta y)$, iff $\exists y(y = x \wedge \neg F\eta y)$, iff $\neg F\eta x$, contradiction.

SuperRussell and Rule (V) are sentences of the language of FA about the existence of extensions every bit as much as Numbers is. Just as Numbers asserts the existence (and uniqueness) of an extension containing just those concepts that are equinumerous with any given concept, so SuperRussell asserts the existence of an extension containing just those concepts that fail to be in some object falling under them and Rule (V) asserts the existence (and uniqueness) of an extension containing just those concepts under which fall the same objects as fall under any given concept. Frege must deny that SuperRussell and Rule (V) are principles of logic—if he maintains that the comprehension axioms are principles of logic. Principles of logic cannot imply falsity. But then Frege cannot maintain both that every predicate of concepts determines a higher level concept and that every higher level concept determines an extension and would thus appear to be deprived of any way at all to distinguish Numbers from SuperRussell and Rule (V) as a principle of logic.

Too bad. The principles Frege *employs* in the *Foundations* are consistent. Arithmetic can be developed on their basis in the elegant manner sketched there. And although Frege couldn't and we can't supply a reason for regarding Numbers (but nothing bad) as a logical truth, Frege was better off than he has been thought to be. After all, the major part of what he was trying to do—develop arithmetic on the basis of consistent, fundamental, and simple principles concerning objects, concepts, and extensions—can be done, in the way he indicated. The threat to the *Foundations* posed by Russell's paradox is to the philosophical significance of the mathematics therein and not at all to the mathematics itself.

It is unsurprising that we cannot regard Numbers as a purely logical principle. Consistent though it is, FA implies the existence of infinitely many objects, in a strong sense: Not only does FA imply $\exists x \exists y (x \neq y)$, $\exists x \exists y \exists z (x \neq y \wedge x \neq z \wedge y \neq z)$, etc., it implies $\exists F(\text{DedInf } F)$, where DedInf F is a formula expressing that F is Dedekind infinite, for example, $\exists x \exists G(\neg Gx \wedge \forall y(Fy \leftrightarrow Gy \vee y = x) \wedge F \text{ eq } G)$. In logic we ban the empty domain as a concession to technical convenience but draw the line there: We firmly believe that the existence of even two objects, let alone infinitely many, cannot be guaranteed by logic alone. After all, logical truth is just *truth no matter what things we may be talking about and no matter what our (nonlogical) words mean*. Since there might be fewer than two items that we happen to be talking about, we cannot take even $\exists x \exists y (x \neq y)$ to be valid.

How then, we might now think, *could* logicism ever have been thought to be a mildly plausible philosophy of mathematics? Is it not obviously demonstrably inadequate? How, for example, could the theorem

$$\forall x (\neg x < x) \wedge \forall x \forall y \forall z (x < y \wedge y < z \rightarrow x < z) \wedge \forall x \exists y (x < y),$$

of (one standard formulation of) arithmetic, a statement that holds in no finite domain but which expresses a basic fact about the standard ordering of the natural numbers, be even a "disguised" truth of logic?[11] The axiom of infinity was soon enough recognized by Russell as both indispensable to his program and as damaging to the claims that could be made on behalf of the program; and it is hard to imagine anyone now taking up even a small cudgel for $\exists x \exists y (x \neq y)$.

I have been arguing for these claims: (1) Numbers is no logical truth; and therefore (2) Frege did not demonstrate the truth of logicism in the *Foundations of arithmetic*. (3) Logic is synthetic if mathematics is, because (4) there are many interesting, logically true

[11]See Paul Benacerraf, "Logicism: some considerations," Ph.D. dissertation, Princeton University, Department of Philosophy, 1960.

conditionals with antecedent Numbers whose mathematical content is not appreciably less than that of their consequents. To these I want to add: (5) Since we have no understanding of the role of logic or mathematics in cognition, the failure of logicism is at present quite without significance for our understanding of mentality. Had Frege succeeded in eliminating the nonlogical residue from his *Foundations*, the question would remain what the information that arithmetic is logic tells us about the cognitive status of arithmetic. But Frege's work is not to be disparaged as a (failed) attempt to inform us about the role of mathematics in thought. It is a powerful mathematical[12] analysis of the notion of natural number, by means of which we can see how a vast body of mathematics can be deduced from one simple and obviously consistent principle, an analysis no less philosophical for its rigor, profundity, and surprise.

A fantasy: After the *Begriffsschrift* Frege writes, not the *Foundations of arithmetic*, but another book with the same title whose main claim is that, since arithmetic is deducible by logic alone from the triviality "the number of F's is the same as the number of G's if and only if the F's can be correlated one-one with the G's," arithmetic is analytic, not synthetic, as Kant supposed. Frege then argues for the analyticity of $NF = NG \leftrightarrow F \text{ eq } G$ on the ground that both halves of the biconditional have the same content, express the same thought. He considers an attempted defense of Kant: Since the existence of an object can be inferred from $NF = NF$, $NF = NF$ must be regarded as synthetic, and therefore so must $NF = NG \leftrightarrow F \text{ eq } G$. Frege replies that $7 + 5 = 7 + 5$ is analytic.

If Frege had abandoned one of his major goals—the quest for an understanding of numbers not as objects but as 'logical' objects— taken as a starting point the self-evident and consistent $\forall F \forall G(NF = NG \leftrightarrow F \text{ eq } G)$, and worked out the consequences of this one axiom in the *Begriffsschrift*, he would have been wholly justified in claiming to have discovered a foundation for arithmetic. To do so would have been to trade a vain philosophical hope for a thoroughgoing mathematical success. Not a bad deal. He could also have plausibly claimed to demonstrate the analyticity of arithmetic. (Of course his own work completely undermines the interest of such a claim.)

---

[12]See Paul Benacerraf, "The last logicist," Chapter 2, above.

Perhaps the saddest effect of Russell's paradox was to obscure from Frege and us the value of Frege's most important work. Frege stands to us as Kant stood to Frege's contemporaries. The *Basic laws of arithmetic* was his *magnum opus*. Are you sure there's nothing of interest in those parts of the *Basic laws* that aren't in prose?

9

The Standard of Equality
of Numbers

One of the strangest pieces of argumentation in the history of logic is found in Richard Dedekind's *Was sind und was sollen die Zahlen?*, where, in the proof of that monograph's theorem 66, Dedekind attempts to demonstrate the existence of infinite systems. Dedekind defines a system S as *infinite* if, as we would now put it, there are a one-one function φ from S to S and an element of S not in the range of φ. Since it is now known that set theory without the axiom of choice does not imply that a set that is infinite in the usual sense is infinite in Dedekind's sense (although it does imply the converse), it is now common to prefix "Dedekind" when speaking of infinity in this stronger sense. The sets with which we shall be concerned are Dedekind infinite if they are infinite at all, however, and I shall therefore omit "Dedekind" before "infinite."

Theorem 66 of *Was sind* reads, "There are infinite systems"; the proof of it Dedekind offered runs:

From *Meaning and method: essays in honor of Hilary Putnam*, Cambridge: Cambridge University Press, 1990, 261–77. Reprinted by kind permission of Cambridge University Press and the author.

I am grateful to Ellery Eells, Dan Leary, Thomas Ricketts, and Gabriel Segal for helpful comments. Research for this essay was carried out under grant no. SES-8808755 from the National Science Foundation.

Proof.[1] The world of my thoughts, i.e., the totality S of all things that can be objects of my thought, is infinite. For if s denotes an element of S, then the thought s', that s can be an object of my thought, is itself an element of S. If s' is regarded as the image $\varphi(s)$ of the element s, then the mapping φ on S determined thereby has the property that its image S' is a part of S; and indeed S' is a proper part of S, because there are elements in S (e.g., my own ego [*mein eigenes Ich*]), which are different from every such thought S' and are therefore not contained in S'. Finally, it is clear that if a, b are different elements of S, then their images a', b' are also different, so that the mapping φ is distinct (similar). Consequently, S is infinite, q.e.d.

It is tempting to think that Dedekind isn't in as deep a hole as his mentioning so wildly nonmathematical an item as his own ego might suggest and to suppose that he has merely chosen a bad example. Wouldn't the sentence *Berlin ist in Deutschland* and the operation of prefixing *Niemand glaubt dass* have been just as good as Dedekind's own ego and the operation that takes any object s in the world of Dedekind's thoughts to that funny thought about s? Instead of the things that can be objects of his thought (whatever these might be) couldn't he have cited (say) the set of German sentences, i.e., sentence-types, as an example of an infinite set? And had he cited that set, wouldn't he have given an obviously correct proof of theorem 66 by giving an obviously correct example of an infinite set?

It is significant that nowhere in the remainder of *Was sind und was sollen die Zahlen?* does Dedekind appeal to theorem 66 in the proof of any other theorem. Why, one might wonder, did not Dedekind simply omit the theorem and its proof, the incongruity of whose argumentation and subject matter Dedekind himself could not have failed to find glaring?

Recall that the aim of *Was sind*, according to Dedekind, was to lay the foundations of that part of logic that deals with the theory of numbers—thus the theory of numbers is a part of logic—and that

---

[1] A similar observation is found in section 13 of Bolzano's, *Paradoxes of the infinite*, Leipzig, 1851. (Footnote by Dedekind.)

his answer to the title question of his monograph was that numbers are "free creations of the human mind." Some of what that saying means emerges in section 73, where he writes:

> If in observing a simply infinite system N, ordered by a mapping φ, the special character of the elements is completely disregarded, only their distinguishability is held fixed, and account is taken of only those relations to one another in which they are placed by the mapping φ that orders them, then these elements are called *natural numbers* or *ordinal numbers* or also simply *numbers*, and the basis element 1 is called the basis-number of the number series N. With regard to this freeing of the elements from all other content (abstraction), one can justifiably call the numbers a free creation of the human mind. The relations or laws which are derived just from the conditions α, β, γ, δ^2 in 71 [these are Dedekind's versions of what have come to be known as the "Peano Postulates"] and therefore are always the same in all ordered simply infinite systems, however the names accidentally given to the individual elements may be pronounced, form the first object of the *Science of Numbers* or *Arithmetic*.

Thus, arithmetic is about certain objects, the numbers, abstracted from simply (we now say "countably") infinite systems, systems satisfying the "Peano" conditions α, β, γ, δ under some appropriate choice of base element and successor operation. Since they have been abstracted from systems satisfying α, β, γ, δ, the numbers too satisfy these conditions. Logic, Dedekind would appear to be claiming, suffices for the derivation of all, or at any rate all familiar, arithmetical facts from the mere assumption that the numbers, together with 1 and successor, are objects that satisfy the "Peano" conditions. (Dedekind proves that the existence of simply infinite systems follows from that of infinite systems.)

The trouble with trying to prove theorem 66 by mentioning the set of sentences of German is that Dedekind would probably have regarded a sentence (or any other abstract object) as as much a free

---

[2]α states that the successor of a number is a number; β, that 1 is a number and that mathematical induction holds of the numbers; γ, that 1 is not the successor of a number; and δ, that successor is one-one.

creation out of ink-tracks or other physical objects as a number is a free creation out of objects. Dedekind did not cite the most obvious infinite system, the system of the natural numbers themselves, in the proof of theorem 66. It would thus appear that he thought that a satisfactory proof of it must mention some infinite set of non-abstract items, out of which the natural numbers could have been freely created, and that he was therefore not at liberty to cite a set of sentences or abstract objects of any other sort as an example of an infinite set.

Dedekind's proof, however, is fallacious if thoughts are taken to be actual physical occurrences. Ignoring worries about opacity, we may grant that if u is a thought that s can be an object of my thought, and likewise for v and t, then $u = v$ if and only if $s = t$. It does not follow, and it is indefensible to assume, that for every object s, or at least for every object s that is an object of my thought, *there is* such a thing as the thought that s can be an object of my thought. There just aren't all those thoughts around. (As Frege, commenting on theorem 66, put it "Now presumably we shall not hurt Dedekind's feelings if we assume that he has not thought infinitely many thoughts.")[3] Dedekind makes this unwarranted assumption in the proof by using the definite article and speaking of "*the* thought that s can be an object of my thought." Of course, without some guarantee that all those thoughts exist, the proof fails: Dedekind hasn't defined a function on the (whole) world of his thoughts. The present king of France strikes again.

Thus Dedekind is in trouble that we do not appear to be in. We, but not he, can use the set of sentences of German as an example of an infinite set. The difficulty for us in doing so will shortly emerge.

Dedekind's notion of free creation[4] raises too many problems for us to find it satisfactory: One somewhat less obvious difficulty it poses is a "third man" difficulty, that of saying why we don't get different systems when we abstract *twice* from a system satisfying conditions α, β, γ, δ. Or *do* we get different, but isomorphic, sys-

---

[3]In "Logic" (1897), in Gottlob Frege, *Posthumous writings*, H. Hermes et al., eds., Chicago: University of Chicago Press, 1979, p. 136. I am grateful to Arnold Koslow for telling me of this quotation.

[4]For a more detailed account of this notion, see Charles Parsons's "The structuralist view of mathematical objects," *Synthese*, **84** (1990) 303–47.

tems? Or can we abstract only *once*? Best not to take him too seriously here.

The view can be made more appealing and more plausible if we forget about abstraction and free creation, and take Dedekind to be saying that statements about the natural numbers can be regarded as logically true statements about all systems satisfying conditions α, β, γ, δ. Charles Parsons, in an illuminating study, "The structuralist view of mathematical objects,"[5] has called this the eliminative reading of *Was sind*. Perhaps we might take Dedekind to be claiming that an arithmetical statement, expressed by a sentence S is in the notation, say, of second-order logic, in which all number quantifiers are relativized to the predicate letter N, 1 denotes one, and s denotes successor, has the logical form: $\forall N, s, 1(\alpha, \beta, \gamma, \delta(N, s, 1) \to S)$. (Addition and multiplication, etc., can be handled by familiar techniques due to Dedekind.) Thus the monadic predicate letter N, the monadic function sign s, and the constant 1 turn into a second-order monadic predicate variable, a second-order monadic function variable, and a (peculiarly shaped) first-order variable, which are then universally quantified upon. To complete the interpretation of the resulting sentence, we might want to add that the first-order variables range over all the things there are.

For any such arithmetical sentence S, let $D(S)$ be the second-order sentence $\forall N, s, 1(\alpha, \beta, \gamma, \delta(N, s, 1) \to S)$. $D(S)$, it will be observed, contains no non-logical constants at all. We now want to inquire into the relation between S and $D(S)$.

Suppose that S is true, i.e., true when interpreted over the natural numbers, together with successor and one and that N', s', and $1'$ satisfy $\alpha, \beta, \gamma, \delta(N, s, 1)$. Then since the natural numbers together with successor and one also satisfy $\alpha, \beta, \gamma, \delta(N, s, 1)$, by a valid second-order argument given by Dedekind, N', s', and $1'$ are isomorphic to the natural numbers, successor, and one and therefore satisfy S. Thus $D(S)$ is a logical truth. Conversely, suppose that $D(S)$ is a logical truth. Let

$$(*) \qquad N', s', \text{ and } 1' \text{ satisfy } \alpha, \beta, \gamma, \delta(N, s, 1).$$

They therefore also satisfy S. Since the natural numbers, successor, and one also satisfy $\alpha, \beta, \gamma, \delta(N, s, 1)$, they are isomorphic to N',

---

[5] Parsons, "The structuralist view of mathematical objects."

s', and $1'$, and therefore also satisfy S; that is, S is true. Thus, it would appear, we have shown that S is true if and only if $D(S)$ is a logical truth. Does not this argument show that arithmetical truths are logical truths disguised only by the omission of an antecedent condition and a few symbols of logic?

Of course—on the assumption $(*)$, true, by our lights, that *there are N', s', and $1'$ together satisfying $\alpha, \beta, \gamma, \delta(N, s, 1)$:* We used this assumption when we argued that if $D(S)$ is a logical truth, then S is true.' We have had to make a true assumption, but one that we have as yet found no reason to regard as logically true, in order to show that we can effectively associate with each sentence of arithmetic, a sentence in the vocabulary of logic in such a way that with each truth and no falsehood of arithmetic there is associated a logical truth. To succeed this far in reducing arithmetic to logic we have had to make an assumption not yet certified as logically true: There are infinite systems.

Is that a difficulty? It might seem not. We make a non-logical assumption to reduce arithmetic to logic. We then throw away the ladder. But ladder or no, we have reduced arithmetic to logic, haven't we?

Parsons has pointed out a difficulty in supposing that we have.[7] He notes that if there are no infinite systems, then $D(S)$ is true, for every arithmetical sentence S " ... both A and $\neg A$ have true canonical forms, which amounts to the inconsistency of arithmetic."

Parsons's observation leads us to the heart of the matter. Logicism is not adequately characterized as the view that arithmetic is reducible to logic if all that is meant thereby is that there is an effective mapping of statements of arithmetic to statements of logic that assigns logical truths to all and only the truths of arithmetic. Nor is it vindicated merely by exhibiting such a mapping E. For E to vindicate logicism, it must show that arithmetic is "reducible to" logic, "really" logic, logic "in disguise," "a part of" logic. Then at least, for any arithmetical sentence S, $E(S)$ must give the content of

---

[6]We didn't need it for the other half, since we are entitled to *assume* that there is a system satisfying $\alpha, \beta, \gamma, \delta(N, s, 1)$. *Was sind* proves that this assumption could have been replaced by the assumption that there is an infinite set.

[7]Parsons, "The structuralist view of mathematical objects."

S, must state, in the language of logic, how matters must be if and only if S is true.[8] But then E must do for falsity what it does for truth and also assign logical falsehoods to the falsehoods of arithmetic; otherwise there will be certain truths S of arithmetic such that $E(S)$ is compatible with $E(\neg S)$, and no mapping that thus violates negation can be regarded as giving the content of statements of arithmetic in logical terms and hence as reducing arithmetic to logic.

For arithmetical sentences, like all others, come in triples: For any two arithmetical sentences there is a third, their conjunction, that, however matters may be, holds when matters are that way if and only if both sentences hold when matters are that way. A mapping E under which $E(S \wedge S')$ is not equivalent in this sense to the conjunction of $E(S)$ and $E(S')$ cannot be thought to give the content of all three of S, S' and $(S \wedge S')$ and cannot therefore count as a reduction of arithmetic to logic.

Similarly for negation: If for some arithmetical sentence S, $E(\neg S)$ is not equivalent to the negation of $E(S)$, then E does not give the content of both S and $\neg S$, and therefore does not show arithmetic reducible to logic. In advance of any possible reduction to logic certain arithmetical statements immediately (logically) imply certain others, and certain arithmetical statements are immediately incompatible with certain others. A reduction of arithmetic to logic, although it may reveal previously unrecognized implications or incompatibilities among the statements of arithmetic, cannot disclose that these immediate implications and incompatibilities actually fail to obtain.

For a mapping E to vindicate logicism, then, E must at the very least respect the truth-functional operators on closed formulae. Since $E(S)$ will always be logically true if S is true, E will respect negation if and only if $E(S)$ is always logically false for false S. It is clear that Dedekind's mapping D respects conjunction.

But D does not respect negation. $\neg \forall N, s, 1(\alpha, \beta, \gamma, \delta(N, s, 1) \to S)$ is not logically equivalent to $\forall N, s, 1(\alpha, \beta, \gamma, \delta(N, s, 1) \to \neg S)$. The latter follows logically from the former, as Dedekind showed.

---

[8]Is it possible to say what logicism is without using intensional notions like *part* or *content*?

But the former follows from the latter in general only under the assumption that infinite systems exist. Indeed, if S is, say, $1 = 1$, then the former is equivalent to $\perp$, the latter to "there are no infinite systems," and the conditional with antecedent the latter and consequent the former is then equivalent to "there are infinite systems."

If S is the statement "there are infinite systems," a truth, but presumably not a logical truth, then $D(S)$ is a logical truth, as desired; but $D(\neg S)$ is equivalent to $\neg S$, and therefore not a logical falsehood.

A third example: The mapping D assigns to "$\neg 17 \times 14 = 228$" a sentence that is (absent logically guaranteed infinite systems) consistent with what it assigns to "$17 \times 14 = 228$." D cannot therefore count as reducing arithmetic to logic in any reasonable sense of the phrase.

All would be well, of course, if, as a matter of logic, there were infinite systems, if theorem 66 had been established as securely as, and in the manner of, theorems 65 and 67. But no purely logical ground has been given for thinking theorem 66 true. That, and not the non-mathematical character of the objects it mentions, is the real problem with Dedekind's proof of theorem 66.

We ought to mention that there can be no effective mapping of sentences of arithmetic to sentences of logic under which truths of arithmetic are mapped to logical truths and falsehoods to logical falsehoods: Otherwise arithmetic would be decidable, since the truth-value of any statement could be ascertained by calculating the truth-value of its image under the mapping in any one-element model. Nor is there a mapping of sentences of arithmetic to sentences of *first*-order logic under which the truths of arithmetic and only those are mapped to logical truths. Otherwise it would be possible to decide effectively whether an arithmetical sentence S is true or false: Effectively enumerate all first-order logical truths; then the image of S under the mapping appears in the enumeration if and only if that of $\neg S$ does not, and S is true if and only if its image appears.

Infinity is cheap. As Dedekind showed, a domain S will be infinite if there are a one-one function φ from S to S and an object in S not in the range of φ. Indeed, it's often easier than one may suspect to show a domain infinite. For example, in conjunction with the trivial truth "$\exists x \exists y\, x \neq y$," the ordered pair axiom, commonly thought to be innocuous, is an axiom of infinity. Any domain in which both hold is infinite, for if $a \neq b$, then the function that assigns to each object x in the domain the ordered pair $\langle a, x \rangle$ will be one-one and omit $\langle b, b \rangle$ from its range.

It is well-known that very weak systems of set theory guarantee that there are infinitely many objects: the conjunction of the null set and unit set axioms supply an object and a one-one function meeting Dedekind's criterion. It is thus not difficult to provide a theory committed to there being infinitely many objects. The difficulty (insuperable, I will urge) is to find a logically true theory with this commitment.

We have seen that in order to be able to claim that the function that assigns $\forall N, s, 1(\alpha, \beta, \gamma, \delta(N, s, 1) \rightarrow S)$ to any sentence S of arithmetic shows that "arithmetic is a part of logic," Dedekind needs a proof from logical truths that there are infinitely many objects. No satisfactory way has yet presented itself.

I want to consider the suggestion that a principle I call Hume's principle can be used to help Dedekind out. As we shall see, Dedekind would probably not like the suggestion. And in the end, I shall argue, we can't accept it, either.

"We are possest of a precise standard," wrote Hume, "by which we can judge of the equality and proportion of numbers; and according as they correspond or not to that standard, we determine their relations, without any possibility of error. When two numbers are so combin'd as that the one has always an unite answering to every unite of the other, we pronounce them equal; and 'tis for want of such a standard of equality in extension, that geometry can scarce be esteem'd a perfect and infallible science."[9] Reflecting on Frege's idea that statements about numbers are assertions about what he called concepts, we may formalize Hume's dictum as a second-order formula. Let "$\#F$" abbreviate "the number of objects falling under the concept F" and "$F \approx G$" express the existence of a one-one correspondence between the objects falling under

---

[9] *Treatise*, I, III, I para. 5.

F and those falling under G.[10] Then Hume's principle may be written: $\forall F \forall G(\#F = \#G \leftrightarrow F \approx G)$.

Frege attempts to prove Hume's principle in § 73 of his *Foundations of arithmetic*. The difficulty with the proof he gives there is that it appeals to the theory of concepts and objects, whose inconsistency Russell pointed out in his first letter to Frege. After having derived Hume's principle from this inconsistent theory, Frege derives the axioms of arithmetic from Hume's principle.

More exactly, in the *Foundations of arithmetic* Frege gives definitions of *zero*, *succeeds* (*"follows directly after"*), and *finite* (*natural*) *number*, and shows, easily enough, that zero is a finite number, that anything that succeeds any finite number is a finite number, that zero succeeds nothing and that if m, m', n, and n' are finite numbers, n succeeds m, and n' succeeds m', then $m = m'$ iff $n = n'$. It is by no means evident, however, that every finite number is succeeded by something, and it was a matter of considerable difficulty for Frege to prove it. The central argument of the *Foundations of arithmetic* is a fairly complete sketch of a proof that every finite number is succeeded by a finite number; in proving this and the other facts about the numbers, Frege makes use only of Hume's principle and the system of logic set forth in his *Begriffsschrift*. The intricacy of his reasoning is astonishing and repays careful attention. See the appendix for a reconstruction. I do not know whether Frege realized that Hume's principle plus the logic of the *Begriffsschrift* was all he used or needed. Perhaps not; *he* would have had no reason to value the observation. In any event, it is a pity that the derivability of arithmetic from Hume's principle isn't known as *Frege's theorem*.

Frege thus succeeds where Dedekind has failed. He has demonstrated the existence of an infinite system. With hard work, he has proved the analogue of what Dedekind simply assumes to be the case for the system of objects x of his thought, that for each x, *there is* a y to which x bears the appropriate relation.

Do not be deceived by the absence of the sort of wallpaper found in the *Begriffsschrift* into thinking that the *Foundations* is not fundamentally a mathematical work.[11] In a letter of September 1882,

---

[10] "$F \approx G$" abbreviates the second-order formula: $\exists\varphi(\forall x[Fx \rightarrow \exists y(Gy \wedge \varphi xy)] \wedge \forall y[Gy \rightarrow \exists(Fx \wedge \varphi xy)] \wedge \forall x \forall x' \forall y \forall y'[\varphi xy \wedge \varphi x'y' \rightarrow (x = x' \leftrightarrow y = y')])$.

[11] Cf. Paul Benacerraf, "Frege: the last logicist," Chapter 2, above.

Carl Stumpf suggested to Frege that it might be "appropriate to explain your line of thought first in ordinary language and then—perhaps separately on another occasion or in the very same book—in conceptual notation. I should think that this would make for a more favorable reception of both accounts."[12] Frege apparently took Stumpf's suggestion, which was that his *mathematical* ideas be first published in ordinary language. At the heart of the *Foundations* there lies a *proof*.

Frege outlines a demonstration in the *Foundations* that arithmetic, i.e., the basic axioms of the second-order arithmetic of zero and successor (from which that of full second-order arithmetic, of addition and multiplication, can be derived, as in *Was sind*; Frege seems never to have been interested in deriving the axioms of addition and multiplication), can be derived from Hume's principle. Second-order arithmetic is consistent, presumably; Frege derives Hume's principle from an inconsistent theory of concepts. Nevertheless, an inconsistent theory may, indeed must, have consistent consequences, and it turns out that in the same sense in which second-order arithmetic may be derived from Hume's principle in the system of logic of the *Begriffsschrift*, Hume's principle may be derived back from second-order arithmetic. (Deriving axioms from theorems has been called "reverse mathematics" by Harvey Friedman.) Hume's principle and second-order arithmetic, which is sometimes called "analysis," are thus equiconsistent, and very effectively so: A proof of an inconsistency from either could easily be turned into a proof of an inconsistency from the other.

In analysis, there are two sorts of variables, one sort ranging over the natural numbers, the other over sets of natural numbers. The axioms are the usual ones: the Peano axioms, together with the usual axioms for addition and multiplication (which, as Dedekind showed, are dispensable), and a comprehension scheme: For any formula of the language of analysis, there is a set of all and only the numbers satisfying the formula.

The *Foundations* of course shows that if analysis is inconsistent, so is Hume's principle. How may the converse be shown? Let α be

---

[12]In Gottlob Frege, *Philosophical and mathematical correspondence*, H. Hermes et al., eds., Chicago: University of Chicago Press, 1980, 172.

a set of natural numbers. Call the natural number n the *grumber* belonging to α if either α has infinitely many members and $n = 0$ or α has $n - 1$ members. The grumber of U.S. senators is 101, the grumber of roots of the equation "$x - 5 = 0$" is 2, and the grumber of even numbers is zero, which is also the grumber of numbers divisible by four. It is then a *theorem* of analysis that the grumber belonging to $\alpha =$ the grumber belonging to β if and only if the members of α and those of β are in one-one correspondence. Any derivation in second-order logic of a contradiction from Hume's principle could thus be turned into a derivation in analysis of a contradiction from the theorem of analysis about grumbers just cited. Thus if analysis is consistent, so is the result of adjoining Hume's principle to second-order logic.

Some trick like the introduction of grumbers is necessary because "the natural number belonging to" is not defined for all sets of natural numbers, indeed not defined for any set containing infinitely many natural numbers. There is no natural number that is the number of members of the set of evens; but the grumber zero belongs to this set.

Frege, then, gave an intricate and mathematically interesting derivation of arithmetic from a simple, consistent, and trivial-seeming principle. Since the principle is as weak as any from which arithmetic can be derived, Frege's derivation was "best possible."

Dedekind, of course, might well have objected to our suggestion that Hume's principle be used to obtain an infinite system on the ground that the arithmetical notion *the number belonging to* that figures in the principle is undefined, and that arithmetic is therefore not shown to be entirely a part of logic. A weak reply can be made: There is a principle, discussed farther on, that deals only with objects and concepts, which licenses a definition of number, from which Hume's principle can be derived. The principle is that for every concept F there is a unique object x such that for all concepts G, G is in x if and only if F and G are equinumerous. Unlike Hume's, this principle does not explicitly mention numbers. The number belonging to F may of course be defined to be the unique x such that for all G, etc.

Even if the objection that the expression "is in" is not a logical "constant" is waived, this principle cannot be held to be a logical principle for a reason we shall consider at length: It commits us to the existence of too many objects. Here we should note that a truth's being couched in purely logical terms is not sufficient for it to count as truth *of* logic, a logical truth, a truth that is true solely in virtue of logic. A distinction must be drawn between truths of logic and truths expressed in the language of logic. I suspect that failure to draw this distinction was largely responsible for there ever being any thought at all that the axiom of infinity might actually count as a logical truth.

According to Hume's principle, for any concepts F and G, there are certain *objects*, namely, the number x belonging to F and the number y belonging to G, such that x is identical to y if and only if F and G are equinumerous. It is the objecthood of numbers that explains why Hume's principle, despite appearances, cannot be considered to be a truth of logic, a definition,[13] an immediate consequence of a definition, analytic, quasi-analytic, or anything of that sort.

The reason that it may appear so is that it can easily be confused with a principle that has a considerably greater claim to the status of truth of logic. Assume that some version of the theory of types, including axioms of comprehension and extensionality, counts as logic. Then matters are as the theory of types has it: There are individuals, sets of individuals, classes of sets of individuals, etc. (The words "set" and "class" are used here just to keep the types straight: We've got classes of sets of individuals.) According to a comprehension axiom, for any set, there will be a class containing all and only the sets that are equinumerous with that set; by extensionality, there will be at most one such class. We may call classes containing all and only the sets that are equinumerous with any one set *Russellnumbers*, and say that the Russellnumber of a set is the Russellnumber that contains the set. The proposition that Russellnumbers are identical if and only if the sets they are Russellnumbers of are equinumerous is then a theorem of (this version of) the theory of types.

Russellnumbers are classes of sets of individuals. One of the (in-

---

[13]On hearing me say what I meant by "Hume's principle," a *very* famous philosopher exclaimed, "But that's just a *definition*!"

effable?) doctrines of the usual formulation of the theory of types
is that the types are disjoint. No set is a class or individual, and no
class is an individual. Russellnumbers are not individuals.

The disadvantage of this way of defining numbers, of course, is
that arithmetic cannot be derived in the theory of types without
postulating that there are infinitely many individuals. With honest
toil, however, Frege succeeds in proving from Hume's principle the
infinity of the natural numbers.

Observe that although the theory of objects and concepts that is
sketched in the *Foundations* is almost certainly inconsistent, there
is a consistent fragment of it that is all Frege needs, or uses, to
derive arithmetic. According to this fragment, there are objects
(or individuals), first level concepts, under which objects may or
may not fall, and second level concepts, under which first level concepts may or may not fall. So far, matters look pretty much as
they do on the theory of types, if one substitutes "object," "first
level concept," and "second level concept" for "individual," "set,"
and "class." Crucially, though, Frege does not analogously define
the numbers as those second level concepts under which fall all and
only those first level concepts that are equinumerous with some one
concept. Call such second level concepts *numerical*. Instead, he introduces a new primitive relation between objects and concepts ("is
the extension of") and then defines a number to be an object that
is the extension of some numerical concept. (Numerical concepts,
to repeat, are second level.) Thus Frege assumes that for every first
level concept F there is a unique object that is the extension of the
second level concept under which fall all and only those first level
concepts that are equinumerous with F.

The introduction of second level concepts is not necessary. All
that Frege need do is introduce a primitive predicate "is in" for a
relation between first level concepts and objects and assume that
for any first level concept F there is exactly one object x such that
for any first level concept G, G is in x if and only if F and G are
equinumerous. Frege may then define the number "belonging to" F
as that object x.

Notice the additional step Frege has taken. Unlike Russell, Frege
has assumed there is a way of associating objects with numerical
concepts so that different objects are associated with different nu-

merical concepts. (Assume that coextensive concepts are identified.) This cannot be done, if there are only finitely many objects; if there are (say) eighteen objects, then there will be nineteen numerical concepts. It is a weighty assumption of Frege's, to put it slightly differently, that the first level concepts can be mapped into objects in such a way that concepts are mapped onto the same object only if they are equinumerous, and *it is a lucky break that the assumption is even consistent*.

The well-known comparison that Frege draws in §§ 64–9 of the *Foundations* between "the direction of line l" and "the number belonging to concept F" is therefore seriously misleading. We do not suspect that lines are made up of directions, that directions are some of the ingredients of lines. Had Frege appended to the direction principle, "The direction of line l is equal to the direction of line k if and only if l and k are parallel," the claim that directions are points, we would never have regarded the principle as anything like a definition, and would perhaps have wondered whether there are enough points to go around. (In fact, there are: There are continuously many points and continuously many directions.) The principle that directions of lines are identical just in case the lines are parallel looks, and is, trivial only because we suppose that directions are one or more types up from, or at any rate are all distinct from, the things of which lines are made.

The principle that numbers belonging to concepts are identical if and only if the concepts are equinumerous, then, should count as a logical truth only if it is supposed that numbers do not do to concepts or sets the corresponding sort of thing, namely, fall under, or be elements of, them. On the theory of types, matters so fall out. But Frege's proof from Hume's principle that every number has a successor cannot be carried out in the theory of types: the proof cannot succeed unless it is supposed that numbers are objects.

For how does Frege show that the number 0 is not identical with the number 1? Frege defines 0 as the number belonging to the concept *not identical with itself*. He then defines 1 as the number belonging to the concept *identical with* 0. Since no object falls under the former concept, and the object 0 falls under the latter, the two concepts are, by logic, not equinumerous, and hence their numbers 0 and 1 are, by Hume's principle, not identical. Notice that for this argument to work it is crucial that 0 be supposed to be an object that falls under the concept *identical with* 0. 2 arises in like manner: Now that 0 and 1 have been defined and shown different, form the concept *identical with* 0 *or* 1, take its number, call it 2, and observe that the new concept is coextensive with neither of these concepts *because the distinct objects* 0 *and* 1 *fall under it*. Conclude by Hume that 2 is distinct from both 0 and 1.

Frege proves that if n is a finite number, then it is succeeded by the number belonging to *being less than or equal to* n; the proof works because n is an object that can be proved not to fall under *being less than* n.

Thus it is only to one who supposes that numbers are not objects that Hume's principle looks analytic or obvious. Frege's proof that every number has a successor depends vitally on the contrary supposition that numbers are indeed objects.

A sentence is a logical truth only if it is true no matter what objects it speaks of and no matter to which of them its predicates or other non-logical words apply. (The vagueness of the consequent, including that as to which words count as logical, is matched by that of the antecedent.) A sentence is not a logical truth if it is false when interpreted over a domain containing infinitely many things, and it is not a logical truth if, like Hume's principle, it is false when only finitely many things belong to the domain.

It is clear that an account of logical truth that attempts to distinguish Hume's principle as a logical truth will have the hard task of explaining why Hume's principle is a logical truth even though two other similar-looking principles are not. These are the principle about extensions embodied in Frege's Rule (V) and a principle about relation numbers that is strikingly analogous to Hume's principle. They read: Extensions of concepts are identical if and only if those concepts are coextensive; and: Relation numbers of relations are identical if and only if those relations are isomorphic. Russell showed the former inconsistent; Harold Hodes has astutely observed that the latter leads to the Burali-Forti paradox.[14]

It will not do to say: Hume's principle, unlike the other two, and like the principle by which we take ourselves to introduce the two

---

[14] *The journal of philosophy*, **81** (1984) 138.

truth-values, is a logical truth because it is consistent.

For say that the concepts F and G *differ evenly* if the number of objects falling under F but not G or under G but not F is even (and finite). The relation between concepts expressed by "F and G differ evenly" is an equivalence relation (exercise), and can of course be defined in purely logical (second-order) vocabulary. Now introduce the term "the parity of" for a function from concepts to objects and consider the parity principle: The parity of F is identical with that of G if and only if F and G differ evenly.

The parity principle is evidently consistent. Let X be any finite domain containing the numbers 0 and 1. Let the parity of a subset of X be 0 if it contains an even number of objects and 1 otherwise. Then with "parity" so defined, the parity principle is true in the domain X, and is therefore consistent.

However, the parity principle is true in no infinite domain. Here's a sketch of the proof. Let X be an infinite set. Then, where Y and Z are subsets of X,

$$|\{\, Y : Y \text{ differs evenly from } Z \,\}|$$
$$= |\{\, (A, B) : A, B \subseteq X,\ A \subseteq Y,\ B \subseteq Z,\ A \cap Z = \emptyset,$$
$$B \cap Y = \emptyset, |A \cup B| \text{ is even, and } Y = (Z \cup A) - B \,\}|$$
$$\leqslant |\{\, (A, B) : A, B \subseteq X \text{ and } |A \cup B| \text{ is even} \,\}| = |X|.$$

Thus $|\{\, Y : Y \text{ differs evenly from } Z \,\}| = |X|$.

Suppose now that $f : PX \to X$ and for all Y, Z, if $fY = fZ$, Y evenly differs from Z. Then for each x in X, $|\{\, Y : fY = x \,\}| \leqslant |X|$, and $|PX| \leqslant |X| \times |X| = |X|$, contradiction.[15]

---

[15]A less natural example to the same purpose, but one that is less heavily dependent on set theory, is the following. Abbreviate: $(\exists x \exists y [x \neq y \wedge Fx \wedge Fy] \vee \exists x \exists y [x \neq y \wedge Gx \wedge Gy]) \to \forall x (Fx \leftrightarrow Gx)$ as: $F \operatorname{Equ} G$. Equ is an equivalence relation. The principle: $\forall F \forall G (\hat{\ }F = \hat{\ }G$ iff $F \operatorname{Equ} G)$ is evidently satisfiable in all domains containing one or two members. It is, however, satisfiable in no domain containing *three* or more members and is therefore inconsistent with Hume's principle. For suppose that $a \neq b \neq c \neq a$. Define Rx as follows: If $x \neq a, b, c$, then Rx iff for some F, $x = \hat{\ }F$ and $\neg Fx$; but if x is one of a, b, c, then let $Ra, Rb, \neg Rc$ hold if none of a, b, c is $\hat{\ }F$ for some F such that $Fa, Fb, \neg Fc$; otherwise let $Ra, \neg Rb, Rc$ hold if none of a, b, c is $\hat{\ }F$, for some F such that $Fa, \neg Fb, Fc$; otherwise let $\neg Ra, Rb, Rc$ hold if none of a,

Consistent principles of the form: The object associated in some manner with the concept F is identical with that associated in the same manner with G if and only if F and G bear a certain equivalence relation to one another, may therefore be inconsistent with each other. Hume's principle is inconsistent with the parity principle. Which is the logical truth?[16]

Indeed, not only do we have no reason for regarding Hume's principle as a truth of logic, it is doubtful whether it is a truth at all. As the existence of a number, 0, belonging to the concept *not-self-identical* is a consequence of Hume's principle, it also follows that there is a number belonging to the concept *self-identical*, a number that is the number of things that there are. Hume's principle is no less dubious than any of its consequences, one of which is the claim, uncertain at best, that there is such a number.

Crispin Wright claims that "there is a programme for the foundations of number theory recoverable from *Grundlagen*."[17] He calls the program "number-theoretic logicism" and characterizes it as the view that "it is possible, using the concepts of higher-order logic with identity to explain a genuinely sortal concept of cardinal number; and hence to deduce appropriate statements of the fundamental truths of number-theory, in particular the Peano Axioms, in an appropriate system of higher-order logic with identity to which a statement of that explanation has been added as an axiom."[18] He adds that he thinks that it would "serve Frege's purpose against the Kantian thesis of the *synthetic a priori* character of number-theoretic truths. For the fundamental truths of number theory would be revealed as consequences of an explanation: a statement whose role is

---

b, c is $\hat{\ }F$ for some F such that $\neg Fa$, Fb, Fc; otherwise let Ra, Rb, Rc hold. Then $\exists x \exists y (x \neq y \wedge Rx \wedge Ry)$. Let $d = \hat{\ }R$. If $d = \hat{\ }F$, then by the principle, $\forall x (Rx \leftrightarrow Fx)$. Thus $d \neq a$, b, c. If Rd, then for some F, $d = \hat{\ }F$ and $\neg Fd$. But then $\forall x (Rx \leftrightarrow Fx)$ and Fd. Thus $\neg Rd$. So $\forall F(d = \hat{\ }F \rightarrow Fd)$, whence Rd, contradiction.

[16]Cf. Allen Hazen's review of Crispin Wright's *Frege's conception of numbers as objects*, cited below. Hazen's review is in *Australasian journal of philosophy*, **63** (1985) 251–54.

[17]*Frege's conception of numbers as objects*, Aberdeen: Aberdeen University Press, 1983, p. 153.

[18]Ibid.

to fix the character of a certain concept."[19]

Wright regards Hume's principle as a statement whose role is to fix the character of a certain concept. We need not read any contemporary theories of the a priori into the debate between Frege and Kant. But Frege can be thought to have carried the day against Kant only if it has been shown that Hume's principle is *analytic*, or a truth of logic. This has not been done. Nor has the view Wright describes been shown to deserve the name "(number-theoretic) logicism." It's logicism only if it's claimed that Hume's principle is a principle of logic. Wright quite properly refrains from calling it one.

We have noted that Dedekind would not have been happy with the suggestion that the existence of infinite systems be derived from Hume's principle. Nor, presumably, would Frege have rested content with it as the foundation of arithmetic. Hume's principle may yield a great deal of information about the natural numbers, but it doesn't tell us how they may be viewed as logical objects, nor even which objects they are. Nor, as Frege noted in § 66 of the *Foundations*, does it enable us to eliminate the phrase "the number belonging to" from all contexts in which it occurs, in particular not from those of the form "$x =$ the number belonging to F."

Well. Neither Frege nor Dedekind showed arithmetic to be part of logic. Nor did Russell. Nor did Zermelo or von Neumann. Nor did the author of *Tractatus* 6.02 or his follower Church. They merely shed light on it.

APPENDIX: ARITHMETIC IN THE *FOUNDATIONS*

Hume's Principle. $\#F = \#G \leftrightarrow F \approx G$.

Def. $0 = \#[x : x \neq x]$. (*Foundations* § 74)

1. $\#F = 0 \leftrightarrow \forall x \neg Fx$. (§ 75)

Proof. Since $0 = \#[x : x \neq x]$, $\#F = 0$ iff $F \approx [x : x \neq x]$. Since $\forall x \neg x \neq x$, $F \approx [x : x \neq x]$ iff $\forall x \neg Fx$. □

Def. mPn iff $\exists F \exists y (Fy \wedge \#F = n \wedge \#[x : Fx \wedge x \neq y] = m)$. (§ 76)

[19]Ibid.

2. mPn and $m'Pn' \to (m = m' \leftrightarrow n = n')$. ($\S$ 75 para. 5)

Proof. Suppose mPn and $m'Pn'$. Let F, y, F', y' be as in the definition of P. Suppose $m = m'$. Then $\#[x : Fx \wedge x \neq y] = \#[x : F'x \wedge x \neq y']$, whence $[x : F'x \wedge x \neq y'] \approx [x : Fx \wedge x \neq y']$ via some φ. Since Fy and $F'y'$, $F \approx F'$ via $\varphi \cup \{\langle y, y'\rangle\}$ and then $n = \#F = \#F' = n'$. Conversely, suppose $n = n'$. Then since $\#F = \#F'$, $F \approx F'$ via some ψ. For some unique x, $x\psi y'$; for some unique x', $y\psi x'$. Let

$$\varphi = ((\psi - \{\langle x, y'\rangle, \langle y, x'\rangle\}) \cup \{\langle x, x'\rangle\}) - \{\langle y, y'\rangle\}.$$

Then $[x : Fx \wedge x \neq y] \approx [x : F'x \wedge x \neq y']$ via φ and $m = m'$. (Since x and x' might be identical with y and y', it is necessary to include "$-\{\langle y, y'\rangle\}$" in the definition of φ.) $\square$

3. $\neg mP0$. ($\S$ 78 para. 6?)

Proof. Otherwise for some y, Fy and $\#F = 0$, contra 1. $\square$

Def. xR^*y iff $\forall F(\forall a \forall b([(a = x \vee Fa) \wedge aRb] \to Fb) \to Fy)$. ($\S$ 79)

Thus to show that $xR^*y \to \ldots y \ldots$, it suffices to let $F = [z : \ldots z \ldots]$, assume $a = x \vee Fa$ and aRb, and show Fb.

4. $xRy \to xR^*y$. (*Begriffsschrift*, Prop. 91)

Proof. Suppose xRy and $\forall a \forall b([(a = x \vee Fa) \wedge aRb] \to Fb)$. Then Fy follows, if we let $a = x$ and $b = y$. $\square$

5. $xR^*y \wedge yR^*z \to xR^*z$. (*Begriffsschrift*, Prop. 98)

Proof. Suppose xR^*y, yR^*z, and $(*)$ $\forall a \forall b([(a = x \vee Fa) \wedge aRb] \to Fb)$. Show Fz. Since yR^*z, it suffices to show $\forall a \forall b([(a = y \vee Fa) \wedge aRb] \to Fb)$. Suppose $(a = y \vee Fa)$ and aRb. Show Fb. Since xR^*y, by $(*)$ Fy. We may suppose Fa. But then by $(*)$ we are done. $\square$

6. $xP^*n \to \exists m\, mPn \wedge \forall m(mPz \to [xP^*m \vee x = m])$.

Proof. Let $F = [z : \exists m\, mPz \wedge \forall m(mPz \to [xP^*m \vee x = m])]$. Suppose $a = x \vee Fa$, aPb. Show Fb. Since aPb, $\exists m\, mPb$. Suppose mPb. By 2, $m = a$. If $a = x$, $x = m$, and we are done. So suppose Fa. Then for some m', $m'Pa$, and xP^*m' or $x = m'$. Since $m'Pa = m$, $m'Pm$. If xP^*m', then $xP^*m'Pm$, whence by 4 and 5, xP^*m; if $x = m'$, xPm, whence by 4, xP^*m. $\square$

7. $0P^*n \rightarrow \neg nP^*n$. (§ 83)

Proof. Let $F = [z : \neg zP^*z]$. Suppose $a = 0 \vee Fa$, aPb. Show Fb. Suppose bP^*b. By 6, $bP^*a \vee b = a$, whence by 4 and 5, aP^*a, contra Fa. Thus $a = 0$, $0P^*0$, and by 6, $\exists m\, mP0$, contra 3.   □

Defs. $m \leqslant n$ iff $mP^*n \vee m = n$. Finite n iff $0 \leqslant n$.

8. $mPn \wedge 0P^*n \rightarrow \forall x(x \leqslant m \leftrightarrow x \leqslant n \wedge x \neq n)$. (§ 83)

Proof. Suppose mPn, $0P^*n$. If $xP^*m \vee x = m$, then xP^*n, by 4 and 5; and by 7, $x \neq n$. If $x \leqslant n$ and $x \neq n$, then xP^*n, and by 6, $x \leqslant m$. ("$0P^*n$" cannot be dropped: if $n =$ Frege's ∞_1, i.e., $\#[x : 0 \leqslant x]$, then nPn but $x \leqslant n$ iff $x = n$.)   □

9. $mPn \wedge 0P^*n \rightarrow \#[x : x \leqslant m]P\#[x : x \leqslant n]$. (§ 82)

Proof. Suppose mPn, $0P^*n$. By 8, $[x : x \leqslant m] \approx [x : x \leqslant n \wedge x \neq n]$; since $n \leqslant n$, $\#[x : x \leqslant m]P\#[x : x \leqslant n]$.   □

10. $mPn \rightarrow (0 \leqslant m \wedge mP\#[x : x \leqslant m] \rightarrow 0 \leqslant n \wedge nP\#[x : x \leqslant n])$. (§ 82)

Proof. Suppose mPn, $0 \leqslant m$. By 4 and 5, $0P^*n$. Thus $0 \leqslant n$. Suppose $mP\#[x : x \leqslant m]$. By 2, $\#[x : x \leqslant m] = n$. By 9, $nP\#[x : x \leqslant n]$.   □

11. $0P\#[x : x \leqslant 0]$. (§ 82)

Proof. Let $F = [x : x \leqslant 0]$. $F0$; but by 6, if xP^*0, $\exists m\, mP0$, contra 3. Thus $\forall x \neg(Fx \wedge x \neq 0)$; and by 1, $\#[x : Fx \wedge x \neq 0] = 0$, whence $0P\#[x : x \leqslant 0]$.   □

12. $0 \leqslant n \rightarrow 0 \leqslant n \wedge nP\#[x : x \leqslant n]$.

Proof. If $0 = n$, done by 11. Suppose $0P^*n$. Let $F = [z : 0 \leqslant z \wedge zP\#[x : x \leqslant z]]$. Suppose $m = 0 \vee Fm$ and mPn. Show Fn. If $m = 0$, by 11, Fm. And then by 10, Fn.   □

13. Finite $n \rightarrow nP\#[x : x \leqslant n]$. (§ 83)

Proof. By 12.   □

PART III

Grundgesetze der Arithmetik

10

The Development of Arithmetic in Frege's *Grundgesetze der Arithmetik*

I. OPENING

In his *Grundlagen der Arithmetik*, Frege explicitly defines numbers as the extensions of concepts of a certain form: The number of Fs is, according to Frege, the extension of the (second-level) concept "is a concept which can be correlated one-one with the concept $F\xi$" (*Gl* § 68).[1] Frege then derives, from his explicit definition, what is known as *Hume's principle* (see *Gl* § 73):

> The number of Fs is the same as the number of Gs just in case the Fs can be correlated one-one with the Gs.

Once the proof of Hume's principle is complete, Frege sketches the proofs of a variety of facts about numbers; this sketch is intended to show that the fundamental laws of arithmetic can be formally derived from his explicit definition. In the course of sketching these

Reprinted with minor revisions and a Postscript from *The journal of symbolic logic*, **58** (1993) 579–601. ©1993, Association for Symbolic Logic. All rights reserved. This reproduction is by special permission only for this publication.

[1]Gottlob Frege, *The foundations of arithmetic*, J. L. Austin, tr., Evanston: Northwestern University Press, 1980. References are in the text, marked by "*Gl*" and a section number.

proofs, however, Frege makes no further use of extensions: He appeals to his explicit definition of numbers only in the proof of Hume's principle.

It is tempting, therefore, to understand Frege's derivation of the laws of arithmetic as consisting of two parts: A derivation of the laws of arithmetic from Hume's principle, and a derivation of Hume's principle from the explicit definition. If this is, indeed, the right way to interpret Frege, this is of significance for our understanding of his philosophy of mathematics. For, first, there is a question what the relation between the explicit definition and Hume's principle should be taken to be; the need to make a sharp separation between these two parts of his derivation of arithmetic constrains the sorts of answers we can give to this question.[2] Secondly, it has, in recent years, been shown that the Dedekind-Peano axioms for arithmetic can indeed be derived, within second-order logic, from Hume's principle; moreover, Fregean arithmetic—second-order logic, with Hume's principle taken as the sole non-logical axiom—is consistent.[3] As is well-known, however, the formal theory in which Frege proves the axioms of arithmetic, in his *Grundgesetze der Arithmetik*,[4] is inconsistent, Russell's paradox being derivable in it: Presumably, any axiom, governing extensions, to which Frege might have implicitly

[2]See here Michael Dummett, *Frege: philosophy of mathematics*, Cambridge: Harvard University Press, 1992, esp. Ch. 14.

[3]For the derivation of the Dedekind-Peano axioms, see Crispin Wright, *Frege's conception of numbers as objects*, Aberdeen: Aberdeen University Press, 1983, Ch. 4. To the best of my knowledge, the derivability of arithmetic from Hume's principle was first noted, in modern times, by Charles Parsons: See his "Frege's theory of number," Chapter 7, above.

More precisely, Fregean arithmetic is equi-consistent with second-order arithmetic: See George Boolos, "The consistency of Frege's *Foundations of arithmetic*," Chapter 8, above. The consistency of Fregean arithmetic was noted, independently, by John Burgess, Allen Hazen, and Harold Hodes in their reviews of Wright.

[4]Gottlob Frege, *Grundgesetze der Arithmetik*, Hildesheim: Georg Olms Verlagsbuchhandlung, 1966. Further references are marked by "*Gg*" and volume and section numbers.

Grundgesetze has never been fully translated into English: For a translation of Part I, see *The basic laws of arithmetic: exposition of the system*, M. Furth, tr., Berkeley: University of California Press, 1964. Parts of Part III are translated in *Translations from the philosophical writings of Gottlob Frege*, M. Black and P. Geach, eds. and trs., Oxford: Blackwell, 1970.

appealed in *Grundlagen* would be similarly inconsistent. Nonetheless, Frege does not appeal to extensions in his sketch of the derivations of the axioms of arithmetic in *Grundlagen*, except during the proof of Hume's principle; his sketch therefore constitutes a sketch of the derivation of the axioms from Hume's principle, within a consistent sub-theory of the formal theory of *Grundgesetze*. It is this which leads George Boolos to remark that "it is a pity that the derivability of arithmetic from Hume's principle isn't known as *Frege's theorem*."[5]

One ought to be struck by a number of questions at this point. First, do Frege's formal proofs of the axioms of arithmetic, in *Grundgesetze*, also depend only upon Hume's principle, except, of course, for the proof of Hume's principle itself? That is to say: Does Frege present a formal proof of Frege's theorem in *Grundgesetze*? And, secondly, is it, as it were, just an accident that his proof can be so construed, or did Frege know that arithmetic could be derived from Hume's principle? And, thirdly, if he did, of what significance for our understanding of his position is this?

My purpose here is to answer the first of these questions, to show that Frege does, indeed, derive the axioms of arithmetic, in *Grundgesetze*, from Hume's principle, within second-order logic. Discussion of the latter two questions, whether Frege himself understood the proofs in this way and what significance this fact has for our understanding of his philosophy, must be reserved for another occasion.

II. AXIOM V IN *GRUNDGESETZE*

Our question is whether Frege derives the axioms of arithmetic from Hume's principle in *Grundgesetze*. We might naturally ask whether, just as in *Grundlagen*, Frege makes no use of extensions after deriving Hume's principle. So formulated, the answer to our question is "No." In *Grundgesetze*, Frege has an axiom, the infamous Axiom V, which governs what he there calls 'value-ranges.' Axiom V, for

---

[5]George Boolos, "The standard of equality of numbers," Chapter 9, p. 241. Emphasis in original.

our purposes, may be taken to be:[6]

$$(\acute{\varepsilon}.F\varepsilon = \acute{\varepsilon}.G\varepsilon) \equiv \forall x(Fx \equiv Gx).$$

So formulated, Axiom V governs terms which refer to the extensions of concepts (see *Gg* I p. vii). As said above, every second-order[7] theory containing Axiom V is inconsistent.

Terms standing for value-ranges are used throughout *Grundgesetze*: There is not a page in Part II on which a term for a value-range does not occur. It is thus just not the case that Frege makes no mention of value-ranges once he has derived Hume's principle. Nonetheless, that does not settle the question whether Frege derives the laws of arithmetic from Hume's principle. There are ways and ways of making use of value-ranges, some of them eliminable, others not. Indeed, what I intend to show is that, with the exception of the use in the proof of Hume's principle itself, *all* uses of value-ranges in Frege's proof of the basic laws of arithmetic can be eliminated in a uniform manner. Moreover, with that exception, Frege uses value-ranges merely for convenience.

Consider the following example, in which '$\Phi_x(\varphi x)$' is a schematic variable for a second-level formula ('φ' a placeholder): Suppose we have proven the propositions "$\Phi_x(Fx)$" and '$\forall x(Fx \equiv Gx)$'; we wish to infer "$\Phi_x(Gx)$." There is, in standard axiomatic second-order logic, no *uniform* way to make this inference: That is not to say that "$\Phi_x(Gx)$" can not be derived from "$\Phi_x(Fx)$" and '$\forall x(Fx \equiv Gx)$,' whatever second-level formula $\Phi_x(\varphi x)$ may be; indeed, it can be

---

[6]Strictly speaking, Axiom V is: $(\acute{\varepsilon}.F\varepsilon = \acute{\varepsilon}.G\varepsilon) = \forall x(Fx = Gx)$. Since, for Frege, the truth-values are objects in the domain of the first-order variables, the Axiom is formulated in terms of the *identity* of the values of the functions $F\xi$ and $G\xi$ and of the identity of the truth-value of $\forall x(Fx \equiv Gx)$ with that of $\acute{\varepsilon}.F\varepsilon = \acute{\varepsilon}.G\varepsilon$; so formulated, it introduces, not just extensions of concepts, but those of functions, the *graphs* of functions.

I shall uniformly translate "*Werthverlauf*" as "value-range"; it is translated, also, as "courses-of-values," "range of values," and so forth.

[7]This caveat is, in fact, needed, since the first-order fragment of the formal theory of *Grundgesetze* is consistent. For a proof, see Terence Parsons, "On the consistency of the first-order portion of Frege's logical system," Chapter 16, below. Indeed, an extension of Parsons's argument shows that the *predicative* second-order fragment of Frege's formal theory is consistent. See my "The consistency of the predicative fragment of Frege's formal theory," draft.

proven, by induction on the complexity of the formula "$\Phi_x(\varphi x)$," that we shall always be able to construct such a proof. But derivations within the formal system will be complicated by the need to prove, for each case in which we wish to make inferences of this sort, specific theorems licensing them.[8]

If we have value-ranges at our disposal, however, matters are much simplified. Given a second-level concept $\Phi_x(\varphi x)$, we define a related, first level concept as follows:

$$\Phi(z) \stackrel{\text{df}}{\equiv} \exists F[z = \acute{\varepsilon}.F\varepsilon \;\&\; \Phi_x(Fx)].$$

The predicate '$\Phi(\xi)$' is thus true of an object if, and only if, that object is the value-range of a concept which falls under $\Phi_x(\varphi x)$. If we then make use, not of second-level concepts, but of their first-order relatives, we can argue as follows:

(1)	$\Phi(\acute{\varepsilon}.F\varepsilon)$	Premise
(2)	$\forall x(Fx \equiv Gx)$	Premise
(3)	$\acute{\varepsilon}.F\varepsilon = \acute{\varepsilon}.G\varepsilon$	(2), Axiom V
(4)	$\Phi(\acute{\varepsilon}.G\varepsilon)$	(1), (3) Identity
(5)	$\Phi(\acute{\varepsilon}.F\varepsilon) \;\&\; \forall x(Fx \equiv Gx) \to \Phi(\acute{\varepsilon}.G\varepsilon)$	(4), Discharge (1), (2).

Given that Frege is appealing to Axiom V, such a use of value-ranges simplifies his formal system.

Now, it is sometimes said that Frege uses value-ranges to replace second-order quantification with first-order quantification: Quantification over concepts can be replaced by quantification over their value-ranges. It is not at all clear, though, what motivation Frege might have for doing this, unless he intended somehow to eliminate second-order quantification from his system. But second-order quantification is not eliminated in this way: At best, it is merely hidden; at worst, it introduces additional (hidden) second-order quantifiers; and it is entirely unclear why Frege should have any interest

---

[8]Charles Parsons has remarked to me that a similar phenomenon prevents the derivation of a version of Axiom V for second-level functions from that for first-level functions. (Axiom V for first-level functions can, on the other hand, be derived from that for second-level functions, via type elevation.)

in hiding second-order quantification when it introduces additional
second-order quantifiers. Moreover, second-order quantification occurs explicitly in Frege's definition of the ancestral (see *Gg* I §§ 45,
108–9), so he has no *general* interest even in hiding second-order
quantification.

The correct account of what is going on here is as follows: Frege
is indeed using value-ranges to represent first-level functions by objects; but he is not doing so in order to replace quantification over
functions by quantification over objects. Rather, he wishes to replace expressions for second-level functions, such as our '$\Phi_x(\varphi x)$,' by
expressions for first-level functions, just as we introduced the first-
level predicate '$\Phi(\xi)$' in place of '$\Phi_x(\varphi x)$.' Frege is explicit about
this:

> ... [I]n further developments, instead of second-level func-
> tions, we may employ first-level functions. ... [T]his is made
> possible through the functions that appear as arguments of
> second-level functions being represented by their value-ranges
> ... (*Gg* I § 34; cf. I § 25).

Exactly why Frege would care to use first-level functions in place of
second-level functions is itself a nice question: Part of the explanation, presumably, is that doing so simplifies his formal system in just
the ways discussed above.[9] Such uses of value-ranges, however, are
inessential to most of Frege's proofs and can be eliminated without
difficulty.

Most of the uses Frege makes of value-ranges, except in the proof
of Hume's principle, are of this sort; there is, however, one other kind
of use to which he puts them. To explain it, we need to introduce
Frege's application-operator. We may write Frege's definition as
follows:[10]

$$a^\frown f \overset{\mathrm{df}}{\equiv} \exists F[f = \grave{\varepsilon}.F\varepsilon \;\&\; Fa].$$

[9]Frege also appears to think that doing so will simplify the meta-theory, since
he need not explain his notation for second-order parameters "in full generality"
(see *Gg* I § 25). Charles Parsons once mentioned to me that one could almost
say that Frege uses value-ranges to avoid the use of *third*-order quantification.

[10]Again, Frege's definition covers not just concepts but functions in general.
Note, too, that it is this definition which introduces *additional* second-order
quantification into Frege's system when he uses value-ranges as discussed above.

It is not difficult to prove Frege's Theorem 1 (using Axiom V):

$$Fa \equiv a \frown \grave{\varepsilon}.F\varepsilon.$$

Now, consider the following sentence:

$$\forall x(Fx \rightarrow (Fx \vee x = c)).$$

Using the application operator, Frege might write this sentence thus:

$$\forall x(x \frown \grave{\varepsilon}.F\varepsilon \rightarrow (x \frown \grave{\varepsilon}.F\varepsilon \vee x = c)).$$

But he might instead write it so:[11]

$$\forall x[x \frown \grave{\varepsilon}.F\varepsilon \rightarrow x \frown \grave{\varepsilon}.(F\varepsilon \vee \varepsilon = c)].$$

The reason is that, in the context of a given proof, our interest may be focused on the concept: ξ is an F or ξ is identical with c. Frege is emphasizing this fact by using value-ranges in the same way one might use predicate-abstraction. We might, that is, achieve the same effect thus:

$$\forall x[Fx \rightarrow \lambda\xi(F\xi \vee \xi = c)(x)].$$

Such uses, either of value-ranges or of lambda-abstraction, are eliminable.[12]

Careful examination of the proofs of the axioms of arithmetic in *Grundgesetze* shows that all uses of value-ranges within those proofs are of one of three types:[13]

(1) The ineliminable use in the proof of Hume's principle.

---

[11]Depending on the context, Frege might write this as: $\forall x[x \frown \grave{\varepsilon}.F\varepsilon \rightarrow x \frown \grave{\varepsilon}.(\varepsilon \frown \grave{\alpha}.F\alpha \vee \varepsilon = c)]$. Or more confusingly: $\forall x[x \frown \grave{\varepsilon}.F\varepsilon \rightarrow x \frown \grave{\varepsilon}.(\varepsilon \frown \grave{\varepsilon}.F\varepsilon \vee \varepsilon = c)]$. According to the rules of his system, the occurrence of "ε" here which replaces the occurrence of "α" in the original is bound by the nearest occurrence of "$\grave{\varepsilon}$."

[12]From a Fregean perspective, we may take the λ–operator to be introduced as a defined second-level relation: $\lambda\xi(F\xi, x) \equiv_{\text{df}} Fx$.

[13]Frege does *not* use Axiom V as a second-order comprehension principle. Comprehension is built into his second-order *rule* of Universal Instantiation (as opposed to his *axiom* of Universal Instantiation), which allows for the uniform replacement of a free variable, of arbitrary type, by any well-formed expression of the appropriate type, containing arbitrarily many other free variables of arbitrary types (subject, of course, to the usual restrictions, which Frege formulates precisely). See *Gg* I § 48, Rule 9.

(2) The use which allows the representation of second-level functions by first-level functions.

(3) The formation of complex predicates to emphasize what is being proven.

And, except for those of the first sort, all uses of value-ranges are therefore easily, and uniformly, eliminable from Frege's proofs. Frege's proof of the axioms of arithmetic, from Hume's principle, therefore require no essential reference to value-ranges.

III. HUME'S PRINCIPLE AND FREGEAN ARITHMETIC

In *Grundlagen*, Frege formulates Hume's principle as follows:[14]

the Number which belongs to the concept F is identical with the Number which belongs to the concept G if the concept F is equinumerous with the concept G (Gl § 72).

According to Frege, a concept F is equinumerous with a concept G just in case

there exists a relation φ which correlates one to one the objects falling under the concept F with the objects falling under the concept G (Gl § 71).

And the correlation is, of course, one-one just in case (see Gl § 72):

$$\forall x\forall y\forall z(\varphi xy \ \& \ \varphi xz \rightarrow y = z) \ \& \ \forall x\forall y\forall z(\varphi xz \ \& \ \varphi yz \rightarrow x = y).$$

Frege explains, further, that the relation φ correlates the Fs with the Gs just in case every F "stands in the relation φ" to some G; and, conversely, for each G, there is some F which "stands in the relation φ" to it. Moreover, each F "stands in the relation φ" to some G just in case "the two propositions 'a falls under F' and 'a does not stand in the relation φ to any object falling under G' cannot, whatever be signified by a, both be true together ... " (Gl § 71). Equivalently:

$$\forall x(Fx \rightarrow \exists y(Gy \ \& \ \varphi xy)) \ \& \ \forall x(Gx \rightarrow \exists y(Fy \ \& \ \varphi yx)).$$

[14] Austin's translation has "equal" for "equinumerous." The German word is "*gleichzahlig*."

Using "$Nx : \Phi x$" as a second-level functional expression, to be read "The number of Φs," we may thus formalize Hume's principle as follows:

$$Nx : Fx = Nx : Gx$$

iff

$$\exists R[\forall x \forall y \forall z(Rxy \ \& \ Rxz \to y = z)$$
$$\& \ \forall x \forall y \forall z(Rxz \ \& \ Ryz \to x = y) \ \& \ \forall x(Fx \to (\exists y)(Gy \ \& \ Rxy)$$
$$\& \ \forall x(Gx \to (\exists y)(Fy \ \& \ Ryx)].$$

This version of Hume's principle is thus that most immediately suggested by Frege's remarks in *Grundlagen*.[15]

Frege's formulation of Hume's principle in *Grundgesetze* may initially strike one as rather different. Frege, however, does *not* conceive of it as a departure from his earlier informal statement of it: On the contrary, when giving his definitions in *Grundgesetze*, Frege himself quotes the passages quoted above (*Gg* I § 38). To state the version of Hume's principle used in *Grundgesetze*, we need a number of these definitions, the first of which is Frege's definition of the *converse* of a relation:[16]

$$\mathrm{Conv}_{\alpha\varepsilon}(R\alpha\varepsilon)(x, y) \overset{\mathrm{df}}{\equiv} Ryx.$$

The second definition is that of a relation's being *functional*:[17]

$$\mathrm{Func}_{\alpha\varepsilon}(R\alpha\varepsilon) \overset{\mathrm{df}}{\equiv} \forall x \forall y(Rxy \to \forall z(Rxz \to y = z)).$$

[15] It is also essentially the version used by Wright (see p. 105).

[16] *Gg* I § 39. Frege writes the converse-operator as a script-U and applies it to the (double) value-range of a relation rather than to the relation itself; its value too is the (double) value-range of a relation. As said earlier, Frege uses value-ranges in those cases where the relation would be an argument of a *second-level* function: "$\mathrm{Conv}_{\alpha\varepsilon}(R\alpha\varepsilon)(\xi, \eta)$" is a second-level function.

"$\mathrm{Conv}_{\alpha\varepsilon}(\Phi\alpha\varepsilon)(\xi, \eta)$" refers to a concept of mixed level, which takes a relation and two objects as arguments and which binds two variables. I shall usually just write: "$\mathrm{Conv}(\Phi)(\xi, \eta)$," when there is no danger of confusion, or, occasionally, "$\mathrm{Conv}(\Phi\alpha\varepsilon)(\xi, \eta)$." I shall use these same conventions throughout.

[17] *Gg* I § 37. Frege writes the operator as "I" and again applies it to the double value-range of a relation. Frege's word here is "*eindeutig*," which is often translated "many-one," but I prefer the translation "functional," especially in light of Frege's definition of mapping, to be mentioned shortly.

The third definition is that of a relation's mapping the objects falling under one concept into those falling under another:[18]

$$\text{Map}_{\alpha\varepsilon xy}(R\alpha\varepsilon)(Fx, Gy) \stackrel{\text{df}}{\equiv} \text{Func}_{\alpha\varepsilon}(R\alpha\varepsilon)$$
$$\& \ \forall x(Fx \rightarrow \exists y(Rxy \ \& \ Gy)).$$

That is: $R\xi\eta$ maps the Fs into the Gs just in case $R\xi\eta$ is functional and each F is related to some G.[19] (Note that we say that $R\xi\eta$ *relates* x to y if, and only if, Rxy.)

We should note two important points about this definition. First, "Map$(R)(F, G)$" states that $R\xi\eta$ is a functional relation which maps the Fs *into* the Gs, not, as might have seemed more natural, one which maps the Fs *onto* the Gs: That would entail that there are at least as many Fs as Gs, whence, if there is a functional relation which maps the Gs onto the Fs, there are just as many Fs as Gs (by the Schröder-Bernstein theorem). The fact that $R\xi\eta$ maps the Fs *into* the Gs says, of itself, nothing whatsoever about the relative cardinalities of the Fs and the Gs: So long as there is at least one G, there will always be a relation which maps the Fs—whatever concept $F\xi$ may be—into the Gs, in Frege's sense.[20]

Secondly, "Map$(R)(F, G)$" says that $R\xi\eta$ is functional and that it relates each F to a G. It says absolutely nothing else about $R\xi\eta$: $R\xi\eta$ may, for all we know, map the entire domain into the Gs; it may relate every non-F to itself; it may not relate non-Fs to anything. In reading Frege's proofs, one must keep this fact in mind. Suppose,

---

[18] *Gg* I § 38. Frege writes "Map" as ")." The argument of ")" is to be the double value-range of a relation; its value is the double value-range of a relation between value-ranges.

[19] Strictly speaking, Frege's definition is: Map$(R)(F, G) \equiv_{\text{df}}$ Func(R) & $\forall x[\forall y(Rxy \rightarrow \neg Gy) \rightarrow \neg Fx]$. Frege works exclusively with this formulation, as the mechanics of his system make it easier for him to do so. There is no rule in it which allows him to infer (an equivalent of) "$Fx \rightarrow \exists y(Rxy \ \& \ Gy)$" from "$Fx \rightarrow Rxt \ \& \ Gt$." He would contrapose to get "$\neg(Rxt \ \& \ Gt) \rightarrow \neg Fx$"; cite "$\forall y.\neg(Rxy \ \& \ Gy) \rightarrow \neg(Rxt \ \& \ Gt)$"; infer "$\forall y.\neg(Rxy \ \& \ Gy) \rightarrow \neg Fx$," by the transitivity of the conditional; and contrapose again. (See Frege's discussion in *Gg* I § 17.) It is somewhat easier for Frege to prove "$(Rxt \rightarrow \neg Gt) \rightarrow \neg Fx$" and infer "$\forall y(Rxy \rightarrow \neg Gy) \rightarrow \neg Fx$," as above.

[20] If Ga, we define: $Rxy \equiv_{\text{df}} y = a$. $R\xi\eta$ then maps the whole domain into the Gs and a fortiori maps the Fs into the Gs.

for example, that the relation $R\xi\eta$ maps the concept $F\xi$ into the concept $G\xi$, that c is not an F and that b is a G; and suppose we wish to show that some relation maps $F\xi \vee \xi = c$ into $G\xi$. What we should like to say is that the relation which is just like $R\xi\eta$, but which relates c to b (that is, $R\xi\eta \vee [\xi = c \ \& \ \eta = b]$), accomplishes this. But we can not proceed so quickly: $R\xi\eta$ may already relate c to something; in particular, it may relate c to something other than b, in which case the relation so defined is not functional and so maps no concept into any other.[21] Clearly, this is not an obstacle which can not be overcome: We shall see an example of the sort of theorem which must be proven in order to overcome it below.

It is easy to formulate Hume's principle in terms of the definitions given, thus:

$$Nx : Fx = Nx : Gx \stackrel{\text{df}}{\equiv} \exists R[\text{Map}(R)(F,G) \ \& \ \text{Map}(\text{Conv}\,R)(G,F)].$$

That is: The number of Fs is the same as the number of Gs if, and only if, there is a relation which maps the Fs into the Gs and whose converse maps the Gs into the Fs. This formulation of Hume's principle is easily seen to be equivalent to that discussed above: However, it has certain technical advantages over that version, an example of which we shall see below.[22]

IV. FREGE'S DERIVATION OF THE AXIOMS OF ARITHMETIC

As in *Grundlagen*, Frege's first task in Part II of *Grundgesetze* is to derive Hume's principle from his explicit definition of numbers. Having done so, he turns to the proofs of a number of basic truths

[21]See Frege's discussion of this point in *Gg* I § 66.

[22]Note that what Frege has done is to group "$\forall x \forall y(Rxy \rightarrow \forall z(Rxz \rightarrow y = z))$" and "$\forall x(Fx \rightarrow \exists y(Rxy \ \& \ Gy))$," on the one hand, and "$\forall x \forall y(Ryx \rightarrow \forall z(Rzx \rightarrow y = z))$" and "$\forall x(Gx \rightarrow \exists y(Ryx \ \& \ Fy))$," on the other. Had he instead grouped "$\forall x \forall y(Rxy \rightarrow \forall z(Rxz \rightarrow y = z))$" and "$\forall x(Gx \rightarrow \exists y(Ryx \ \& \ Fy))$," he would have had the alternative definition of 'Map,' which states that $R\xi\eta$ is functional and that it maps the Fs onto the Gs. As said, formulating Hume's principle in terms of either Frege's or the alternate definition of 'Map' has technical advantages over the more modern construal, which groups the conjuncts "$\forall x \forall y(Rxy \rightarrow \forall z(Rxz \rightarrow y = z))$" and "$\forall x \forall y(Ryx \rightarrow \forall z(Rzx \rightarrow y = z))$," on the one hand, and "$\forall x(Fx \rightarrow \exists y(Rxy \ \& \ Gy))$" and "$\forall x(Gx \rightarrow \exists y(Ryx \ \& \ Fy))$," on the other.

about numbers. He says, at the beginning of *Grundgesetze*, that, in *Grundlagen*, he "sought to make it plausible that arithmetic is a branch of logic" and that "this shall now be confirmed, by the derivation of the simplest laws of Number by logical means alone" (*Gg* I § 0). One might object, rather facetiously, that the derivation of the *simplest* laws of number hardly confirms that arithmetic is a branch of logic: It would more interesting were some really *complicated* laws of number derivable. Presumably, however, Frege meant, by "the simplest laws," not the simplest laws in any syntactic or psychological sense, but those laws from which all other laws of arithmetic follow. One might suggest, indeed, that, to show that "arithmetic is a branch of logic," Frege must show that *every* law of arithmetic can be derived within logic.[23] There is surely no way to do so without isolating some (presumably, finitely many) principles, *the* basic laws of arithmetic, from which all laws of arithmetic follow, and deriving these basic laws within logic.

These basic laws are just axioms for arithmetic. Thus, Frege's demonstration that arithmetic is a branch of logic must ultimately rest upon some axiomatization of arithmetic: For the moment, we take arithmetic to be axiomatized by the Dedekind-Peano axioms. Where '$N\xi$' is a predicate to be read 'ξ is a natural number,' and 'Pred(ξ, η)' as 'ξ immediately precedes η in the number-series,' we may formulate the Dedekind-Peano axioms as follows, using the definitions introduced above:

(1) $N0$
(2) $\forall x(Nx \rightarrow \exists y(Ny \ \& \ \text{Pred}(x, y))$
(3) $\neg\exists x.\,\text{Pred}(x, 0)$
(4) (a) Func(Pred)
      (b) Func[Conv(Pred)]

---

[23]I say "one might suggest" because, so framed, the requirement can not be met within any formal theory, so long as we take the "laws" of arithmetic to be the *truths* of arithmetic: The incompleteness theorem precludes any such demonstration. For present purposes, this may be ignored, since the point is to argue that Frege is committed to providing some axiomatization of arithmetic. A logicist may not so characterize her project as to make it self-fulfilling, by stipulating that the theory of arithmetic is the smallest deductively closed set of sentences which contains the axioms of her theory. Some independently plausible characterization of arithmetic is required: Frege has one. See below.

(5) $\forall x \{ \mathrm{N}x \rightarrow \forall F[F0 \ \& \ \forall x(\mathrm{N}x \ \& \ Fx \rightarrow \forall y(\mathrm{Pred}(x,y) \rightarrow Fy)) \rightarrow Fx] \}$.

Frege derives each of these axioms, from Hume's principle, in *Grundgesetze*.

To prove the axioms, we obviously need some additional definitions. We begin with the definitions of '0' and the relation-sign 'Pred(ξ, η).' The definitions given in *Grundgesetze* are essentially those given in *Grundlagen*. Frege defines zero as the number of objects which are not self-identical (*Gg* I § 41; see *Gl* § 74):

$$0 = \mathrm{N}x : x \neq x.$$

His definition of 'Pred(ξ, η)' is as follows (*Gg* I § 43):

$$\mathrm{Pred}(m,n) \overset{\mathrm{df}}{\equiv} \exists F \exists x [Fx \ \& \ n = \mathrm{N}z : Fz$$
$$\& \ m = \mathrm{N}z : (Fz \ \& \ z \neq x)].$$

That is: m precedes n if "there exists a concept F, and an object falling under it x, such that the Number which belongs to the concept F is n and the Number which belongs to the concept 'falling under F but not identical with x' is m" (*Gl* § 74). We shall return to the definition of '$\mathrm{N}\xi$.'

Frege's proofs of the Dedekind-Peano axioms follow, for the most part, the sketch given in *Grundlagen* (and subsequently worked out by Wright). The proof that Pred(ξ, η) is functional, Axiom (4)(a), occupies section B(eta); that its converse is functional, Axiom (4)(b), section Γ.[24] I shall not discuss the latter proof here. I shall, however, say a few things about the proof that Pred(ξ, η) is functional, primarily to illustrate some of the technical points about Frege's use of value-ranges and his definition of "Map" discussed

---

[24]Frege's proofs show that Pred(ξ, η) is one-one, not that it is one-one if it is restricted to natural numbers. At this point in Part II of the *Grundgesetze*, the concept of a natural number has not been defined: The proof shows that zero, $\aleph_0$ and all other cardinal numbers have at most one predecessor and at most one successor. Note, however, that Frege's notion of succession, as applied to infinite cardinals, does not coincide with that now common in set-theory: On Frege's definition, the successor of $\aleph_0$ is $\aleph_0$. (It then follows, from Theorem 145, to be proven below, that $\aleph_0$ is not a finite number.)

above. To show that $\text{Pred}(\xi, \eta)$ is functional, we must prove that, if $\text{Pred}(x, y)$ and $\text{Pred}(x, w)$, then $y = w$. Assume the antecedent. Then, by the definition of '$\text{Pred}(\xi, \eta)$,' there is a concept $F\xi$ and an object c such that Fc, $Nz : Fz = y$, and $Nz : (Fz \ \& \ z \neq c) = x$. Similarly, there is a concept $G\xi$ and an object b such that Gb, $Nz : Gz = w$, and $Nz : (Gz \ \& \ z \neq b) = x$. So the Theorem will follow from Frege's Theorem 66:

$$Fc \ \& \ Gb \ \& \ Nz : (Gz \ \& \ z \neq b) = Nz : (Fz \ \& \ z \neq c)$$
$$\rightarrow Nz : Fz = Nz : Gz.$$

For $Nz : (Gz \ \& \ z \neq b)$ and $Nz : (Fz \ \& \ z \neq c)$ are both x and, hence, are identical; so, $Nz : Fz = Nz : Gz$; hence, $y = w$.

The proof of Theorem 66 requires two lemmas, the first of which is Theorem 63:

$$\neg \exists z. Qbz \ \& \ \text{Map}(Q)(G\xi \ \& \ \xi \neq b, F\xi \ \& \ \xi \neq c) \ \& \ Fc$$
$$\rightarrow \text{Map}(Q\xi\eta \vee (\xi = b \ \& \ \eta = c))(G, F).$$

In words: If there is no object to which $Q\xi\eta$ relates b, if $Q\xi\eta$ maps the Gs, other than b, into the Fs, other than c, and if c is an F, then the relation which is just like $Q\xi\eta$, except that it relates b to c, maps the Gs into the Fs. The second lemma is (a contrapositive of) Theorem 56:

$$\neg Fc \ \& \ \neg Gb \ \& \ Nz : Fz = Nz : Gz$$
$$\rightarrow \exists Q[\neg \exists z.(\text{Conv}\, Q)(c, z) \ \& \ \neg \exists z. Qbz$$
$$\& \ \text{Map}(\text{Conv}\, Q)(F, G) \ \& \ \text{Map}(Q)(G, F)].$$

In words: If c is not an F and b is not a G and the number of Fs is the same as the number of Gs, then there is a relation $Q\xi\eta$, which relates b to no object and whose converse relates c to no object, which correlates the Fs one-one with the Gs.[25] Theorem 56 is an

[25] When we have $\text{Map}(Q)(G, F)$ and $\text{Map}(\text{Conv}\, Q)(F, G)$, we say that $Q\xi\eta$ correlates the Gs one-one with the Fs. Note the order here. Again, Frege works with a contrapositive because the mechanics of his system make it easier for him to do so.

example of the sort of result which is required if we are to allow that a relation may map the Gs into the Fs, yet not relate non-Gs to other objects.

The proof of Theorem 66, from Theorems 63 and 56, provides us with an example of the sort of purely technical advantages which Frege's version of Hume's principle has: Frege is able to substitute into Theorem 63 to prove results about the *converse* of $Q\xi\eta$. Substituting 'Conv$(Q)(\xi,\eta)$' for '$Q\xi\eta$,' swapping '$F\xi$' and '$G\xi$,' and swapping 'b' and 'c' in Theorem 63, we have:

$$\neg\exists z.\,\mathrm{Conv}(Q)(c,z)$$
$$\&\;\;\mathrm{Map}(\mathrm{Conv}\,Q)(F\xi\;\&\;\xi\neq c, G\xi\;\&\;\xi\neq b)\;\&\;Gb$$
$$\rightarrow\mathrm{Map}[(\mathrm{Conv}(Q)(\xi,\eta)\vee(\xi=c\;\&\;\eta=b)](F,G).$$

But we have also Frege's Theorem 64, ι:[26]

$$\forall x\forall y\{[\mathrm{Conv}(Q)(x,y)\vee(x=c\;\&\;y=b)]$$
$$\equiv\mathrm{Conv}[Q\xi\eta\vee(\xi=b\;\&\;\eta=c)](x,y)\}.$$

That is: $\mathrm{Conv}(Q)(\xi,\eta)\vee(\xi=c\;\&\;\eta=b)$ is the converse of $Q\xi\eta\vee(\xi=b\;\&\;\eta=c)$, as may easily be verified. Hence, Theorem 64, λ:[27]

$$\neg\exists z.\,\mathrm{Conv}(Q)(c,z)$$
$$\&\;\;\mathrm{Map}(\mathrm{Conv}\,Q)(F\xi\;\&\;\xi\neq c, G\xi\;\&\;\xi\neq b)\;\&\;Gb$$
$$\rightarrow\mathrm{Map}[\mathrm{Conv}(Q\xi\eta\vee(\xi=c\;\&\;\eta=b))](F,G).$$

Putting this together with Theorem 63 and applying Hume's principle, then, we have Theorem 64, ν:

$$Gb\;\&\;Fc\;\&\;\neg\exists z.\,\mathrm{Conv}(Q)(c,z)\;\&\;\neg\exists z.Qbz$$
$$\&\;\;\mathrm{Map}(\mathrm{Conv}\,Q)(F\xi\;\&\;\xi\neq c, G\xi\;\&\;\xi\neq b)$$
$$\&\;\;\mathrm{Map}(Q)(G\xi\;\&\;\xi\neq b, F\xi\;\&\;\xi\neq c)\rightarrow Nx:Gx=Nx:Fx.$$

[26]Theorem 64, ι is the theorem marked "*iota*" which occurs during, as opposed to after, the proof of Theorem 64.

[27]This is the sort of point at which Axiom V would be used to ease the transition: We should need here to prove a theorem which allows substitution of co-extensive relational expressions in the relevant argument place of "Map" (not that it would be difficult to prove such a theorem).

Finally, substituting, in Theorem 56, "$F\xi$ & $\xi \neq c$" for "$F\xi$"; "$G\xi$ & $\xi \neq b$," for "$G\xi$"; we have:

$$\neg[Fc \ \& \ c \neq c] \ \& \ \neg[Gb \ \& \ b \neq b]$$
$$\& \ Nx : (Fx \ \& \ x \neq c) = Nx : (Gx \ \& \ x \neq b)$$
$$\rightarrow \exists Q[\neg\exists z.(\mathrm{Conv}\,Q)(c, z) \ \& \ \neg\exists z.Qbz$$
$$\& \ \mathrm{Map}(\mathrm{Conv}\,Q)(F\xi \ \& \ \xi \neq c, G\xi \ \& \ \xi \neq b)$$
$$\& \ \mathrm{Map}(Q)(G\xi \ \& \ \xi \neq b, F\xi \ \& \ \xi \neq c)].$$

The first two conjuncts are obvious, so Theorem 66 follows immediately from this and Theorem 64, ν.

We shall leave the proof that $\mathrm{Pred}(\xi, \eta)$ is functional here: The proofs of the two lemmas, Theorems 63 and 56, pose no difficulty.

V. FREGE'S DERIVATION OF THE
AXIOMS OF ARITHMETIC, CONTINUED

We turn now to Frege's proofs of the other axioms. Axiom 3, that zero has no predecessor, is perhaps the easiest of all to prove; Frege proves it in section E as Theorem 108. Suppose that $\mathrm{Pred}(a, 0)$; by definition, there is a concept $F\xi$ and an object x falling under it, such that the number of Fs is 0 and the number of Fs, other than x, is a:

$$\exists F \exists x[Fx \ \& \ a = Nz : (Fx \ \& \ x \neq z) \ \& \ 0 = Nx : Fx].$$

A fortiori, there is a concept $F\xi$, under which some object falls, whose number is 0:

$$\exists F \exists x[Fx \ \& \ 0 = Nz : Fz].$$

But that yields a contradiction, for it is easy to show that, if something is F, the number of Fs is not zero, Theorem 93:

$$\exists x.Fx \rightarrow 0 \neq Nz : Fz.$$

Proving the contrapositive, if 0 is the number of Fs, there is a relation $R\xi\eta$ which maps the Fs into the non-self-identicals (and whose converse maps the non-self-identicals into the Fs). But then, by definition:

$$\forall x[Fx \rightarrow \exists y(y \neq y \ \& \ Rxy)].$$

But nothing is non-self-identical, so nothing is F.

To make any further progress, Frege must define the predicate '$N\xi$,' that is, give a definition of the predicate 'ξ is a natural number.' His definition is again the same as that given in *Grundlagen*. First, Frege introduces the ancestral: Given a relation $Q\xi\eta$, we say that a concept $F\xi$ is *hereditary in the Q-series* just in case, whenever x is F, each object to which $Q\xi\eta$ relates it is F; i.e.:

$$\forall x[Fx \rightarrow \forall y(Qxy \rightarrow Fy)].$$

We now say that an object b *follows* an object a in the Q-series just in case b falls under every concept which is hereditary in the Q-series and under which each object to which $Q\xi\eta$ relates a falls. Formally, writing "$\mathcal{F}(Q)(a,b)$" for "b follows a in the Q-series," Frege's definition of the strong (or proper) ancestral[28] is:

$$\mathcal{F}(Q)(a,b) \overset{\text{df}}{\equiv} \forall F[\forall x(Fx \rightarrow \forall y(Qxy \rightarrow Fy))$$
$$\& \ \forall x(Qax \rightarrow Fx) \rightarrow Fb].$$

Frege then defines the *weak* ancestral as:[29]

$$\mathcal{F}^{=}(Q)(a,b) \overset{\text{df}}{\equiv} \mathcal{F}(Q)(a,b) \vee a = b.$$

Frege reads "$\mathcal{F}^{=}(Q)(a,b)$" as "b is a member of the Q-series beginning with a"; or, equivalently, "a is a member of the Q-series ending with b."

---

[28] So-called because we need not have that $\mathcal{F}(Q)(a,a)$; take $Q\xi\eta$, for example, to be the empty relation. Following Russell, Boolos and Wright use "$Q^{*}(\xi,\eta)$" as notation for the (weak) ancestral of $Q\xi\eta$. In the actual formal development of Frege's theory, however, we need to consider the ancestrals of relations such as: $Q\xi\eta \ \& \ \mathcal{F}(Q)(a,\eta)$. In such cases, the '*' notation can become rather confusing unless some special notation for concept-abstraction is introduced.

[29] See *Gg* I §§ 45–6. It will come as no surprise that, on Frege's definition, the arguments of the strong and weak ancestral are the double value-ranges of relations, as are their values. The weak ancestral is written as "⌐"; the strong, as "⌣."

The concept, $\mathbb{N}\xi$, that is, ξ is a finite (or natural) number, is then definable as $\mathcal{F}^=(\text{Pred})(0,\xi)$: An object is a natural number just in case it belongs to the Pred-series beginning with 0.[30] Axiom 1, which states that zero is a natural number, follows immediately from the definition of the weak ancestral. Famously, induction too follows almost immediately from this definition. Matters are slightly more complicated than one might have thought, however. The hypothesis of induction is not that, *whenever* x is F, its successor is F; it is only that, whenever x is a natural number which is F, its successor is F. However, this is not a substantial difficulty; for, as Frege shows, the following, his Theorem 152, is provable:[31]

$$\mathcal{F}^=(Q)(a,b) \ \& \ Fa$$
$$\& \ \forall x[\mathcal{F}^=(Q)(a,x) \ \& \ Fx \to \forall y(Qxy \to Fy)] \to Fb.$$

That is: If b is a member of the Q-series beginning with a, if a is F, and if $F\xi$ is hereditary in the Q-series, as we say, *restricted* to members of the Q-series beginning with a, then b is F. Induction is the generalization of an instance of Theorem 152: Take $Q\xi\eta$ to be $\text{Pred}(\xi,\eta)$; take a to be zero; generalize.[32]

VI. AN ELEGANT PROOF THAT EVERY NUMBER HAS A SUCCESSOR

a. The Strategy of the Proof.

To complete the discussion of Frege's proofs of the axioms of arithmetic, we have now only to discuss Axiom 2, which states that every natural number has

---

[30] Frege has no special symbol for our predicate '$\mathbb{N}\xi$': He does, however, read '$\mathcal{F}^=(\text{Pred})(0,\xi)$' as '$\xi$ is a finite number.' See *Gg* I § 108.

[31] For the proof of Theorem 152, we use Theorem 144, to be mentioned below: $\mathcal{F}^=(Q)(a,b) \ \& \ Fa \ \& \ \forall x(Fx \to \forall y(Qxy \to Fy)) \to Fb$. We take $F\xi$ as $F\xi \ \& \ \mathcal{F}^=(Q)(a,\xi)$ and must show, on the suppositions that $\mathcal{F}^=(Q)(a,b)$, that Fa, and that $F\xi$ is hereditary in the Q-series restricted to members of the Q-series beginning with a, that: (i) $Fa \ \& \ \mathcal{F}^=(Q)(a,a)$; (ii) $\forall x[Fx \ \& \ \mathcal{F}^=(Q)(a,x) \to \forall y[Qxy \to Fy \ \& \ \mathcal{F}^=(Q)(a,y)]]$. The former is obvious, since $\mathcal{F}^=(Q)(a,a)$, by definition. Suppose, then, that $Fx \ \& \ \mathcal{F}^=(Q)(a,x)$ and Qxy. Since $F\xi$ is hereditary in the Q-series restricted to members of the Q-series beginning with a, Fy; moreover, we have $\mathcal{F}^=(Q)(a,x)$ and Qxy, so $\mathcal{F}^=(Q)(a,y)$, by Frege's Theorem 137, to be mentioned below. Hence, by (144), Fb.

[32] Oddly enough, Frege never actually writes this instance of Theorem 152 down, though it is, of course, applied.

a successor. The basic idea behind the proof is that each natural number is succeeded by the number of numbers less than or equal to it; more precisely, what we want to prove is Frege's Theorem 155:

$$\mathcal{F}^=(\text{Pred})(0, b) \to \text{Pred}[b, Nx : \mathcal{F}^=(\text{Pred})(x, b)].$$

That is: If b is a natural number, then b precedes the number of members of the Pred-series ending with b. Now, in *Grundlagen*, Frege's sketch of the proof is roughly as follows: The proof is to proceed by induction, the induction justified by Theorem 152, mentioned above. The relevant concept for the induction (i.e., what we substitute for '$F\xi$') is:

$$\text{Pred}[\xi, Nx : \mathcal{F}^=(\text{Pred})(x, \xi)].$$

So the object is to show that this concept is hereditary in the Pred-series, restricted to natural numbers, and that zero falls under it. To prove the concept hereditary, according to Frege, we must show that each natural number a is the number of numbers which belong to the Pred-series ending with a, but are not identical with a; that is, we must show that:

$$\mathcal{F}^=(\text{Pred})(0, a) \to a = Nx : [\mathcal{F}^=(\text{Pred})(x, a) \ \& \ x \neq a].$$

(See *Gl* §§ 82–3.)

The theorem can, indeed, be proven in this way: Wright's proof follows Frege's sketch closely. As is clear even from a cursory glance at Wright's proof, however, the proof of the theorem along these lines is long and difficult. In *Grundgesetze*, however, Frege gives a shorter and more elegant proof. The outlines of the proof are the same: The proof is by induction; the relevant concept is $\text{Pred}[\xi, Nx : \mathcal{F}^=(\text{Pred})(x, \xi)]$. We need, therefore, to prove that zero falls under this concept, which is Theorem 154:

$$\text{Pred}[0, Nx : \mathcal{F}^=(\text{Pred})(x, 0)].$$

And we need to prove it hereditary in the number-series, restricted to natural numbers, which is Theorem 150:

$$\forall y \{\mathcal{F}^=(\text{Pred})(0, y) \ \& \ \text{Pred}[y, Nx : \mathcal{F}^=(\text{Pred})(x, y)]$$
$$\to \forall z [\text{Pred}(y, z) \to \text{Pred}(z, Nx : \mathcal{F}^=(\text{Pred})(x, z)]\}.$$

We shall return to Theorem 150 shortly.

The proof of Theorem 154 is relatively easy. Its proof relies only upon the fact that nothing *ancestrally* precedes zero in the Pred-series, Frege's Theorem 126:

$$\neg\mathcal{F}(\mathrm{Pred})(x,0).$$

This lemma follows immediately from Frege's Theorem 124:[33]

$$\mathcal{F}(Q)(a,b) \rightarrow \exists x.Qxb.$$

Since nothing immediately precedes zero in the Pred-series, nothing ancestrally precedes it; therefore, by the definition of the weak ancestral, the only member of the Pred-series ending with zero is zero itself (Theorem 154, β):

$$\mathcal{F}^{=}(\mathrm{Pred})(x,0) \rightarrow x = 0.$$

Hence, nothing is a member of the Pred-series ending with zero, other than zero:

$$\neg\exists x[\mathcal{F}^{=}(\mathrm{Pred})(x,0) \ \& \ x \neq 0].$$

And so, the number of members of the Pred-series ending with zero, other than zero, is zero (since, by Frege's Theorem 97, if nothing is F, the number of Fs is zero):[34]

$$Nx : [\mathcal{F}^{=}(\mathrm{Pred})(x,0) \ \& \ x \neq 0] = 0.$$

---

[33] The proof of this theorem is entirely analogous to the usual proof, in first-order arithmetic, that every natural number other than zero has a predecessor. In this case, the role of induction is played by Frege's Theorem 123, which is immediate from the definition of the ancestral: $\mathcal{F}(Q)(a,b) \ \& \ \forall x(Qax \rightarrow Fx) \ \& \ \forall x[Fx \rightarrow \forall y(Qxy \rightarrow Fy)] \rightarrow Fb$. The relevant concept, for the induction, is $\exists z.Qz\xi$. The proofs that $\forall x(Qax \rightarrow \exists z.Qzx)$ and that $\forall x[\exists z.Qzx \rightarrow \forall y(Qxy \rightarrow \exists z.Qzy)]$ are then near trivial.

[34] Theorem 97 follows from Theorem 96, which states that if the Fs are the Gs, the number of Fs is the same as the number of Gs. If nothing is F, the Fs are the non-self-identicals: $\neg\exists x.Fx \rightarrow \forall x[Fx \equiv x \neq x]$. Hence, if nothing is F, the number of Fs is the number of non-self-identicals, which is zero, by definition.

Now, zero is a member of the Pred-series ending with zero (by the definition of the weak ancestral), so we have:

$$\mathcal{F}^=(\text{Pred})(0,0) \ \& \ Nx : [\mathcal{F}^=(\text{Pred})(x,0) \ \& \ x \neq 0] = 0$$
$$\& \ Nx : \mathcal{F}^=(\text{Pred})(x,0) = Nx : \mathcal{F}^=(\text{Pred})(x,0).$$

Generalizing ($F\xi$ is here $\mathcal{F}^=(\text{Pred})(\xi,0)$; y is 0):

$$\exists F \exists y [Fy \ \& \ Nx : (Fx \ \& \ x \neq y) = 0$$
$$\& \ Nx : Fx = Nx : \mathcal{F}^=(\text{Pred})(x,0)].$$

Hence, $\text{Pred}[0, Nx : \mathcal{F}^=(\text{Pred})(x,0)]$, by definition.

The proof of Theorem 150 is of more interest, for its proof does not in fact require a proof that each natural number is the number of numbers which precede it. Frege's proof in fact shows the theorem to be a consequence of certain quite general facts about the ancestral, together with the fact that $\text{Pred}(\xi, \eta)$ is one-one, and I shall present the theorem in this way. Proofs of the relevant theorems concerning the ancestral may be found in the footnotes.

b. An Important Lemma?   A central lemma in the proof of Theorem 150 is Theorem 145, to whose proof section Zeta is devoted:

$$\mathcal{F}^=(\text{Pred})(0,b) \rightarrow \neg \mathcal{F}(\text{Pred})(b,b).$$

That is: No natural number follows itself in the Pred-series. For the proof, we need the following version of induction, Theorem 144:[35]

$$\mathcal{F}^=(Q)(a,b) \ \& \ Fa \ \& \ \forall x(Fx \rightarrow \forall y(Qxy \rightarrow Fy)) \rightarrow Fb.$$

To prove Theorem 145, Frege takes $F\xi$ to be $\neg\mathcal{F}(\text{Pred})(\xi,\xi)$; $Q\xi\eta$, to be $\text{Pred}(\xi, \eta)$; a, to be 0. Hence:

$$\mathcal{F}^=(\text{Pred})(0,b) \ \& \ \neg\mathcal{F}(\text{Pred})(0,0)$$
$$\& \ \forall x\{\neg\mathcal{F}(\text{Pred})(x,x) \rightarrow \forall y[\text{Pred}(x,y)$$
$$\rightarrow \neg\mathcal{F}(\text{Pred})(y,y)]\} \rightarrow \neg\mathcal{F}(\text{Pred})(b,b).$$

---

[35]Theorem 144 is a consequence of Theorem 128: $\mathcal{F}(Q)(a,b) \ \& \ Fa \ \&$ $\forall x[Fx \rightarrow \forall y(Qxy \rightarrow Fy)] \rightarrow Fb$. For suppose that $\mathcal{F}^=(Q)(a,b)$, that Fa, and that $\forall x[Fx \rightarrow \forall y(Qxy \rightarrow Fy)]$. Then either $a = b$ or $\mathcal{F}(Q)(a,b)$. If the latter, then Fb, by (128); if the former, then Fb, by identity.

Theorem 128 follows from (123). For, if Fa and $Q\xi\eta$ is hereditary, then surely $\forall x(Qax \rightarrow Fx)$.

Again, the second conjunct follows from Frege's Theorem 126. What needs to be established is therefore the following, which is Frege's Theorem 145, α, from which the third conjunct will follow by contraposition and generalization:

$$\mathcal{F}(\text{Pred})(y, y) \ \& \ \text{Pred}(x, y) \rightarrow \mathcal{F}(\text{Pred})(x, x).$$

This Theorem Frege derives from the following two propositions:

(i) $\mathcal{F}(\text{Pred})(y, y) \ \& \ \text{Pred}(x, y) \rightarrow \mathcal{F}^=(\text{Pred})(y, x).$
(ii) $\mathcal{F}^=(\text{Pred})(y, x) \ \& \ \text{Pred}(x, y) \rightarrow \mathcal{F}(\text{Pred})(x, x).$

The latter is an instance of Frege's Theorem 132:[36]

$$\mathcal{F}^=(Q)(y, x) \ \& \ Qzy \rightarrow \mathcal{F}(Q)(z, x).$$

The former is an instance of Frege's Theorem 143:

$$\mathcal{F}(\text{Pred})(y, z) \ \& \ \text{Pred}(x, z) \rightarrow \mathcal{F}^=(\text{Pred})(y, x).$$

Theorem 143, in turn, is a consequence of the following more general one, which I shall call Theorem 143, *:

$$\text{Func}(\text{Conv}\, Q) \ \& \ \mathcal{F}(Q)(y, z) \ \& \ Qxz \rightarrow \mathcal{F}^=(Q)(y, x).$$

(Substitute; note that Func(Conv Pred).) Frege is perfectly aware that this more general proposition is provable. He does not give a formal derivation of it, but he does give an informal proof of it during his discussion of Theorem 143:[37]

---

[36]Theorem 132 follows from Frege's Theorem 129: $\mathcal{F}(Q)(b, c) \ \& \ Qab \rightarrow \mathcal{F}(Q)(a, c)$. For suppose that $\mathcal{F}^=(Q)(b, c)$ and Qab. Then either $\mathcal{F}(Q)(b, c)$ or $b = c$. If the former, then $\mathcal{F}(Q)(a, c)$, by (129); if the latter, then $\mathcal{F}(Q)(a, c)$, by identity.

Theorem 129, in turn, follows from the previously mentioned Theorem 128 and Theorem 127: $\forall F[\forall x(Qax \rightarrow Fx) \ \& \ \forall x[Fx \rightarrow \forall y(Qxy \rightarrow Fy)] \rightarrow Fc] \rightarrow \mathcal{F}(Q)(a, c)$. (This is immediate from the definition.) For suppose that $\mathcal{F}(Q)(b, c)$ and Qab. Let $F\xi$ be hereditary and such that $\forall x(Qax \rightarrow Fx)$. Then Fb; hence, by Theorem 128, Fc. Done.

[37]I have translated Frege's formalism into modern notation. The translation of the text is due to myself and Jason Stanley.

Evidently, the corresponding proposition would not hold, in general, in an arbitrary series. It is here essential that predecession in the number-series takes place functionally (Theorem 88). We rely upon the proposition that, if in some $(Q\text{-})$series an object (b) follows after a second object (a), there is an object which belongs to the $(Q\text{-})$series beginning with the second object (a) and stands to the first object (b) in the series-forming $(Q\text{-})$relation. In signs [Theorem 141]:

$$\mathcal{F}(Q)(a,b) \to \exists x[Qxb \ \& \ \mathcal{F}^=(Q)(a,x)].$$

Now, if one knows that there is no more than one object which stands to the first object (b) in the $(Q\text{-})$relation, then this object must also belong to the $(Q\text{-})$series beginning with the second object (a) (Gg I § 112).

That is: Suppose that $\text{Func}(\text{Conv}\,Q)$, that $\mathcal{F}(Q)(y,z)$, and that Qxz. By Theorem 141, there is some object, call it w, such that Qwz and $\mathcal{F}^=(Q)(y,w)$. But, since the converse of $Q\xi\eta$ is functional and Qxz, we have $x = w$, and hence $\mathcal{F}^=(Q)(y,x)$. That establishes Theorem 143, $\ast$. The proof of Theorem 141, in turn, poses no great difficulty.[38]

Frege's proof of Theorem 145 thus consists essentially in its derivation from a much more general result, which I shall call Theorem 145, $\ast$, from which it immediately follows:[39]

$$\text{Func}(\text{Conv}\,Q) \ \& \ \neg\mathcal{F}(Q)(a,a) \to [\mathcal{F}^=(Q)(a,b) \to \neg\mathcal{F}(Q)(b,b)].$$

That is: If the converse of $Q\xi\eta$ is functional and a does not follow itself in the Q-series, then no member of the Q-series beginning with

---

[38] The proof is, *of course*, by induction, the induction justified by (123). The relevant concept is $\exists z[Qz\xi \ \& \ \mathcal{F}^=(Q)(a,\xi)]$. We must thus establish that: (i) $\forall x(Qax \to \exists z[Qzx \ \& \ \mathcal{F}^=(Q)(a,z)])$; (ii) $\forall x\{\exists z[Qzx \ \& \ \mathcal{F}^=(Q)(a,x)] \to \forall y[Qxy \to \exists z[Qzy \ \& \ \mathcal{F}^=(Q)(a,y)]]\}$. The former is obvious. For the latter, suppose that $\exists z[Qzx \ \& \ \mathcal{F}^=(Q)(a,x)]$ and Qxy; obviously, $\exists z.Qzy$, so we need only show that $\mathcal{F}^=(Q)(a,y)$. Now, we have that $\mathcal{F}^=(Q)(a,x)$ and Qxy; hence $\mathcal{F}(Q)(a,y)$, by Theorem 134; so, $\mathcal{F}^=(Q)(a,y)$, by definition.

[39] Substituting gives: $\text{Func}(\text{Conv}\,\text{Pred}) \ \& \ \neg\mathcal{F}(\text{Pred})(0,0) \to [\mathcal{F}^=(\text{Pred})(0,b) \to \neg\mathcal{F}(\text{Pred})(b,b)]$. But the converse of $\text{Pred}(\xi,\eta)$ is functional; and, as mentioned earlier, 0 does not follow itself in the Pred-series. Theorem 145 then follows by *modus ponens*.

a follows itself in the Q-series. For note that, if we take $F\xi$ in (144) to be $\neg\mathcal{F}(Q)(\xi, \xi)$, then we have the following, Theorem 144, $*$:

$\mathcal{F}^=(Q)(a, b)$ & $\neg\mathcal{F}(Q)(a, a)$

& $\forall x\{\neg\mathcal{F}(Q)(x, x) \to \forall y[Qxy \to \neg\mathcal{F}(Q)(y, y)]\} \to \neg\mathcal{F}(Q)(b, b)$.

We have, then, the following instances of Theorems 132 and 143, $*$, just as in the proof of Theorem 145, α:

(i) $\mathcal{F}^=(Q)(y, x)$ & $Qxy \to \mathcal{F}(Q)(x, x)$.
(ii) $\text{Func}(\text{Conv } Q)$ & $\mathcal{F}(Q)(y, y)$ & $Qxy \to \mathcal{F}^=(Q)(y, x)$.

Hence:

$\qquad \text{Func}(\text{Conv } Q)$ & $\mathcal{F}(Q)(y, y)$ & $Qxy \to \mathcal{F}(Q)(x, x)$.

Theorem 145, $*$ then follows easily from this and Theorem 144, $*$.

c. Another Important Lemma.   Given Theorem 145, our next goal is Theorem 149:

$\mathcal{F}^=(\text{Pred})(0, a)$ & $\text{Pred}(d, a)$

$\qquad \to Nx : \mathcal{F}^=(\text{Pred})(x, d) = Nx : [\mathcal{F}^=(\text{Pred})(x, a)$ & $x \neq a]$.

That is: If a is a natural number and d precedes a, then the number of numbers less than or equal to d is the same as the number of numbers less than or equal to a, other than a. This theorem follows from Theorem 149, α:

$\mathcal{F}^=(\text{Pred})(0, a)$ & $\text{Pred}(d, a)$

$\qquad \to \forall x\{\mathcal{F}^=(\text{Pred})(x, d) \equiv [\mathcal{F}^=(\text{Pred})(x, a)$ & $x \neq a]\}$.

That is: If a is a natural number and d precedes a, then the members of the Pred-series ending with d just are the members of the Pred-series ending with a, other than a. If so, then the number of members of the Pred-series ending with d surely is the same as the number of members of the Pred-series ending with a, other than a.

Theorem 149, α, in turn, follows immediately from the following general fact about the ancestral, which we may call Theorem 149, $*$:

$\text{Func}(\text{Conv } Q)$ & $\neg\mathcal{F}(Q)(a, a)$ & Qda

$\qquad\qquad \to \forall x\{\mathcal{F}^=(Q)(x, d) \equiv [\mathcal{F}^=(Q)(x, a)$ & $x \neq a]\}$.

In words: If the converse of $Q\xi\eta$ is functional, a does not follow itself in the Q-series, and Qda, then the members of the Q-series ending with d are the members of the Q-series ending with a, other than a. Substituting, we have:

$$\text{Func}(\text{Conv Pred}) \ \& \ \neg\mathcal{F}(\text{Pred})(a,a) \ \& \ \text{Pred}(d,a)$$
$$\to \forall x\{\mathcal{F}^=(\text{Pred})(x,d) \equiv [\mathcal{F}^=(\text{Pred})(x,a) \ \& \ x \neq a]\}.$$

Again, the converse of $\text{Pred}(\xi,\eta)$ is functional, and, if a is a natural number, it does not follow itself in the Pred-series, by Theorem 145 (which is used only at this point).

Frege does not prove Theorem 149, $*$, but, as in the case of Theorem 145, $*$, his proof can easily be generalized to yield it. He derives Theorem 149, α, from the following two results:

$(148, \alpha)$ $\text{Pred}(d,a)$
$$\to \{[\mathcal{F}^=(\text{Pred})(x,a) \ \& \ x \neq a] \to \mathcal{F}^=(\text{Pred})(x,d)\}$$

$(148, \zeta)$ $\text{Pred}(d,a) \ \& \ \mathcal{F}^=(\text{Pred})(0,a)$
$$\to \{\mathcal{F}^=(\text{Pred})(x,d) \to [\mathcal{F}^=(\text{Pred})(x,a) \ \& \ x \neq a]\}.$$

For the former: If $\mathcal{F}^=(\text{Pred})(x,a)$ and $x \neq a$, then $\mathcal{F}(\text{Pred})(x,a)$, by the definition of '$\mathcal{F}^=$.' Hence, if $\text{Pred}(d,a)$, then $\mathcal{F}^=(\text{Pred})(x,d)$, by Theorem 143.[40] For the latter: If $\text{Pred}(d,a)$ and $\mathcal{F}^=(\text{Pred})(x,d)$, then $\mathcal{F}(\text{Pred})(x,a)$, by Theorem 137, which is:

$$\mathcal{F}^=(Q)(x,d) \ \& \ Qda \to \mathcal{F}(Q)(x,a).$$

Since $\mathcal{F}(\text{Pred})(x,a)$, $\mathcal{F}^=(\text{Pred})(x,a)$, by definition; and, if $x = a$, then $\mathcal{F}(\text{Pred})(a,a)$, contradicting Theorem 145, since a is a natural number.[41]

---

[40] For the proof of Theorem 149, $*$, we prove: $\text{Func}(\text{Conv }Q) \ \& \ Qda \to \{[\mathcal{F}^=(Q)(x,a) \ \& \ x \neq a] \to \mathcal{F}^=(Q)(x,d)\}$. As before, we have $\mathcal{F}(Q)(x,a)$ and Qda; hence, $\mathcal{F}^=(Q)(x,d)$, by Theorem 143, $*$.

[41] For the proof of Theorem 149, $*$, we prove: $Qda \ \& \ \neg\mathcal{F}(Q)(a,a) \to \{\mathcal{F}^=(Q)(x,d) \to [\mathcal{F}^=(Q)(x,a) \ \& \ x \neq a]\}$. Since Qda and $\mathcal{F}^=(Q)(x,d)$, $\mathcal{F}(Q)(x,a)$; hence $\mathcal{F}^=(Q)(x,a)$. If $x = a$, then $\mathcal{F}(Q)(a,a)$.

d. Completion of the Proof.   We now complete the proof of Theorem 150. (This explanation of the proof closely follows that given by Frege: See Gg I § 114.) Recall that Theorem 150 is:

$$\forall y \{\mathcal{F}^=(\mathrm{Pred})(0, y) \ \& \ \mathrm{Pred}[y, Nx : \mathcal{F}^=(\mathrm{Pred})(x, y)]$$
$$\rightarrow \forall z[\mathrm{Pred}(y, z) \rightarrow \mathrm{Pred}(z, Nx : \mathcal{F}^=(\mathrm{Pred})(x, z))]\}.$$

Much of the elegance of Frege's proof lies in the ease with which he derives Theorem 150 from Theorem 149, which is, again:

$$\mathcal{F}^=(\mathrm{Pred})(0, a) \ \& \ \mathrm{Pred}(d, a)$$
$$\rightarrow Nx : \mathcal{F}^=(\mathrm{Pred})(x, d) = Nx : [\mathcal{F}^=(\mathrm{Pred})(x, a) \ \& \ x \neq a].$$

Indeed, the formal derivation, in *Grundgesetze*, takes just six lines. Theorem 150 follows, by generalization, from Theorem $150, \varepsilon$:

$$\mathcal{F}^=(\mathrm{Pred})(0, d) \ \& \ \mathrm{Pred}[d, Nx : \mathcal{F}^=(\mathrm{Pred})(x, d)]$$
$$\& \ \mathrm{Pred}(d, a) \rightarrow \mathrm{Pred}[a, Nx : \mathcal{F}^=(\mathrm{Pred})(x, a)].$$

For the proof, suppose that d is a natural number, that d precedes the number of members of the Pred-series ending with d, and that d precedes a. We must show that a precedes the number of members of the Pred-series ending with a. To do so, we must find some concept $F\xi$ and some object x falling under $F\xi$ such that the number of Fs, other than x, is a and the number of Fs is the same as the number of members of the Pred-series ending in a. That is, we must show that:

$$\exists F \exists x \{a = Nz : [Fz \ \& \ z \neq x] \ \& \ Fx$$
$$\& \ Nz : \mathcal{F}^=(\mathrm{Pred})(z, a) = Nz : Fz\}.$$

The concept in question is to be $\mathcal{F}^=(\mathrm{Pred})(\xi, a)$; the object in question is to be a itself. Hence, we must show that:

$$a = Nz : [\mathcal{F}^=(\mathrm{Pred})(z, a) \ \& \ z \neq a] \ \& \ \mathcal{F}^=(\mathrm{Pred})(a, a)$$
$$\& \ Nz : \mathcal{F}^=(\mathrm{Pred})(z, a) = Nz : \mathcal{F}^=(\mathrm{Pred})(z, a).$$

Now, the latter two conjuncts are trivial. The former, we may derive, by the transitivity of identity, from:

(i) $a = Nx : \mathcal{F}^=(\mathrm{Pred})(x, d)$.
(ii) $Nx : \mathcal{F}^=(\mathrm{Pred})(x, d) = Nx : [\mathcal{F}^=(\mathrm{Pred})(x, a) \ \& \ x \neq a]$.

The former follows from the functionality of $\text{Pred}(\xi, \eta)$, since, by hypothesis, d precedes both a and the number of members of the Pred-series ending with d. The latter, in turn, is the consequent of Theorem 149: And, since $\mathcal{F}^=(\text{Pred})(0, d)$ and $\text{Pred}(d, a)$, $\mathcal{F}^=(\text{Pred})(0, a)$, by Frege's Theorem 137, mentioned above. Since we are assuming that $\text{Pred}(d, a)$, that establishes the antecedent of (149).

That completes the proof of (150) and so Frege's proof that every number has a successor.

VII. FREGE'S AXIOMATIZATION OF ARITHMETIC

We have, literally, only begun a discussion of Part II of *Grundgesetze*: The derivation of the axioms of arithmetic occupies just one third of Part II. We are plainly not going to complete such a discussion here, but one theorem proven in the remainder deserves special mention, namely, Frege's Theorem 263. Where '∞,' read "*Endlos*," is a name for the number of natural numbers,[42] Theorem 263 is essentially:

$$\text{Func}(Q) \ \& \ \forall x[Gx \equiv \mathcal{F}^=(Q)(a, x)] \ \& \ \neg \exists x.\mathcal{F}(Q)(x, x)$$
$$\& \ \forall x[Gx \rightarrow \exists y.Qxy] \rightarrow Nx : Gx = \infty.$$

That is: If $Q\xi\eta$ is functional, the Gs are the members of the Q-series beginning with a, no object follows itself in the Q-series, and each G is related, by $Q\xi\eta$, to some object, then the number of Gs is Endlos.

It is essential, if we are to understand *Grundgesetze*, that we ask ourselves, concerning such theorems, *why they are here*: *Grundgesetze* is not a random collection of results, and Frege did not just include whatever came to mind; each of the results has a purpose. Unfortunately, he rarely stops to tell us what purposes the various results have. In the case of this theorem, its significance is not at all apparent from its statement; to understand its significance, we must look at its proof. What Frege proves is that, if $Q\xi\eta$ is functional, if the Gs are the members of the Q-series beginning with a, if no member of the Q-series beginning with a follows itself in the Q-series,[43] and if each G is related, by $Q\xi\eta$, to some object, then the

---

[42] That is: $\infty \equiv_{\text{df}} Nx : \mathcal{F}^=(\text{Pred})(0, x)$.

[43] The theorem, with the stronger condition, that no object follow itself in the Q-series, is inferred from one with this weaker condition. Obviously, if $\neg \exists x.\mathcal{F}(Q)(x, x)$, then $\neg \exists x[\mathcal{F}^=(Q)(a, x) \ \& \ \mathcal{F}(Q)(x, x)]$.

number of Gs is Endlos because the Gs, ordered by $\mathcal{F}(Q)(\xi, \eta)$, are *isomorphic to the natural numbers*, ordered by the $\mathcal{F}(\mathrm{Pred})(\xi, \eta)$, that is, by less-than. If we write '$Q\xi\eta$' as '$S\xi\eta$,' '$G\xi$' as '$\mathbb{N}\xi$,' and 'a' as '0', then what Frege proves is that the following determine a structure isomorphic to the natural numbers:[44]

(1) $\mathrm{Func}(S\xi\eta)$.

(2) $\neg\exists x[\mathcal{F}^=(S)(0, x) \;\&\; \mathcal{F}(S)(x, x)]$.

(3) $\forall x[\mathbb{N}x \rightarrow \exists y.Sxy]$.

(4) $\forall x[\mathbb{N}x \equiv \mathcal{F}^=(S)(0, x)]$.

That is to say, Theorem 263 establishes that (1)-(4) are axioms for arithmetic, which are different from, though closely related to, those due to Dedekind and Peano. The accusation considered earlier, that Frege's argument that "arithmetic is a branch of logic" suffers from a failure to isolate *the* basic laws, *the* axioms, of arithmetic, would thus be completely unjustified. Frege does not explicitly say so, but it is clear from his proof of Theorem 263 that he conceived of (1)-(4) as the basic laws of arithmetic. It is for this reason that Theorem 145 is as heavily emphasized by Frege as it is (see his discussion in *Gg* I § 108): For Theorem 145 is his second axiom for arithmetic.

Frege's axioms capture, at least as well as do the Dedekind-Peano Axioms, our intuitions about the structure of the natural numbers. The natural numbers are the members of a series beginning with the number zero (Axiom (4)). Each natural number is immediately followed in that series by one, and only one, natural number (Axioms (1) and (3)). And, finally, each natural number is followed by a *new* natural number, one which has not previously occurred in the series; that is, no natural number follows itself in the series (Axiom (2)).[45] These general principles are, it seems to me, entirely consonant with our intuitive notion of a natural number, as well

---

[44]Of course, Dedekind had published a similar result some years before Frege: See his *The nature and meaning of numbers*, in *Essays on the theory of numbers*, W. W. Beman, tr., New York: Dover, 1963, section X; *Was sind?* was first published in 1887. It seems unlikely that Frege knew of Dedekind's proof before he composed his own, but it is impossible to be certain of this.

Frege's proof is of substantial independent interest. It includes, for example, a proof of the recursion theorem for ω. See my "Definition by induction in Frege's *Grundgesetze der Arithmetik*," which appears here as Chapter 11.

[45]Note that Axiom (2) is equivalent to: $\forall x[\mathbb{N}x \rightarrow \neg\exists y(Sxy \,\&\, \mathcal{F}^=(S)(y, x))]$. If $\exists y(Sxy \;\&\; \mathcal{F}^=(S)(y, x))$, then, by a theorem related to Theorem 134,

they should be: After all, Frege intends to derive the basic laws of *arithmetic*, not the basic laws of some formal theory which bears no obvious relation to arithmetic as we ordinarily understand it. It is therefore essential that his axioms not only characterize the right kind of structure, but that they successfully capture our ordinary understanding of the structure of the natural numbers.

Not that Frege would have cared, but it should be said that his axioms are, compared with Dedekind's, more involved with second-order notions. Of course, the Induction axiom, as formulated by Frege and Dedekind, is second-order as it stands: But, as is well-known, it can, for many purposes, be weakened and written as a first-order axiom schema. Frege's Axiom (2), on the other hand, also contains a second-order universal quantifier in negative position. It is, however, not difficult to see that it follows from "$\neg \exists x.Sx0$" and "Func(Conv $S\xi\eta$)," the two missing axioms from the Dedekind-Peano axiomatization: Indeed, it is just this which is established by the proof of Theorem 145.

VIII. CLOSING

Frege's derivation of the axioms of arithmetic in *Grundgesetze* has been unjustly neglected. Not only *can* his proofs be re-constructed in second-order logic, but the proofs he gives are themselves derivations of the axioms for arithmetic—both his and the Dedekind-Peano axioms—from Hume's principle, within second-order logic. Frege proved Frege's theorem, which is as it should be.

I said earlier that this fact has no significance for our understanding of Frege's philosophy unless Frege *knew* that the axioms could be proven from Hume's principle. Surely, however, he did: For he says, in *Grundlagen*, that he "attach[es] no decisive importance even to bringing in the extensions of concepts at all" (*Gl* § 107). This would be a strange remark for Frege to make if extensions were required, not just for the formulation of the explicit definition and the derivation of Hume's principle from it, but for his proofs of the axioms

---

$\mathcal{F}(S)(x,x)$; and, conversely, a theorem closely related to Frege's Theorem 141 implies that, if $\mathcal{F}(S)(x,x)$, then $\exists y(Sxy \ \& \ \mathcal{F}^=(S)(y,x))$.

See also Frege's discussion of the significance of Theorem 145 in *Gg* I § 108.

from Hume's principle itself.[46]

That said, there are a variety of questions which an appreciation of the role Axiom V plays in Frege's derivation of the axioms of arithmetic might raise. The most important of these is: How can an axiom which plays such a limited *formal* role be of such fundamental importance to Frege's philosophy of mathematics? Rarely, so far as I know, is it said that Frege's abandonment of the logicist project was due to some realization that, with Axiom V refuted and no suitable weakening forthcoming, he could no longer derive the laws of arithmetic within logic: It is said so rarely because it can seem so little worth saying; it can seem so much common sense, and that, in some sense, it must be. But there was, in a clear sense, no *formal* obstacle to the logicist program, even after Russell's discovery of the contradiction, and Frege knew it. The explanation, that Frege realized he could no longer derive the laws of arithmetic *within logic* is surely right: The question, however, is why Frege was prepared to accept Axiom V, but not Hume's principle, as a fundamental truth of logic, as a "primitive truth," one which "neither need[s] nor admit[s] of proof" (*Gl* § 3).

A particularly nice statement of the philosophical significance of Axiom V is made in a letter Frege wrote to Russell in 1902:[47]

> ... [T]he question is, How do we apprehend logical objects? And I have found no other answer to it that this, We apprehend them as extensions of concepts, or more generally, as value-ranges of functions. I have always been aware that there were difficulties with this, and your discovery of the contradiction has added to them; but what other way is there?

It is to answer this question, first raised in the famous § 62 of *Grundlagen*, that Frege introduces extensions of concepts in the first place: And the difficulty which demands their introduction is the so-called Julius Caesar problem, the problem how numbers (or abstract ob-

---

[46]See also Frege's letter to Russell of 28 July 1902, Gottlob Frege, *Philosophical and mathematical correspondence*, H. Hermes, et al., eds., H. Kaal, tr., Chicago: University of Chicago Press, 1980, 141; the letter is number xxxvi/6, in the German edition. He here suggests the use of Hume's principle, as an axiom, as a way of avoiding the use of Axiom V.

[47]Ibid.

jects of other kinds) are to be differentiated from persons and objects of other sorts (see *Gl* §§ 56, 66–8).

The questions to which we *really* need answers are thus: What, exactly, does Frege mean by his question, how we apprehend logical objects? and what is the real point of the Caesar problem? In what, for Frege, does it consist that a truth is a "primitive" truth? one which "neither need[s] nor admit[s] of proof?" Or, if the notion of a primitive truth is not absolute, but relative to a particular systematization of logic (or arithmetic), in what does it consist that some truth is even a *candidate* for being a "primitive" truth? And, if Frege would not have been opposed to accepting Hume's principle as a primitive truth, perhaps there was some obstacle, in his view, to accepting it as a primitive *logical* truth. So, in what does it consist that a "primitive" truth *is* a primitive logical truth?—I could hazard a guess at answers to such questions, but it would, at this point, only be a guess.

What I do know is that the questions lately raised need answers: For, until we have such answers, we shall not understand the significance Axiom V had for Frege, since we shall not understand why he could not abandon it in favor of Hume's principle; and that is to say that we shall not understand how he conceived the logicist project. We shall thus not fully understand Frege's philosophy until we understand the enormous significance the question how we apprehend logical objects, and the Caesar problem, had for him, until, that is, we understand how he could consistently be so hostile to psychologism, yet hold his own philosophy of mathematics hostage to such epistemological considerations.[48]

[48]I should like to thank George Boolos for asking the question which led to my study of the *Grundgesetze* as well as for the numerous discussions we have had about these issues. I have also benefited from discussions with Michael Dummett, Michael Glanzberg, Warren Goldfarb, Kathrin Koslicki, and Charles Parsons. I owe an immeasurable debt to Jason Stanley, whose enthusiasm about Frege has made my work on *Grundgesetze* both better and more enjoyable.

A version of this paper was read to a conference on the foundations of mathematics, held at Boston University, in April 1992. Parts of the paper were also read to the conference *Philosophy of mathematics today* held in Munich, in June 1993, and to the joint session of the Mind Association and British Society for the Philosophy of Science held in St. Andrews, in September 1993.

POSTSCRIPT: A RE-CONSIDERATION OF THE PROOF OF
THE INFINITY OF THE NUMBER-SERIES
IN *GRUNDLAGEN*

What is said in the foregoing concerning *Grundlagen* now strikes me
as incorrect; given the opportunity presented by the re-printing of
this paper, I should like to add a few remarks concerning the proof
of the infinity of the number-series given in *Grundlagen*. Above, I
wrote that this proof requires a proof of:

$$\mathcal{F}^=(\text{Pred})(0, a) \rightarrow a = Nx : [\mathcal{F}^=(\text{Pred})(x, a) \ \& \ x \neq a].$$

And I went on to say: "The theorem can, indeed, be proven in this
way .... [H]owever, the proof of the theorem along these lines is
long and difficult. In *Grundgesetze*, however, Frege gives a shorter
and more elegant proof." This now seems to me to be wrong: If a
coherent proof is suggested by sections 82–3 of *Grundlagen* at all,[49]
the proof there outlined is only slightly different from that given in
Grundgesetze.

To see this, let us reproduce the relevant portions of *Grundlagen*
§§ 82–3 as follows, introducing some indices and substituting the
logical symbols used above for some of Frege's prose:

§ 82. It is now to be shown that—subject to a condition still
to be specified—

(0) $\text{Pred}[n, Nx : \mathcal{F}^=(\text{Pred})(x, n)]$.

... It is to be proved that

(1)

$$\text{Pred}(d, a) \ \& \ \text{Pred}[d, Nx : \mathcal{F}^=(\text{Pred})(x, d)]$$
$$\rightarrow \text{Pred}[a, Nx : \mathcal{F}^=(\text{Pred})(x, a)].$$

---

[49] For discussion of this question, see George Boolos and Richard G. Heck,
Jr., "*Die grundlagen der arithmetik* §§ 82–3," forthcoming in M. Schirn, ed.,
Philosophy of mathematics today, Oxford: Oxford University Press. As what
is under discussion here is considered in more detail in that paper, I have here
merely sketched the relevant line of thought.

Just in case it is not clear from what is to follow: I am even more heavily
indebted to George Boolos in this Postscript than I am in the main body of the
paper.

It is then to be proved, secondly, that

(2) $\qquad\qquad \mathrm{Pred}[0, Nx : \mathcal{F}^=(\mathrm{Pred})(x, 0)].$

And finally, (0) is to be deduced. The argument here is an application of the definition [of the ancestral].

$83. In order to prove (1), we must show that

(3) $\qquad a = Nx : [\mathcal{F}^=(\mathrm{Pred})(x, a) \ \& \ x \neq a].$

And for this, again, it is necessary to prove that

(4) $\qquad \forall x\{[\mathcal{F}^=(\mathrm{Pred})(x, a) \ \& \ x \neq a] \equiv \mathcal{F}^=(\mathrm{Pred})(x, d)\}.$

For this we need

(5) $\qquad\qquad \forall x[\mathcal{F}^=(\mathrm{Pred})(0, x) \rightarrow \neg\mathcal{F}(\mathrm{Pred})(x, x)].$

Now, these sections have often been read as suggesting a proof of (0) which requires a proof of

(3′) $\qquad \mathcal{F}^=(\mathrm{Pred})(0, a) \rightarrow a = Nx : [\mathcal{F}^=(\mathrm{Pred})(x, a) \ \& \ x \neq a].$

One might reason as follows. The last proposition, (5), is provable by induction. Since (4) is obviously not provable unless some relation between d and a is assumed, the supposition that $\mathrm{Pred}(d, a)$, made in (1), must still be in force. Nor is (4) provable unless $\neg\mathcal{F}(\mathrm{Pred})(a, a)$, so the point of (5) must be to allow us to conclude this from the additional hypothesis that $\mathcal{F}^=(\mathrm{Pred})(0, d)$. So, what is to be proven is certainly not (4) but

(4′) $\mathcal{F}^=(\mathrm{Pred})(0, d) \ \& \ \mathrm{Pred}(d, a)$
$\qquad\qquad \rightarrow \forall x\{[\mathcal{F}^=(\mathrm{Pred})(x, a) \ \& \ x \neq a] \equiv \mathcal{F}^=(\mathrm{Pred})(x, d)\}.$

Similarly, (3) is not provable without the assumption that a is finite, so what is to be proven is surely not (3) but

(3′) $\qquad \mathcal{F}^=(\mathrm{Pred})(0, a) \rightarrow a = Nx : [\mathcal{F}^=(\mathrm{Pred})(x, a) \ \& \ x \neq a].$

And, similarly, what is to be proven is not (1), but

(1') $\mathcal{F}^=(\text{Pred})(0, d)$ & $\text{Pred}(d, a)$

                 & $\text{Pred}[d, Nx : \mathcal{F}^=(\text{Pred})(x, d)]$

                   $\rightarrow \text{Pred}[a, Nx : \mathcal{F}^=(\text{Pred})(x, a)].$

And, finally, what is to be proven is not (0) but

(0') $\mathcal{F}^=(\text{Pred})(0, n) \rightarrow \text{Pred}[n, Nx : \mathcal{F}^=(\text{Pred})(x, n)],$

which follows by induction, using Gg Thm 152, from (2)—which is Gg Thm 154—and (1').

As George Boolos has pointed out, the proof, so read, suffers from redundancy.[50] For (0') can be made to follow from (3') extremely easily, so that the derivation of (1') from (3'), the proof of (2), and the derivation of (0') from (1') and (2) fall away as unnecessary. For note that we have, by Gg Thm 102, that

$\mathcal{F}^=(\text{Pred})(a, a)$

    $\rightarrow \text{Pred}\{Nx : [\mathcal{F}^=(\text{Pred})(x, a)$ & $x \neq a], Nx : \mathcal{F}^=(\text{Pred})(x, a)\}.$

But (0') follows from this, Gg Thm 140, and (3') immediately.

To read the proof sketched in *Grundlagen* as requiring a proof of (3'), however, is to misread it; it is this misreading of which I myself am guilty above. Part of what makes the misreading natural is Frege's insistence, in other contexts, that there is no distinctive mode of 'reasoning from a supposition.' It is easy to be misled into supposing that, according to Frege, there is *no such thing* as reasoning from an hypothesis; if so, then, in an informal argument such as that of *Gl* §§ 82–3, Frege ought always be read as sketching the derivation of *one theorem*—e.g., (1')—from other theorems—e.g., (3'). But, for one thing, Frege's view is only that reasoning from an hypothesis *amounts to* or *should be formalized as* reasoning concerning conditionals whose tacit antecedents contain the 'hypothesis' in question; and, furthermore, he frequently engages in informal

---

[50]In a paper read to the conference *Philosophy of mathematics today* held in Munich in June 1993.

reasoning from suppositions, even in *Grundgesetze*. The sketch being given in the relevant sections of *Grundlagen* thus is, or so it now seems to me, precisely an informal argument in which Frege is reasoning from hypotheses. The best way to see this is probably to imagine *Gl* § 82–3 re-written so as to emphasize this aspect of Frege's presentation. To simplify the exposition, we here incorporate our supposition that he is proposing to derive (0′) from (1′) and (2), by means of *Gg* Thm 152, rather than to derive it from (1) and (2), by means of *Gg* Thm 144.[51] Viz.:

§ 82. It is now to be shown that . . .

$$(0') \qquad \mathcal{F}^=(\mathrm{Pred})(0,n) \rightarrow \mathrm{Pred}[n, Nx : \mathcal{F}^=(\mathrm{Pred})(x,n)].$$

. . . It is to be proved that, if $\mathcal{F}^=(\mathrm{Pred})(0,d)$, $\mathrm{Pred}(d,a)$ and $\mathrm{Pred}[d, Nx : \mathcal{F}^=(\mathrm{Pred})(x,d)]$, then

$$(1^*) \qquad \mathrm{Pred}[a, Nx : \mathcal{F}^=(\mathrm{Pred})(x,a)].$$

It is then to be proved, secondly, that

$$(2) \qquad \mathrm{Pred}[0, Nx : \mathcal{F}^=(\mathrm{Pred})(x,0)].$$

[And then (0′) gets proven by induction, etc., etc.]

§ 83. In order to derive (1*) from the suppositions, we must show that, *given these assumptions*,

$$(3) \qquad a = Nx : [\mathcal{F}^=(\mathrm{Pred})(x,a) \ \& \ x \neq a].$$

And for this, again, it is necessary to prove that, *on these same assumptions*,

$$(4) \qquad \forall x \{[\mathcal{F}^=(\mathrm{Pred})(x,a) \ \& \ x \neq a] \equiv \mathcal{F}^=(\mathrm{Pred})(x,d)\}.$$

[51]On this way of reading *Gl* §§ 82–3, Frege is proceeding as he does in order to explain how the finiteness condition comes to be attached to (0), so that when he writes, at the end of § 83, "We are obliged hereby attach to the proposition [(0)] the condition that [*n* is finite]," what he means is that the need for the finiteness condition in the proof of (4) is, in some way, transferred to (1). There are reasons, however, to be dissatisfied with this reading. For discussion, and for a suggested (if radical) alternative, see Boolos and Heck, op. cit.

For this we need

(5) $\forall x[\mathcal{F}^=(\text{Pred})(0, x) \rightarrow \neg \mathcal{F}(\text{Pred})(x, x)]$.

Thus, were we to wish to formalize this proof in an *axiomatic* system, such as that of *Grundgesetze*, we should proceed as follows. We should first prove (5). Secondly, we should derive

(4′) $\mathcal{F}^=(\text{Pred})(0, d)$ & $\text{Pred}(d, a)$
    $\rightarrow \forall x\{[\mathcal{F}^=(\text{Pred})(x, a)$ & $x \neq a] \equiv \mathcal{F}^=(\text{Pred})(x, d)\}$,

from (5), the third supposition not being needed. Thirdly, we should derive

(3*) $\mathcal{F}^=(\text{Pred})(0, d)$ & $\text{Pred}(d, a)$
      & $\text{Pred}[d, Nx : \mathcal{F}^=(\text{Pred})(x, d)]$
      $\rightarrow a = Nx : [\mathcal{F}^=(\text{Pred})(x, a)$ & $x \neq a]$

from (4′). Fourthly, we should derive (1′) from (3*). And, finally, we should complete the proof by proving (2) and concluding (0′) by induction.

This proof of (0′) is closely related to that given in *Grundgesetze*. In fact, (5) is *Gg* Thm 145, and (4′) is only inessentially different from *Gg* Thm 149, α. (3*) is not proven in *Grundgesetze*, but (1′) is there derived from (4′) by means of an argument which is but a permutation of that required to derive (1′) from (4′) via (3*). Both of these proofs begin by concluding

(145) $\mathcal{F}^=(\text{Pred})(0, a)$ & $\text{Pred}(d, a)$
    $\rightarrow Nx : [\mathcal{F}^=(\text{Pred})(x, a)$ & $x \neq a] = Nx : \mathcal{F}^=(\text{Pred})(x, d)$

from (4′) or (149, α), by Hume's principle.[52] It is only in how (1′) is derived from *Gg* Thm 149 that they differ.

---

[52]I am here ignoring the niceties regarding whether the finiteness condition concerns a or d. It is in this respect that (4′) and *Gg* Thm 149, α differ.

The formal derivation of $(1')$ via (3^*) might proceed as follows:

(i) $\text{Pred}(d, a)$ $(Gg\ 71)$
   & $\text{Pred}[d, Nx : \mathcal{F}^=(\text{Pred})(x, d)]$
   $\rightarrow a = Nx : \mathcal{F}^=(\text{Pred})(x, d).$

(ii) $Nx : [\mathcal{F}^=(\text{Pred})(x, a) \ \& \ x \neq a]$ Identity
     $= Nx : \mathcal{F}^=(\text{Pred})(x, d)$
   & $\text{Pred}(d, a)$
   & $\text{Pred}[d, Nx : \mathcal{F}^=(\text{Pred})(x, d)]$
   $\rightarrow a = Nx : [\mathcal{F}^=(\text{Pred})(x, a) \ \& \ x \neq a].$

(3^*) $\mathcal{F}^=(\text{Pred})(0, a) \ \& \ \text{Pred}(d, a)$ (ii, Gg 149) PC
   & $\text{Pred}[d, Nx : \mathcal{F}^=(\text{Pred})(x, d)]$
   $\rightarrow a = Nx : [\mathcal{F}^=(\text{Pred})(x, a) \ \& \ x \neq a].$

*   *   *

(iii) $\mathcal{F}^=(\text{Pred})(a, a)$ $(Gg\ 102)$
   & $a = Nx : [\mathcal{F}^=(\text{Pred})(x, a) \ \& \ x \neq a]$
   $\rightarrow \text{Pred}[a, Nx : \mathcal{F}^=(\text{Pred})(x, a)].$

(iv) $a = Nx : [\mathcal{F}^=(\text{Pred})(x, a) \ \& \ x \neq a]$ $(Gg\ 140)$
   $\rightarrow \text{Pred}[a, Nx : \mathcal{F}^=(\text{Pred})(x, a)].$

(v) $\mathcal{F}^=(\text{Pred})(0, a) \ \& \ \text{Pred}(d, a)$ $(3^*, \text{iv})$ PC
   & $\text{Pred}[d, Nx : \mathcal{F}^=(\text{Pred})(x, d)]$
   $\rightarrow \text{Pred}[a, Nx : \mathcal{F}^=(\text{Pred})(x, a)].$

$(1')$ $\mathcal{F}^=(\text{Pred})(0, d) \ \& \ \text{Pred}(d, a)$ $(Gg\ 137)$
   & $\text{Pred}[d, Nx : \mathcal{F}^=(\text{Pred})(x, d)]$
   $\rightarrow \text{Pred}[a, Nx : \mathcal{F}^=(\text{Pred})(x, a)].$

The *Grundgesetze* version of the proof, on the other hand, is as follows:

$(\#)$ $\quad \mathcal{F}^=(\mathrm{Pred})(a,a)$ $\hspace{3cm}$ $(Gg\ 102)$
$\quad\quad$ & $Nx : [\mathcal{F}^=(\mathrm{Pred})(x,a)$ & $x \neq a]$
$\quad\quad\quad = Nx : \mathcal{F}^=(\mathrm{Pred})(x,d)$
$\quad\quad\quad \to \mathrm{Pred}[Nx : \mathcal{F}^=(\mathrm{Pred})(x,d), Nx : \mathcal{F}^=(\mathrm{Pred})(x,a)].$

$(150,\alpha)$ $\quad \mathcal{F}^=(\mathrm{Pred})(a,a)$ $\hspace{3cm}$ $(149, \#)$ PC
$\quad\quad$ & $\mathcal{F}^=(\mathrm{Pred})(0,a)$ & $\mathrm{Pred}(d,a)$
$\quad\quad\quad \to \mathrm{Pred}[Nx : \mathcal{F}^=(\mathrm{Pred})(x,d), Nx : \mathcal{F}^=(\mathrm{Pred})(x,a)].$

$(150,\beta)$ $\quad \mathcal{F}^=(\mathrm{Pred})(0,a)$ & $\mathrm{Pred}(d,a)$ $\hspace{2cm}$ $(Gg\ 140)$
$\quad\quad\quad \to \mathrm{Pred}[Nx : \mathcal{F}^=(\mathrm{Pred})(x,d), Nx : \mathcal{F}^=(\mathrm{Pred})(x,a)].$

$(150,\gamma)$ $\quad a = Nx : \mathcal{F}^=(\mathrm{Pred})(x,d)$ $\hspace{2.5cm}$ Identity
$\quad\quad$ & $\mathcal{F}^=(\mathrm{Pred})(0,a)$ & $\mathrm{Pred}(d,a)$
$\quad\quad\quad \to \mathrm{Pred}[a, Nx : \mathcal{F}^=(\mathrm{Pred})(x,a)].$

$(150,\delta)$ $\quad \mathrm{Pred}[d, Nx : \mathcal{F}^=(\mathrm{Pred})(x,d)]$ $\hspace{2cm}$ $(Gg\ 71)$
$\quad\quad$ & $\mathcal{F}^=(\mathrm{Pred})(0,a)$ & $\mathrm{Pred}(d,a)$
$\quad\quad\quad \to \mathrm{Pred}[a, Nx : \mathcal{F}^=(\mathrm{Pred})(x,a)].$

$(150,\varepsilon)$ $\quad \mathcal{F}^=(\mathrm{Pred})(0,d)$ $\hspace{3.5cm}$ $(Gg\ 137)$
$\quad\quad$ & $\mathrm{Pred}(d,a)$ & $\mathrm{Pred}[d, Nx : \mathcal{F}^=(\mathrm{Pred})(x,d)]$
$\quad\quad\quad \to \mathrm{Pred}[a, Nx : \mathcal{F}^=(\mathrm{Pred})(x,a)].$

Note, of course, that Gg Thm $150,\varepsilon$ is what we have been calling
$(1')$. Note, also, that the very same steps occur in each of these
proofs; the difference lies only in a slight re-arrangement of the proof,
which obviates the need to stop the chain of inferences after the proof
of (3^*). The version of the proof given in *Grundgesetze* thus proves
to be *slightly* shorter and *slightly* more elegant than that we have
extracted from *Grundlagen*.

RICHARD G. HECK, JR.

11

Definition by Induction in Frege's
Grundgesetze der Arithmetik

I. OPENING

Dedekind's *Was sind und was sollen die Zahlen?* has long been celebrated for the proofs of two important theorems. The first of these is
the proof of the validity of the definition, by induction, of a function
defined on the natural numbers, that is, of the recursion theorem
for ω. The second is the proof that all "simply infinite systems"—
that is, structures which satisfy the Dedekind-Peano axioms—are
isomorphic.[1] Dedekind's proofs of these theorems are carried out
set-theoretically, the set theory in question being unformalized. It
is well-known that the proofs of these theorems can be carried out in
second-order arithmetic or in standard set-theories, and Dedekind
is usually credited with having been the first to give a set-theoretic
proof of these results.

It is almost unknown, however, that, in his *Grundgesetze der
Arithmetik*,[2] Frege gives formal proofs of both of these results. Of

---

[1] Richard Dedekind, "The nature and meaning of numbers," in *Essays on the
theory of numbers*, W. W. Beman, tr., New York: Dover, 1963. *Was sind?* was
first published in 1887; the second edition, in 1893. Of the results mentioned,
the former is Theorem 126; the latter, Theorem 132.

[2] Gottlob Frege, *Grundgesetze der Arithmetik*, Hildesheim: Georg Olms,

course, the formal system of *Grundgesetze* is inconsistent, Russell's paradox being derivable, in second-order logic, from Frege's infamous Axiom V. For present purposes, Axiom V may be taken to be:

$$(\grave{\varepsilon}.F\varepsilon = \grave{\varepsilon}.G\varepsilon) \equiv \forall x(Fx \equiv Gx).$$

Axiom V, so formulated, governs terms which purport to stand for the extensions of concepts,[3] and such terms occur throughout *Grundgesetze*. As I have shown elsewhere, however, Frege makes reference to value-ranges, for the most part, only for convenience, and such reference may be eliminated, uniformly, from the great majority of his proofs.[4]

Frege makes essential use of Axiom V only in his proof of what has come to be known as *Hume's principle*.[5] Hume's principle may be formalized in second-order logic as:

$$Nx : Fx = Nx : Gx$$

iff

$$(\exists R)[\forall x \forall y \forall z (Rxy \ \& \ Rxz \rightarrow y = z)$$
$$\& \ \forall x \forall y \forall z (Rxz \ \& \ Ryz \rightarrow x = y) \ \& \ \forall x(Fx \rightarrow \exists y(Rxy \ \& \ Gy))$$
$$\& \ \forall x(Gx \rightarrow \exists y(Ryx \ \& \ Fy))].$$

The second-order theory whose sole "non-logical" axiom is Hume's principle we may call *Fregean arithmetic*: Fregean arithmetic is equiconsistent with second-order arithmetic and is thus almost certainly consistent.[6] Frege's proofs of the axioms of arithmetic, in *Grundgesetze*, can thus be reconstructed as proofs in Fregean arithmetic:

1966. The first volume was, of course, published in 1893; the second, in 1903. References are in the text, marked "*Gg*" with volume and section numbers.

[3]Note that the *first-order* theory which contains Axiom V as its sole nonlogical axiom is actually consistent. See Terence Parsons, "On the consistency of the first-order portion of Frege's logical system," Chapter 16, below.

[4]See my "The development of arithmetic in Frege's *Grundgesetze der Arithmetik*," Chapter 10, above.

[5]Of course, "Hume's principle" is not a definite description. It is so called because Frege quotes Hume as a way of introducing the principle.

[6]See George Boolos, "The consistency of Frege's *Foundations of arithmetic*," Chapter 8, above.

Indeed, it can be argued that Frege knew full well that the axioms of arithmetic are derivable, in second-order logic, from Hume's principle. That is to say: The main theorem of *Grundgesetze*, which George Boolos has rightly urged us to call *Frege's theorem*, is that Hume's principle implies the axioms of second-order arithmetic.

That Frege offered proofs of the axioms of arithmetic in *Grundgesetze* is well-known, even if the fact that he proved Frege's theorem has not been. However, only one-third of Part II of *Grundgesetze*, entitled "Proofs of the basic laws of number," is concerned with the proofs of these axioms. The remainder of Part II contains proofs of a number of additional results, among them Theorem 167, that there is an infinite cardinal (namely, the number of natural numbers); Theorem 359, of which the least number principle is an instance; and Theorem 469, the main theorem required for the definition of addition. Of interest to us here is Theorem 263.

To state Theorem 263 and the form of Hume's principle which Frege employs in his proof of it, we need a number of definitions. These definitions, and the theorems themselves, I give in translation into second-order logic. The first is that of the *converse of a relation* (*Gg* I § 39):[7]

$$\text{Conv}_{\alpha\varepsilon}(R\alpha\varepsilon)(x,y) \overset{\text{df}}{\equiv} Ryx.$$

Thus, x stands in the converse of the relation $R\xi\eta$ to y if y stands in the relation $R\xi\eta$ to x. The second definition is the familiar one of the *functionality* of a relation (*Gg* I § 37):

$$\text{Func}_{\alpha\varepsilon}(R\alpha\varepsilon) \overset{\text{df}}{\equiv} \forall x \forall y \forall z(Rxy \ \& \ Rxz \to y = z).$$

The third is that of a relation's *mapping* one concept into another (*Gg* I § 38):

$$\text{Map}_{\alpha\varepsilon xy}(R\alpha\varepsilon)(Fx, Gy) \overset{\text{df}}{\equiv} \text{Func}_{\alpha\varepsilon}(R\alpha\varepsilon) \ \& $$
$$\forall x[Fx \to \exists y(Rxy \ \& \ Gy)].$$

Thus, a relation $R\xi\eta$ maps a concept $F\xi$ into a concept $G\xi$ just in case $R\xi\eta$ is functional and every F stands in it to some (and

---

[7]I insert the bound variables in the definitions but will drop them when it causes no confusion.

therefore exactly one) G.[8] Using these definitions, Frege concisely formulates Hume's principle as:

$$Nx : Fx = Nx : Gx \text{ iff } (\exists R)[\text{Map}(R)(F, G) \ \& \ \text{Map}(\text{Conv } R)(G, F)].$$

This formulation is equivalent to the more familiar formulation given above, but it has certain technical advantages over it.[9] Additionally, we need Frege's famous definition of the *strong ancestral* of a relation (*Gg* I § 45):

$$\mathcal{F}_{\alpha\varepsilon}(Q\alpha\varepsilon)(a, b) \stackrel{\mathrm{df}}{\equiv} \forall F[\forall x(Qax \rightarrow Fx)$$
$$\& \ \forall x \forall y(Fx \ \& \ Qxy \rightarrow Fy) \rightarrow Fb].$$

That is, b *follows after* a *in the Q-series* if, and only if, b falls under every concept (i) under which all objects to which a stands in the Q-relation fall and (ii) which is *hereditary in the Q-series*, i.e., under which every object to which an F stands in the Q-relation falls. Frege defines the *weak ancestral* as follows (*Gg* I § 46):

$$\mathcal{F}^{=}_{\alpha\varepsilon}(Q\alpha\varepsilon)(a, b) \stackrel{\mathrm{df}}{\equiv} \mathcal{F}_{\alpha\varepsilon}(Q\alpha\varepsilon)(a, b) \vee a = b.$$

Thus, b *belongs to* (or *is a member of*) *the Q-series beginning with* a if, and only if, either b follows after a in the Q-series or b is identical with a.

We now turn to the definitions of more particularly arithmetical notions. The number zero is defined by Frege as the number of objects which are not self-identical (*Gg* I § 41; see *Gl* § 74):[10]

$$0 = Nx : x \neq x$$

The relation of predecession is defined as (*Gg* I § 43; see *Gl* § 74):

$$\text{Pred}(m, n) \stackrel{\mathrm{df}}{\equiv} \exists F \exists y[n = Nx : Fx \ \& \ Fy$$
$$\& \ m = Nx : (Fx \ \& \ x \neq y)].$$

---

[8]Note that the definition does *not* say that $R\xi\eta$ is functional and is *onto* the Gs.

[9]See the previous chapter for discussion of these. See also *Gg* I § 66.

[10]Gottlob Frege, *The foundations of arithmetic*, J. L. Austin, tr., Evanston: Northwestern University Press, 1980. References are in the text, marked by "*Gl*" and a section number.

That is, m *immediately precedes n in the number-series* if, and only if, there is some concept $F\xi$, whose number is n, and an object y falling under $F\xi$, such that m is the number of Fs other than y. The concept of a *finite* or *natural number* may then be defined as:[11]

$$\mathbb{N}x \overset{\text{df}}{\equiv} \mathcal{F}^{=}(\text{Pred})(0, x).$$

So a number is a natural number, is finite, if, and only if, it belongs to the Pred-series (the number-series) beginning with zero. (Famously, induction is a near immediate consequence of this definition.) And, finally, the first transfinite number, which Frege calls "Endlos," may be defined as (Gg I § 122):

$$\infty \overset{\text{df}}{\equiv} \mathbb{N}x : \mathcal{F}^{=}(\text{Pred})(0, x).$$

Thus, Endlos is the number of natural numbers.

Theorem 263 may then be formulated as:

$$\exists Q[\text{Func}(Q) \ \& \ \neg\exists x.\mathcal{F}(Q)(x, x)$$
$$\& \ \forall x(Gx \to \exists y.Qxy) \ \& \ \exists x\forall y(Gy \equiv \mathcal{F}^{=}(Q)(x, y))]$$
$$\to \mathbb{N}x : Gx = \infty.$$

Suppose that there is a relation $Q\xi\eta$ which satisfies the following conditions: First, it is functional; second, no object follows after itself in the Q-series; thirdly, each G stands in the Q-relation to some object; and, finally, the Gs are the members of the Q-series beginning with some object. Then, says Theorem 263, the number of Gs is Endlos.

II. FREGE'S PROOF OF THEOREM 263

It is worth quoting Frege's initial explanation of Theorem 263 in full:[12]

---

[11] Frege does not have any special symbol for this concept, though he reads "$\mathcal{F}^{=}(\text{Pred})(0, \xi)$" as "$\xi$ is a finite number [*endliche Anzahl*]." See Gg I § 108.

[12] The following is from Gg I § 144. The translation is due to myself and Jason Stanley.

We now prove ... that Endlos is the number which belongs
to a concept, if the objects falling under this concept may
be ordered in a series, which begins with a particular object
and continues without end, without coming back on itself and
without branching.

By an "unbranching series," Frege means one whose determining
relation is functional; by a series which does not "come back on
itself," he means one in which no object follows after itself; by a
series which "continues without end," he means one every member
of which is immediately followed by some object.

What it is essential to show is that Endlos is the number which
belongs to the concept *member of such a series* ... . For this
purpose, we use proposition (32) and have to establish that
there is a relation which maps the number-series into the Q-
series beginning with x, and whose converse maps the latter
into the former.

Proposition (32) is one direction of Hume's principle, namely: If
there is a relation which maps the Fs into the Gs and whose converse
maps the Gs into the Fs, then the number of Fs is the same as the
number of Gs.

It suggests itself to associate 0 with x, 1 with the next member of the Q-series following after x, and so always to associate
the number following next with the member of the Q-series following next. Each time, we combine a member of the number-
series and a member of the Q-series into a pair, and we build
a series from these pairs.

That is, the theorem is to be proven by defining, by induction, a
relation[13] between the numbers and the members of the Q-series
beginning with x: The number, n, which is the immediate successor
of a given number, m, will be related to that member of the Q-
series, call it x_n, which follows immediately after the member of the
Q-series to which m is related, say, x_m:

$$0 \rightarrow 1 \rightarrow 2 \rightarrow \cdots \rightarrow m \rightarrow n \rightarrow \cdots$$

$$x_0 \rightarrow x_1 \rightarrow x_2 \rightarrow \cdots \rightarrow x_m \rightarrow x_n \rightarrow \cdots$$

---

[13]In connection with Frege's use of the word "associate," see *Gg* I § 66.

The proof of the theorem will require a proof of the validity of such definitions. The idea is to define the relation by defining a *series of ordered pairs*: Namely, the series $\langle 0; x_0 \rangle$, $\langle 1; x_1 \rangle$, etc. The relation will then hold between objects x and y just in case $\langle x; y \rangle$ is a member of this series of ordered pairs; as one might put it, the members of this series will be the extension of the relation to be defined. To define this series of ordered pairs, Frege thus needs to introduce ordered pairs into his system and to define the relation in which a given member of the series stands to the next member of the series.

Unfortunately, Frege's definition of ordered pairs is, as George Boolos once put it, extravagant and can not be consistently reconstructed, either in second-order logic or in set-theory. According to Frege's definition, the ordered pair $(a; b)$ is the class to which all and only the extensions of relations in which a stands to b belong. Obviously, this is a (super-)proper class. Frege's proof can, however, be carried out if we take ordered pairs as primitive and subject to the usual ordered pair axiom:

(OP) $(a; b) = (c; d) \equiv [a = c \ \& \ b = d].$

Indeed, Frege derives OP from his definition,[14] and the fact that ordered pairs, so defined, satisfy OP is all he really uses (just as the fact that numbers satisfy Hume's principle is all he uses).[15]

After introducing ordered pairs, Frege continues by defining the relation in which a given member of his series of pairs stands to the next:[16]

[14] From left-to-right, the axiom is proven as Theorem 218; from right-to-left, as Theorem 251.

[15] Frege does, it should be said, also use the following trick, which explains why he defines ordered pairs as he does. Indeed, it is really too bad that his definition does not work. For certain purposes, when making use of ordered pairs, one needs, given a relation $R\xi\eta$, to define a concept $F\xi$ such that: $F[(a; b)] \equiv Rab$. (See the definition of the 'collapse' of a relation, below.) Frege's definition makes this extremely easy: For Rab just in case the extension of the relation $R\xi\eta$ is a *member of* the ordered pair $\langle a; b \rangle$, since the ordered pair is the class of all extensions of relations in which a stands to b. That is, where $\grave{\alpha}\acute{\varepsilon}.R\varepsilon\alpha$ is the extension of $R\xi\eta$: $[\grave{\alpha}\acute{\varepsilon}.R\varepsilon\alpha \in (a; b)] \equiv Rab$. ($\grave{\alpha}\acute{\varepsilon}.R\varepsilon\alpha$ is the *double value-range* of the relation $R\xi\eta$. Note that Frege does *not* use ordered pairs to define the extension of a relation. See *Gg* I § 36.)

[16] I have translated Frege's definition into second-order logic, eliminating the reference to value-ranges.

The series-forming relation is thereby determined: a pair stands in it to a second pair, if the first member of the first pair stands in the Pred-relation to the first member of the second pair, and the second member of the first pair stands in the Q-relation to the second member of the second pair. If, then, the pair $(n; y)$ belongs to our series beginning with the pair $(0; x)$, then n stands to y in the mapping relation to be exhibited.

That is to say: $(m; n)$ will stand in the "series-forming relation" to $(x; y)$ just in case $\text{Pred}(m, n)$ and Qxy. Frege goes on to define this relation for the general case:

> ... For the ... relation, which, in the way given above, is, as I say, *coupled* from the R-relation and the Q-relation, I introduce a simple sign, by defining:

$$\Sigma_{\xi\eta\zeta\tau}(R\xi\eta, Q\zeta\tau)(a, b) \overset{\text{df}}{\equiv} \exists x \exists y \exists z \exists w [a = (x; y)$$
$$\& \ b = (z; w) \ \& \ Rxz \ \& \ Qyw)].$$

If, then, we have a series x_0, x_1, etc., where $Qx_m x_{m+1}$, we have also that $\Sigma(\text{Pred}, Q)[(0; x_0), (1; x_1)]$, since $\text{Pred}(0, 1)$ and $Qx_0 x_1$; similarly, $\Sigma(\text{Pred}, Q)[(1; x_1), (2; x_2)]$. Note, however, that we also have $\Sigma(\text{Pred}, Q)[(0; x_{16}), (1; x_{17})]$, since $\text{Pred}(0, 1)$ and $Qx_{16} x_{17}$. To define the wanted relation, we therefore need to restrict attention to members of the *series* $(0; x_0)$, $(1; x_1)$, $(2; x_2)$, etc. As always, Frege employs the ancestral for this purpose:

> Accordingly,

$$\mathcal{F}^=[\Sigma(\text{Pred}, Q)][(0; x_0), (\xi; \eta)]$$

> indicates our mapping relation ... .

The relation in question is thus that in which ξ stands to η just in case the ordered pair $(\xi; \eta)$ belongs to the $\Sigma(\text{Pred}, Q)$-series beginning with $(0; x)$. It can indeed be proven that, under the hypotheses of Theorem 263, this relation maps the natural numbers into the members of the Q-series beginning with x and that its converse maps the latter into the former. Thus, in terms of ordered pairs and his definition of the coupling of two relations, Frege is able *explicitly to define* a relation which correlates the Gs one-to-one with the natural numbers.

III. FREGE'S USE OF ORDERED PAIRS

As said, Frege's proof of Theorem 263 can be carried out in Fregean arithmetic, if we add the ordered pair axiom to it. More interestingly, however, the proof can also be carried out in Fregean arithmetic itself.

Frege introduces ordered pairs for two reasons. First, he uses them to give his definition of the coupling of two relations. The use of ordered pairs is obviously inessential to this definition, which can instead be given as:[17]

$$(R\,\pi\,Q)(a, b; c, d) \overset{\text{df}}{\equiv} Rac \ \& \ Qbd.$$

Secondly, as we saw, the relation which is to correlate the natural numbers one-one with the Gs is defined by Frege as:

$$\mathcal{F}^=[\Sigma(\text{Pred}, Q)][(0; x), (\xi; \eta)].$$

Thus, Frege uses ordered pairs in order to be able to use the ancestral —which is the ancestral of a *two-place* relation—to define this new relation. Given our definition of $(R\,\pi\,Q)(\xi, \eta; \zeta, \tau)$, it is a *four-place relation*, so we can not apply Frege's definition of the ancestral.

As mentioned, Frege essentially uses nothing about ordered pairs in his proofs other than that they satisfy the ordered pair axiom: Indeed, much of his effort is devoted to *eliminating* reference to ordered pairs from certain of his theorems. A particularly nice example is an instance of induction for series determined by the couplings of relations. The definition of the strong ancestral yields the following, Frege's Theorem 123:

$$\mathcal{F}(Q)(a, b) \ \& \ \forall x \forall y (Fx \ \& \ Qxy \rightarrow Fy) \ \& \ \forall x (Qax \rightarrow Fx) \rightarrow Fb.$$

Taking $Q\xi\eta$ to be $\Sigma(R, Q)(\xi, \eta)$; a to be $(a; b)$; b to be $(c; d)$; we have:

$$\mathcal{F}[\Sigma(R, Q)][(a; b), (c; d)] \ \& \ \forall x \forall y (Fx \ \& \ \Sigma(R, Q)(x, y) \rightarrow Fy]$$
$$\& \ \forall x [\Sigma(R, Q)[(a; b), x] \rightarrow Fx] \rightarrow F[(c; d)].$$

---

[17] This is derived by Frege from his definition: From right-to-left, it is Theorem 208; from left-to-right, Theorem 223.

Here, the bound variables will range, for all intents and purposes, over ordered pairs, since the domain and range of $\Sigma(R, Q)(\xi, \eta)$ consist only of ordered pairs; $F\xi$, in turn, will be a concept under which ordered pairs fall. Let us define the concept $\mathrm{Col}_{\eta\zeta}(F\eta\zeta)(\xi)$—the collapse of $F\xi\eta$—to be that concept under which an ordered pair $(x; y)$ stands just in case Fxy. Formally:

$$\mathrm{Col}_{\alpha\varepsilon}(F\alpha\varepsilon)(a) \stackrel{\mathrm{df}}{\equiv} \exists x\exists y[a = (x; y) \ \& \ Fxy].$$

We then substitute $\mathrm{Col}(F)(\xi)$ for $F\xi$ to get (∗):

$$\mathcal{F}[\Sigma(R, Q)][(a; b), (c; d)]$$
$$\& \ \forall x\forall y(\mathrm{Col}(F)(x) \ \& \ \Sigma(R, Q)(x, y) \to \mathrm{Col}(F)(y)]$$
$$\& \ \forall x[\Sigma(R, Q)[(a; b), x] \to \mathrm{Col}(F)(x)] \to \mathrm{Col}(F)[(c; d)].$$

Consider, now, the third conjunct of (∗):

$$\forall x[\Sigma(R, Q)[(a; b), x] \to \mathrm{Col}(F)(x)].$$

What we wish to show is that this conjunct follows from:

$$\forall y\forall z[Ray \ \& \ Qbz \to Fyz].$$

We suppose that $\Sigma(R, Q)[(a; b), x]$ and must show that $\mathrm{Col}(F)(x)$. Since $\Sigma(R, Q)[(a; b), x]$, we have, by the definition of coupling:

$$\exists y\exists z\exists u\exists v[(a; b) = (u; v) \ \& \ x = (y; z) \ \& \ Ruy \ \& \ Qvz].$$

By the ordered pair axiom, $u = a$ and $v = b$, so by the laws of identity: $\exists y\exists z[x = (y; z) \ \& \ Ray \ \& \ Qbz]$. But, by hypothesis, if $Ray \ \& \ Qbz$, then Fyz; so: $\exists y\exists z[x = (y; z) \ \& \ Fyz]$. But, therefore, $\mathrm{Col}(F)(x)$, by definition.

Similarly, the second conjunct of (∗) follows from:

$$\forall x\forall y\forall z\forall w(Fxy \ \& \ Rxz \ \& \ Qyw \to Fzw).$$

Thus, applying these two results and the definition of '$\mathrm{Col}(F)(\xi)$' to (∗), we have:

$$\mathcal{F}[\Sigma(R, Q)][(a; b), (c; d)]$$
$$\& \ \forall x\forall y\forall z\forall w(Fxy \ \& \ Rxz \ \& \ Qyw \to Fzw)$$
$$\& \ \forall y\forall z[Ray \ \& \ Qbz \to Fyz] \to Fcd.$$

This is Frege's Theorem 231, and it is one of the forms of induction he uses in his proof of Theorem 263. Note that all reference to ordered pairs has been eliminated, except in the first conjunct, where it is needed for the application of the ancestral.

Frege is thus using ordered pairs to define a two-place relation, $\Sigma(R,Q)(\xi,\eta)$, from a four-place relation, $R\xi\eta$ & $Q\zeta\tau$, so that he can use the ancestral and the theorems proven about it (e.g., Theorem 123) to prove results about series determined by such relations. That is to say, just as Frege's definition of ordered pairs is used, essentially, only in the proof of the ordered pair axiom, what is important about ordered pairs is that one can use the ancestral to define *series* of ordered pairs; and what, in turn, is important about series of ordered pairs is that they satisfy theorems such as Theorem 231, from which reference to ordered pairs has been almost entirely eliminated. It therefore seems natural to abandon the use of ordered pairs entirely and to formulate a definition of the ancestral, for *four-place* relations, on the model of Frege's Theorem 231.

We thus define the *strong 2-ancestral* as follows:

$$\mathcal{F}_2(R\xi,\eta;\zeta,\tau)(a,b;c,d)$$
$$\overset{\mathrm{df}}{\equiv} \forall F[\forall x\forall y\forall z\forall w(Fxy \ \& \ Rxyzw \rightarrow Fzw)$$
$$\& \ \forall y\forall z[Rabxy \rightarrow Fxy] \rightarrow Fcd.$$

We similarly define the *weak* 2-ancestral as:

$$\mathcal{F}_2^=(R\xi,\eta;\zeta,\tau)(a,b;c,d) \overset{\mathrm{df}}{\equiv} \mathcal{F}_2(R)(a,b;c,d) \vee (a=c \ \& \ b=d).$$

These definitions are plainly analogous to Frege's definition of the ancestral for two-place relations.

As we shall see, Theorem 263 can be proven using this definition of the 2-ancestral and so without the use of ordered pairs. Of course, if we are to use the 2-ancestral to prove Theorem 263, we must prove the analogues of those theorems about the ancestral which Frege uses in his proof. As an example, consider Frege's Theorem 131:

$$Qad \rightarrow \mathcal{F}(Q)(a,d).$$

The analogue, for the 2-ancestral, we may call Theorem 131_2:

$$Qab;cd \rightarrow \mathcal{F}_2(Q)(a,b;c,d).$$

To prove this, we use Theorem 127_2, which is immediate from the definition of the 2-ancestral:

$$\forall F\{\forall x\forall y[Qab; xy \rightarrow Fxy]$$
$$\&\ \forall x\forall y\forall z\forall w[Fxy\ \&\ Qxy; zw \rightarrow Fzw] \rightarrow Fcd\}$$
$$\rightarrow \mathcal{F}_2(Q)(a, b; c, d).$$

For the proof of (131_2), suppose that $F\xi\eta$ is hereditary in the Q-series and that, if $Qab; xy$, then Fxy; we must show that Fcd. But, by hypothesis, $Qab; cd$; so Fcd. Done.

This proof of Theorem 131_2 simply mirrors Frege's proof of Theorem 131. Indeed, it will *always* be possible to prove analogues of Frege's theorems concerning the ancestral by following his proofs of the original theorems, making use of analogues whenever he makes use of a theorem about the ancestral. The reason for this is, of course, that the definition of the 2-ancestral is itself a precise analogue of his definition of the ancestral. To see exactly in what sense this is so, define the n-ancestral, on the same model, as follows:

$$\mathcal{F}_n(R\xi_1 \ldots \xi_n; \eta_1 \ldots \eta_n)(a_1 \ldots a_n; b_1 \ldots b_n)$$
$$\overset{df}{\equiv}\forall F\{\forall x_1 \ldots \forall x_n[R(a_1 \ldots a_n; x_1 \ldots x_n) \rightarrow Fx_1 \ldots x_n]$$
$$\&\ \forall x_1 \ldots \forall x_n\forall y_1 \ldots \forall y_n[R(x_1 \ldots x_n; y_1 \ldots y_n)$$
$$\&\ Fx_1 \ldots x_n \rightarrow Fy_1 \ldots y_n] \rightarrow Fb_1 \ldots b_n\}.$$

We now write '**x**' for '$x_1 \ldots x_n$' and '**x** = **y**' for '$x_1 = y_1\ \&\ \cdots\ \&\ x_n = y_n$'; the definition of the n-ancestral can then be written as:

$$\mathcal{F}_n(R\xi; \eta)(\mathbf{a}; \mathbf{b}) \overset{df}{\equiv} \forall F[\forall \mathbf{x}(R\mathbf{ax} \rightarrow F\mathbf{x})$$
$$\&\ \forall \mathbf{x}\forall \mathbf{y}(R\mathbf{xy}\ \&\ F\mathbf{x} \rightarrow F\mathbf{y}) \rightarrow F\mathbf{b}].$$

And, of course, we define the weak n-ancestral as:

$$\mathcal{F}_n^=(R\xi; \eta)(\mathbf{a}; \mathbf{b}) \overset{df}{\equiv} \mathcal{F}_n(R\xi; \eta)(\mathbf{a}; \mathbf{b}) \vee \mathbf{a} = \mathbf{b}.$$

To say that this definition schema was reminiscent of Frege's definition of the ancestral would be an understatement. It is because of the similarity which this familiar notational trick reveals that Frege's proofs of theorems concerning the ancestral can be immediately transformed into proofs of theorems concerning each of the various n-ancestrals.

Theorem 231, of course, is *not* an analogue of a theorem concerning the ordinary ancestral, since it makes reference to ordered pairs. Nevertheless, if we employ the definition of coupling introduced above which makes no use of ordered pairs, together with the 2-ancestral, we can formulate a precise analogue of Theorem 231 and prove it essentially as Frege proves Theorem 231 itself. The proof of the analogue shows it to be a nearly immediate consequence of Theorem 123_2:

$$\mathcal{F}_2(R\xi, \eta; \zeta, \tau)(a, b; c, d) \ \& \ \forall x \forall y \forall z \forall w(Fxy \ \& \ Rxyzw \to Fzw)$$
$$\& \ \forall y \forall z[Rabxy \to Fxy] \to Fcd.$$

If we take $R\xi, \eta; \zeta, \tau$ to be a coupling of two relations—that is, if we substitute '$(R\pi Q)(\xi, \eta; \zeta, \tau)$' for '$R\xi, \eta; \zeta, \tau$'—then we have:

$$\mathcal{F}_2[R\pi Q](a, b; c, d)$$
$$\& \ \forall x \forall y \forall z \forall w(Fxy \ \& \ (R\pi Q)(x, y; z, w) \to Fzw)$$
$$\& \ \forall y \forall z[(R\pi Q)(a, b; x, y) \to Fxy] \to Fcd.$$

As above, we have, by the definition of coupling:

$$\mathcal{F}_2[R\pi Q](a, b; c, d)$$
$$\& \ \forall x \forall y \forall z \forall w(Fxy \ \& \ Rxz \ \& \ Qyw \to Fzw)$$
$$\& \ \forall y \forall z[Rax \ \& \ Qby \to Fxy] \to Fcd.$$

This is the mentioned analogue of Frege's Theorem 231.

It seems to me that Frege almost certainly knew that his uses of ordered pairs were inessential to his proof. He uses a number of other forms of induction, for series determined by the couplings of relations. In each of these cases (e.g., Theorem 257), reference to ordered pairs is similarly eliminated from the theorem, except in terms in which the ancestral itself occurs. Surely, it would have been *obvious* to Frege that the use of ordered pairs was unnecessary, if not at the outset, then at least by reflection on the pattern which these theorems display, namely, the pattern used above to motivate the definition of the 2-ancestral. For this reason, I shall reproduce Frege's proof of Theorem 263 using the definition of the 2-ancestral, rather than ordered pairs. One might wonder, however, why, if Frege knew he could do without ordered pairs, he made use of them. One can only speculate about such a question, but the following two part answer has some plausibility: Firstly, ordered pairs were in wide use in mathematics, no suitable definition of them had at that time (1893) been given, and the provision of such a definition contributes to Frege's claim to be able to formalize, in the language of *Grundgesetze*, all of mathematics. Secondly, using ordered pairs in this context relieves Frege both of having to introduce a new definition of the ancestral and of having to prove a number of theorems which are simple analogues of ones he had already proven. More generally, the use of ordered pairs *unifies* the treatment of the 2-ancestral and the usual ancestral (not to mention the various other n-ancestrals). The use of ordered pairs thus has not insignificant advantages: If one can define them, why not use them?

We, however, are not in the same position Frege was. To use ordered pairs in the proof of Theorem 263, we should have to add the ordered pair axiom to Fregean arithmetic: So far as I know, ordered pairs are *not* definable in FA. Indeed, it is an interesting and open question whether FA plus the ordered pair axiom is even a conservative extension of FA.[18]

---

[18] An affirmative answer would imply the definability of addition for transfinite cardinals in FA, which is, for reasons I shall not discuss, equivalent to the following: $\exists F[Nx : Fx = Nx : \neg Fx \ \& \ Nx : Fx = Nx : x = x]$. That is: It is equivalent to the theorem that the domain can be partitioned into two equinumerous classes each of which is equinumerous with the whole domain. This is problematic only if we do not assume Choice, whence the domain can not be well-ordered.

It is, however, easy to prove this theorem in FA+OP: Let $F\xi$ be defined as $\exists x[\xi = (x; 0)]$; $G\xi$ as $\exists x[\xi = (x; 1)]$. Plainly, $F\xi$ and $G\xi$ are equinumerous with $\xi = \xi$ and so with each other. Since $\forall x(Gx \rightarrow \neg Fx)$, by the Schröder–Bernstein theorem (which is provable in second-order logic), $\neg F\xi$ is equinumerous with $\xi = \xi$.

IV. DEFINITION BY INDUCTION

We now return to the proof of Theorem 263. I shall discuss only parts of Frege's proof here, namely, those parts which are of some conceptual interest. Those parts of the proof which proceed primarily by brute force (i.e., by repeated applications of the ancestral) will be omitted.

Recall that Theorem 263 is:

$$\exists Q[\text{Func}(Q) \ \& \ \neg\exists x.\mathcal{F}(Q)(x,x)$$
$$\& \ \forall x(\neg\exists y.Qx,y \to \neg Gx) \ \& \ \exists x\forall y(Gy \equiv \mathcal{F}^=(Q)(x,y))]$$
$$\to Nx : Gx = \infty.$$

The theorem follows immediately from Theorem 262, which is:

$$\text{Func}(Q) \ \& \ \neg\exists x.\mathcal{F}(Q)(x,x) \ \& \ \forall x[\mathcal{F}^=(Q)(a,x) \to \exists y.Qxy]$$
$$\to Nx : \mathcal{F}^=(Q)(a,x) = \infty.$$

Two pieces of terminology. If $Q\xi\eta$ is functional and no object follows after itself in the Q-series, let us say that the Q-series is *simple*. If every member of the Q-series beginning with a stands in the Q-relation to some object, let us say that the Q-series beginning with a is *endless*. Theorem 262 thus says that, if the Q-series is simple and if the Q-series beginning with a is endless, then the number of members of the Q-series beginning with a is Endlos.—Theorem 263 then follows, since the number of Gs is certainly Endlos if the Gs are the members of the Q-series beginning with a.

Hence, the theorem follows from the definition of Endlos and Theorem 262, η:

$$\text{Func}(Q) \ \& \ \neg\exists x.\mathcal{F}(Q)(x,x) \ \& \ \forall x[\mathcal{F}^=(Q)(a,x) \to \exists y.Qxy]$$
$$\to Nx : \mathcal{F}^=(Q)(a,x) = Nx : \mathcal{F}^=(\text{Pred})(0,x).$$

This theorem is to be proven, as Frege indicates in the remarks quoted above, by showing that, if the antecedent holds, then there is a relation which maps the natural numbers into the members of the Q-series beginning with a and whose converse maps the latter concept into the former. Recall that $(R \pi Q)(\xi, \eta)$ is the coupling of the relations $R\xi\eta$ and $Q\xi\eta$, which is defined as:

$$(R \pi Q)(a, b; c, d) \overset{\text{df}}{\equiv} Rac \ \& \ Qbd.$$

The mapping relation is then to be:

$$\mathcal{F}_2^{=}[\text{Pred} \ \pi \ Q](0, a; \xi, \eta).$$

Because relations defined in this way are of such importance in this connection, Frege introduces an abbreviation, which we reproduce as (see *Gg* I § 144):

$$\mathcal{F}^\frown {}_{\alpha\varepsilon\gamma\delta}[R(\alpha, \varepsilon; \gamma, \delta); b, c](x, y) \overset{\text{df}}{\equiv} \mathcal{F}_2^{=}(R)(b, c; x, y).$$

Thus, an object x stands in the $\mathcal{F}^\frown(R; b, c)(\xi, \eta)$-relation to y just in case the pair of x and y follows after the pair of b and c in the R-series. The relation which correlates the natural numbers one-to-one with the members of the Q-series beginning with a is thus to be: $\mathcal{F}^\frown[(\text{Pred} \ \pi \ Q); 0, a](\xi, \eta)$.

We turn, then, to the proof of Theorem 262, η, which is, again:

$$\text{Func}(Q) \ \& \ \neg\exists x.\mathcal{F}(Q)(x, x) \ \& \ \forall x[\mathcal{F}^{=}(Q)(a, x) \to \exists y.Qxy]$$
$$\to Nx : \mathcal{F}^{=}(Q)(a, x) = Nx : \mathcal{F}^{=}(\text{Pred})(0, x).$$

Frege's proof of this theorem requires three lemmas, the first of which is Theorem 254:

$$\text{Func}(R) \ \& \ \text{Func}(Q) \ \& \ \neg\exists y(\mathcal{F}^{=}(R)(m, y) \ \& \ \mathcal{F}(R)(y, y))$$
$$\& \ \forall x[\mathcal{F}^{=}(Q)(a, x) \to \exists y.Qxy]$$
$$\to \text{Map}[\mathcal{F}^\frown(R \pi Q; m, a)](\mathcal{F}^{=}(R)(m, \xi), \mathcal{F}^{=}(Q)(a, \xi)).$$

Another piece of terminology. If $R\xi\eta$ is functional and no member of the R-series beginning with m follows after itself, then we say that the R-series *beginning with* m is simple. Thus, Theorem 254 says that, if the R-series beginning with m is simple, and if $Q\xi\eta$ is functional and the Q-series beginning with a is endless, then $\mathcal{F}^\frown(R \pi Q; m, a)(\xi, \eta)$ maps the members of the R-series beginning with m into the members of the Q-series beginning with a. The second lemma is Theorem 256:

$$\text{Func}(Q) \ \& \ \forall y (\mathcal{F}^{=}(Q)(a,y) \rightarrow \exists z. Qyz)$$
$$\rightarrow \text{Map}[\mathcal{F}^{\wedge}(\text{Pred } \pi \, Q; 0, a)](\mathcal{F}^{=}(\text{Pred})(0, \xi), \mathcal{F}^{=}(Q)(a, \xi)).$$

That is: If $Q\xi\eta$ is functional and the Q-series beginning with a is endless, then $\mathcal{F}^{\wedge}(\text{Pred } \pi \, Q; 0, a)(\xi, \eta)$ maps the natural numbers into the members of the Q-series beginning with a. The third, and last, lemma is Theorem 259:

$$\forall y \forall z \{ \mathcal{F}^{\wedge}(Q \pi R; a, m)(y, z) \equiv \text{Conv}[\mathcal{F}^{\wedge}(R \pi Q; m, a)](y, z) \},$$

i.e., $\mathcal{F}^{\wedge}(Q \pi R; a, m)(\xi, \eta)$ is the converse of $\mathcal{F}^{\wedge}(R \pi Q; m, a)(\xi, \eta)$.

To prove Theorem $262, \eta$, we assume the antecedent and show that $\mathcal{F}^{\wedge}[\text{Pred } \pi \, Q; 0, a](\xi, \eta)$ correlates the natural numbers one-to-one with the members of the Q-series beginning with a. Now, $Q\xi\eta$ is functional and the Q-series beginning with a is endless. So, $\mathcal{F}^{\wedge}(\text{Pred } \pi \, Q; 0, a)(\xi, \eta)$ maps the natural numbers into the members of the Q-series beginning with a, by Theorem 256. Moreover, if, in Theorem 254, we take $Q\xi\eta$ as $\text{Pred}(\xi, \eta)$, $R\xi\eta$ as $Q\xi\eta$, m as a, and a as 0, then we have:

$$\text{Func}(Q) \ \& \ \text{Func}(\text{Pred}) \ \& \ \neg\exists y (\mathcal{F}^{=}(Q)(a, y) \ \& \ \mathcal{F}(Q)(y, y))$$
$$\& \ \forall y (\mathcal{F}^{=}(\text{Pred})(0, y) \rightarrow \exists z. \text{Pred}(y, z))$$
$$\rightarrow \text{Map}[\mathcal{F}^{\wedge}(Q \pi \text{Pred}; a, 0)](\mathcal{F}^{=}(Q)(a, \xi), \mathcal{F}^{=}(\text{Pred})(0, \xi)).$$

The antecedent holds, since, by hypothesis, $Q\xi\eta$ is functional and no object follows after itself in the Q-series (a fortiori, no member of the Q-series beginning with a does); moreover, $\text{Pred}(\xi, \eta)$ is functional (Theorem 71) and every natural number is immediately succeeded by a natural number (Theorem 156); hence, $\mathcal{F}^{\wedge}(Q \pi \text{Pred}; a, 0)(\xi, \eta)$ maps the members of the Q-series beginning with a into the natural numbers. Since, by Theorem 259, this relation is the converse of the relation $\mathcal{F}^{\wedge}(\text{Pred } \pi \, Q; 0, a)(\xi, \eta)$, $\mathcal{F}^{\wedge}(\text{Pred } \pi \, Q; 0, a)(\xi, \eta)$ maps the natural numbers into the members of the Q-series beginning with a and its converse maps the latter into the former. Done.

Most of the interest of Frege's proof of Theorem 263 lies in these three lemmas. Recall that Theorem 256 is:

$$\text{Func}(Q) \ \& \ \forall x[\mathcal{F}^=(Q)(a, x) \to \exists y.Qxy]$$
$$\to \text{Map}[\mathcal{F}^\smallfrown(\text{Pred}\,\pi\,Q; 0, a)](\mathcal{F}^=(\text{Pred})(0, \xi), \mathcal{F}^=(Q)(a, \xi)).$$

It is this theorem which justifies the definition of a function(al relation), defined on the natural numbers, by induction. For, eliminating 'Map' via the definition, we have:

$$\text{Func}(Q) \ \& \ \forall x[\mathcal{F}^=(Q)(a, x) \to \exists y.Qxy]$$
$$\to \{\text{Func}(\mathcal{F}^\smallfrown(\text{Pred}\,\pi\,Q; 0, a)) \ \& \ \forall x[\mathcal{F}^=(\text{Pred})(0, x)$$
$$\to \exists y\{\mathcal{F}^=(Q)(a, y) \ \& \ \mathcal{F}^\smallfrown(\text{Pred}\,\pi\,Q; 0, a)(x, y)\}]\}.$$

If $Q\xi\eta$ is functional and the Q-series beginning with a is endless, then $\mathcal{F}^\smallfrown(\text{Pred}\,\pi\,Q; 0, a)(\xi, \eta)$ *is functional and every natural number is in its domain*. Moreover, it is not difficult to see that its range consists entirely of members of the Q-series beginning with a.[19]

Theorem 256 is thus a version of what is known as the recursion theorem for ω. The usual set-theoretic statement of this theorem is:

Let $g(\xi)$ be a function, $g : A \to A$; let $a \in A$. Then there is a unique function $\varphi : \mathbb{N} \to A$ such that $\varphi(0) = a$ and $\varphi(n + 1) = g(\varphi(n))$.

Assume the antecedent and define $Q\xi\eta \equiv [\eta = g(\xi)]$. $Q\xi\eta$ is then functional, since $g(\xi)$ is a function, and the Q-series beginning with a is endless, as can easily be seen. Thus, the antecedent of Theorem 256 is satisfied, so $\mathcal{F}^\smallfrown(\text{Pred}\,\pi\,Q; 0, a)(\xi, \eta)$ is functional and its domain contains all the natural numbers. Uniqueness is obvious.

It is not hard to see that this relation satisfies the recursion equations. We must show that $\mathcal{F}^\smallfrown(\text{Pred}\,\pi\,Q; 0, a)(0, a)$, that is, that $\varphi(0) = a$. And, we must show that $\mathcal{F}^\smallfrown(\text{Pred}\,\pi\,Q; 0, a)(n + 1, g(x))$, if $\mathcal{F}^\smallfrown(\text{Pred}\,\pi\,Q; 0, a)(n, x)$, that is, that $\varphi(n + 1) = g(\varphi(n))$. But, by definition, $\mathcal{F}^\smallfrown(\text{Pred}\,\pi\,Q; 0, a)(0, a)$ if, and only if, $\mathcal{F}^=(\text{Pred}\,\pi\,Q)(0, a; 0, a)$, which follows from Theorem 140_2.[20] Suppose, then, that

---

[19] See Theorem 232, to be mentioned below.

[20] For proofs of the various theorems concerning the 2- and n-ancestrals, see Appendix II.

$\mathcal{F}^\frown(\mathrm{Pred}\,\pi\,Q; 0, a)(n, x)$. Then, by definition, $\mathcal{F}^=(\mathrm{Pred}\,\pi\,Q)(0, a; n, x)$; furthermore, $\mathrm{Pred}(n, n+1)$ and $Qx, g(x)$, this last by the definition of $Q\xi\eta$; hence, $(\mathrm{Pred}\,\pi\,Q)(n, x; n+1, g(x))$, by the definition of coupling. We have, then, Theorem 137_2:

$$\mathcal{F}_2^=(R)(b, c; y, z) \ \& \ Ryz; uv \rightarrow \mathcal{F}_2^=(R)(b, c; u, v).$$

Since $\mathcal{F}^=(\mathrm{Pred}\,\pi\,Q)(0, a; n, x)$ and $(\mathrm{Pred}\,\pi\,Q)(n, x; n+1, g(x))$, we have that $\mathcal{F}^=(\mathrm{Pred}\,\pi\,Q)(0, a; n+1, g(x))$; hence, $\mathcal{F}^\frown(\mathrm{Pred}\,\pi\,Q; 0, a)$ $(n+1, g(x))$, by definition.

As an example, consider the recursion equations:[21]

$$\varphi(0) = a$$
$$\varphi(Sm) = S(\varphi(m)).$$

($S\xi$ here is the successor-function.) As above, we define $Q\xi\eta$ as $\eta = S\xi$; thus, $Q\xi\eta$ is just $\mathrm{Pred}(\xi, \eta)$. We have, then, by the above, $\mathcal{F}^\frown[\mathrm{Pred}\pi\mathrm{Pred}; 0, a](\xi; \eta)$ is functional and satisfies these equations. We may write them, in more familiar form, as:

$$a + 0 = a$$
$$a + Sm = S(a + m).$$

These are, of course, the standard recursion equations for addition. Thus, $\mathcal{F}^\frown[\mathrm{Pred}\,\pi\,\mathrm{Pred}; 0, a](\xi; \eta)$ holds just in case $\eta = a + \xi$; so

$$\mathcal{F}^\frown[(\mathrm{Pred}\,\pi\,\mathrm{Pred}); 0, \zeta](\xi; \eta)$$

defines $\eta = \zeta + \xi$, for natural numbers.[22]

---

[21]Note that not all functions intuitively given by induction can be defined in this way. Coupling can not be used to define functions when the recursion equations are of the form: $\varphi(0) = a$; $\varphi(n+1) = g(n, \varphi(n))$. However, let $T\xi\eta; \zeta\tau$ be defined as: $\mathrm{Pred}(\xi, \zeta) \ \& \ \tau = g(\xi, \eta)$. It is not hard to show that $\mathcal{F}^\frown[T; 0, a](\xi, \eta)$ is functional and satisfies the recursion equations.

[22]Frege was aware of this. See Gg II § 33: " 'The sum of two numbers is determined by them.' The thought of the proposition [in question here] is most easily recognized in this expression, and it may be cited for this purpose, although ... the word 'sum' here is used differently than we will later use it in connection with [real] numbers. Namely, we here call $[Nx : (Fx \vee Gx)]$ the sum of $[Nx : Fx]$ and $[Nx : Gx]$, if no object falls both under the $[F$-$]$ and under the $[G$-$]$concept. Also, infinite numbers hereby come into consideration. If we wanted to prove the proposition only for finite numbers, then another method would be more sensible." I presume he means this one.

Frege's proof of Theorem 256 can be carried out in Fregean arithmetic. The *general* theorem licensing the definition of a function by induction is, however, Theorem 254, from which Theorem 256 easily follows. Substituting 'Pred(ξ, η)' for '$R\xi\eta$' and '0' for 'm' in (254), we have:

$$\mathrm{Func}(\mathrm{Pred}) \ \& \ \mathrm{Func}(Q) \ \& \ \forall x[\mathcal{F}^=(Q)(a, x) \rightarrow \exists y. Qxy]$$
$$\& \ \neg \exists y(\mathcal{F}^=(\mathrm{Pred})(0, y) \ \& \ \mathcal{F}(\mathrm{Pred})(y, y))$$
$$\rightarrow \mathrm{Map}(\mathcal{F}^\frown(\mathrm{Pred} \,\pi\, Q; 0, x))(\mathcal{F}^=(\mathrm{Pred})(0, \xi), \mathcal{F}^=(Q)(x, \xi)).$$

But Pred(ξ, η) is provably functional, and no natural number follows after itself in the Pred-series, by Frege's Theorem 255 (which follows easily from Frege's Theorem 145). Theorem 256 follows immediately.

The validity of the definition of a function on the natural numbers, by induction, thus follows from Theorem 254, the functionality of Pred(ξ, η), and the fact that no finite number follows after itself in the Pred-series: Note that the *endlessness* (infinity) of the series of natural numbers is *not* needed for the proof of Theorem 256. Theorem 254 itself states that, if $R\xi\eta$ is functional and no member of the R-series beginning with m follows after itself in the R-series, then, if $Q\xi\eta$ is functional and the Q-series beginning with a is endless, $\mathcal{F}^\frown(R \,\pi\, Q; m, a)(\xi, \eta)$ is a functional relation whose domain contains the whole of the R-series beginning with m: Thus, Theorem 254 justifies the definition of a functional relation, by induction, on *any* "unbranching" series which does not "turn back on itself"; it is for this reason that I referred to it earlier as the general theorem licensing definition by induction. What is remarkable about Theorem 254 is that it is itself a theorem of *second-order logic*: The general theory of definition by induction can be developed entirely within second-order logic, and it was Frege who first showed that it can be (modulo his use of ordered pairs).

It is worth proving Theorem 262, η, and so Theorem 263, from the lemmas, in a slightly different way, to see just what Frege has done here. First, we exchange '$R\xi\eta$' with '$Q\xi\eta$', and 'm' with 'a,' in Theorem 254, yielding:

$$\mathrm{Func}(R) \ \& \ \mathrm{Func}(Q) \ \& \ \neg \exists y(\mathcal{F}^=(Q)(a, y) \ \& \ \mathcal{F}(Q)(y, y))$$
$$\& \ \forall y(\mathcal{F}^=(R)(m, y) \rightarrow \exists z. Ryz)$$
$$\rightarrow \mathrm{Map}[\mathcal{F}^\frown(Q \,\pi\, R; a, m)](\mathcal{F}^=(Q)(a, \xi), \mathcal{F}^=(R)(m, \xi)).$$

By Theorem 259, then:

$$\text{Func}(R) \ \& \ \text{Func}(Q) \ \& \ \neg\exists y(\mathcal{F}^=(Q)(a,y) \ \& \ \mathcal{F}(Q)(y,y))$$
$$\& \ \forall y(\mathcal{F}^=(R)(m,y) \to \exists z.Ryz)$$
$$\to \text{Map}[\text{Conv}[\mathcal{F}^\frown(R\,\pi\,Q; m, a)]](\mathcal{F}^=(Q)(a,\xi), \mathcal{F}^=(R)(m,\xi)).$$

Putting this together with Theorem 254, we have:

$$\text{Func}(Q) \ \& \ \neg\exists y(\mathcal{F}^=(Q)(a,y) \ \& \ \mathcal{F}(Q)(y,y))$$
$$\& \ \forall y(\mathcal{F}^=(Q)(a,y) \to \exists z.Qyz)$$
$$\& \ \text{Func}(R) \ \& \ \neg\exists y(\mathcal{F}^=(R)(m,y) \ \& \ \mathcal{F}(R)(y,y))$$
$$\& \ \forall y(\mathcal{F}^=(R)(m,y) \to \exists z.Ryz)$$
$$\to \text{Map}(\mathcal{F}^\frown(R\,\pi\,Q; m, a))(\mathcal{F}^=(R)(m,\xi), \mathcal{F}^=(Q)(a,\xi))$$
$$\& \ \text{Map}[\text{Conv}(\mathcal{F}^\frown(R\,\pi\,Q; m, a))](\mathcal{F}^=(Q)(a,\xi), \mathcal{F}^=(R)(m,\xi)).$$

This result, which we may call the *Isomorphism theorem*, thus says that, if the R-series beginning with m and the Q-series beginning with a both are simple and endless (or, *simply endless*), $\mathcal{F}^\frown(R\,\pi\,Q; m, a)(\xi, \eta)$ maps the former series into the latter and its converse maps the latter series into the former. Since, as it is not particularly difficult to see, these relations are order-preserving—where the ordering relations are the strong ancestrals of $R\xi\eta$ and $Q\xi\eta$—this theorem implies that all simply endless series are isomorphic.[23] Hence the name.

Theorem 262, η then follows by substituting 'Pred(ξ, η)' for '$R\xi\eta$,' and '0' for 'm,' and noting that the Pred-series beginning with 0 is simply endless. Frege thus in effect proceeds by *first* proving that all simply endless series are isomorphic and *then* concluding that, since the series of natural numbers is a simply endless series, every simply endless series is isomorphic to (and hence equinumerous with) it. The Isomorphism theorem is an immediate and trivial consequence of Frege's Theorems 254 and 259, and these two theorems are, as

---

[23] The fact that $\mathcal{F}^\frown(R\,\pi\,Q; m, a)(\xi, \eta)$ is an order-isomorphism is a relatively obvious consequence of the way it is defined. But see Appendix I for a proof, which is not given by Frege. The proof is easy, given certain theorems he does prove. It is perhaps worth mentioning that Dedekind does not explicitly prove this result either.

has been said, theorems of second-order logic. The Isomorphism theorem is thus *itself* a theorem of second-order logic, and Frege's proof in effect shows it so to be. Facts about the natural numbers, derived from Hume's principle, are used only in the derivation of Theorem 256 from Theorem 254 and of Theorem 263 from the Isomorphism theorem.

V. THE DEFINITION OF RELATIONS BY INDUCTION

We now turn to the proof of the central lemma in Frege's proof of Theorem 263, Theorem 254. (The proof of Theorem 259 is not difficult and will be omitted.) Recall that Theorem 254 is:

$$\text{Func}(R) \ \& \ \text{Func}(Q) \ \& \ \neg\exists y[\mathcal{F}^=(R)(m,y) \ \& \ \mathcal{F}(R)(y,y)]$$
$$\& \ \forall y(\mathcal{F}^=(Q)(a,y) \to \exists z.Qyz)$$
$$\to \text{Map}(\mathcal{F}^\frown(R\,\pi\,Q; m, a))(\mathcal{F}^=(R)(m,\xi), \mathcal{F}^=(Q)(a,\xi)).$$

Theorem 254 is derived from two lemmas, the first of which is Theorem 253:

$$\text{Func}(R) \ \& \ \text{Func}(Q) \ \& \ \neg\exists y[\mathcal{F}^=(R)(m,y) \ \& \ \mathcal{F}(R)(y,y)]$$
$$\to \text{Func}(\mathcal{F}^\frown[R\,\pi\,Q; m, a]).$$

That is: If $R\xi\eta$ and $Q\xi\eta$ are functional and if no member of the R-series beginning with m follows after itself in the R-series, then $\mathcal{F}^\frown[(R\,\pi\,Q); m, a](\xi, \eta)$ is functional. The second needed lemma is Theorem 241:

$$\text{Func}[\mathcal{F}^\frown(R\,\pi\,Q; m, a)] \ \& \ \forall y(\mathcal{F}^=(Q)(a,y) \to \exists z.Qyz)$$
$$\to \text{Map}(\mathcal{F}^\frown[R\,\pi\,Q; m, a])(\mathcal{F}^=(R)(m,\xi), \mathcal{F}^=(Q)(a,\xi)).$$

That is: If $\mathcal{F}^\frown(R\,\pi\,Q; m, a)(\xi, \eta)$ is functional and the Q-series beginning with a is endless, then $\mathcal{F}^\frown(R\,\pi\,Q; m, a)(\xi, \eta)$ maps the R-series beginning with m into the Q-series beginning with a. Theorem 254 is an easy consequence of these two lemmas.

Theorem 241 is an immediate consequence of the definition of 'Map' and Theorem 241, ζ:

$$\forall x[\mathcal{F}^=(Q)(a,x) \to \exists y.Qxy] \to \forall x\{\mathcal{F}^=(R)(m,x)$$
$$\to \exists y[\mathcal{F}^\frown(R\,\pi\,Q; m, a)(x,y) \ \& \ \mathcal{F}^=(Q)(a,y)]\}.$$

For the consequent states that, if x belongs to the R-series beginning with m, then x stands in the $\mathcal{F}^\frown(R\,\pi\,Q;m,a)(\xi,\eta)$-relation to some member of the Q-series beginning with a; hence, by definition, if $\mathcal{F}^\frown(R\,\pi\,Q;m,a)(\xi,\eta)$ is functional, it maps the R-series beginning with m into the Q-series beginning with a. Thus: If the Q-series beginning with a is endless, the *domain of* $\mathcal{F}^\frown(R\,\pi\,Q;m,a)(\xi,\eta)$ *contains the whole of the R-series beginning with m*. Now, Theorem 232 is:

$$\mathcal{F}(R\,\pi\,Q)(m,a;x,y) \rightarrow [\mathcal{F}(R)(m,x) \ \& \ \mathcal{F}(Q)(a,y)].$$

It follows easily from this theorem that all and only members of the R-series beginning with m are in the domain and that the range contains only members of the Q-series beginning with a. Thus, the proof of the Theorem 241, ζ amounts to a proof of the validity of the definition, not of a function, but of a *relation*, by induction, on the members of the R-series beginning with m, so long as the Q-series beginning with a does not end.[24] Frege thus derives the validity of the inductive definition of a *functional relation* from this general theorem about the validity of inductive definitions of relations by proving that, under certain conditions (namely, those mentioned in Theorem 253), the relation so defined will be functional.[25]

An example should help to explain the theorem. Let $Q\xi\eta$ relate m to n just in case n is a prime multiple of m; that is, define:

$$Qmn \stackrel{\text{df}}{\equiv} \exists p[\text{Prime}(p) \ \& \ n = pm].$$

Now, it is a theorem of arithmetic that, for every natural number $m \geqslant 1$, there is some n such that Qmn:

$$\mathcal{F}^=(\text{Pred})(1,m) \rightarrow \exists y.Qmy.$$

---

[24]For this reason, certain relations so defined are not defined inductively, in any natural sense, e.g., if the relation $R\xi\eta$ is dense. But let us ignore such complications.

[25]Thus, by substitution: $\forall x[\mathcal{F}^=(\text{Pred})(0,x) \rightarrow \exists y.\,\text{Pred}(x,y)] \rightarrow \forall x\{\mathcal{F}^=(R)$ $(m,x) \rightarrow \exists y[\mathcal{F}^\frown(R\,\pi\,\text{Pred};m,0)(x,y) \ \& \ \mathcal{F}^=(\text{Pred})(0,y)]$. Since the antecedent is a theorem of FA, we have, as a theorem of FA: $\forall x\{\mathcal{F}^=(R)(m,x)$ $\rightarrow \exists y[\mathcal{F}^\frown(R\,\pi\,\text{Pred};m,0)(x,y) \ \& \ \mathcal{F}^=(\text{Pred})(0,y)]$. That is: For *any* relation $R\xi\eta$ and *any* object m, $\mathcal{F}^\frown(R\,\pi\,\text{Pred};m,0)(\xi,\eta)$ is a relation whose domain is exactly the R-series beginning with m and whose range is contained in the natural numbers.

Hence, the Q-series beginning with 1 is endless. By Theorem 241, ζ, then:

$$\forall x\{\mathcal{F}^=(\text{Pred})(0, x) \to \exists y[\mathcal{F}^{\frown}(\text{Pred } \pi Q; 0, 1)(x, y) \ \& \ \mathcal{F}^=(Q)(1, y)]\}.$$

That is, $\mathcal{F}^{\frown}\{\text{Pred}(\xi, \eta) \ \pi \ \exists p[\text{Prime}(p) \ \& \ \tau = p\zeta]; 0, 1\}(\alpha, \beta)$ is a relation whose domain consists of all natural numbers and whose range is wholly contained in the Q-series beginning with 1. Moreover,

$$\mathcal{F}^{\frown}[(\text{Pred}(\xi, \eta) \ \pi \ \exists p(\text{Prime}(p) \ \& \ \tau = p\zeta)); 0, 1](m, x)$$

holds just in case x is a product of m (not necessarily distinct) primes, as can easily be seen.[26]

The proof of Theorem 241, ζ itself is of no great interest and will be omitted.

VI. FUNCTIONALITY AND THE n-ANCESTRAL

We turn then to the proof of Theorem 253, for which we need two lemmas. Recall that Theorem 253 is:

$$\text{Func}(R) \ \& \ \text{Func}(Q) \ \& \ \neg\exists y[\mathcal{F}^=(R)(m, y) \ \& \ \mathcal{F}(R)(y, y)]$$
$$\to \text{Func}(\mathcal{F}^{\frown}[R \pi Q; m, a]).$$

The first of the lemmas, Theorem 252, is essentially:

$$\text{Func}(R) \ \& \ \text{Func}(Q) \to \text{Func}(R \pi Q).$$

That is: If $R\xi\eta$ and $Q\xi\eta$ are functional, then $(R \pi Q)(\xi, \eta; \zeta, \tau)$ is functional. However, since the notion of *functionality* has so far been

---

[26]It is worth mentioning, too, that $\mathcal{F}^{\frown}(Q \pi \text{Pred}; 1, 0)(\xi, \eta)$—the converse of $\mathcal{F}^{\frown}(\text{Pred} \pi Q; 0, 1)(\xi, \eta)$—is also, by (241, η), a relation whose domain consists of all members of the Q-series beginning with 1 and whose range consists wholly of natural numbers, since every natural number has a successor. Note, too, that its range consists of *all* natural numbers and that $\mathcal{F}^{\frown}(Q \pi \text{Pred}; 1, 0)(\xi, \eta)$ is functional. Thus, $\mathcal{F}^{\frown}(Q \pi \text{Pred}; 1, 0)(\xi, \eta)$ is a counterexample to a converse of Theorem 248: $\mathcal{F}^{\frown}(Q \pi R; m, n)(\xi, \eta)$ may be a function, even if $(Q \pi R)(\xi, \eta; \zeta, \tau)$ is not a 2, 2-function. It is a nice question whether some general, informative theorem can be formulated concerning the conditions $Q\xi\eta$ and $R\xi\eta$ must satisfy if $\mathcal{F}^{\frown}(R \pi Q; m, a)(\xi, \eta)$ is to be functional.

defined only for two-place relations, the consequent is not yet well-formed. Now, for Frege, the coupling of two relations is a relation between ordered pairs, whence it is indeed a two-place relation: To say that $(R\,\pi\,Q)[\xi, \eta]$ is functional is then to say that:

$$\forall x \forall y \forall z [(R\,\pi\,Q)(x, y) \,\&\, (R\,\pi\,Q)(x, z) \to y = z].$$

Since 'x' and 'y' here range over ordered pairs, we may, as earlier, reduce this proposition to:

$$\forall x \forall y \forall z \forall w \forall u \forall v [(R\,\pi\,Q)[(x; y), (z; w)]$$
$$\&\ (R\,\pi\,Q)[(x; y), (u; v)] \to z = u \,\&\, w = v].$$

We therefore define:

$$\text{Func}_{2,2}(R\xi, \eta; \zeta, \tau) \overset{\text{df}}{\equiv} \forall x \forall y \forall z \forall u \forall v [Rx, y; z, w$$
$$\&\ Rx, y; u, v \to z = u \,\&\, w = v].$$

That is: A four-place relation $R\xi, \eta; \zeta, \tau$ is $2, 2$-functional if, and only if, whenever the pair of x and y stands in it to the pair of z and w and to the pair of u and v, the pair of z and w is the pair of u and v. We may then write Theorem 252 as:

$$\text{Func}(R) \,\&\, \text{Func}(Q) \to \text{Func}_{2,2}(R\,\pi\,Q).$$

The second needed lemma is Theorem 248:

$$\text{Func}_{2,2}(R\,\pi\,Q) \,\&\, \neg\exists x[\mathcal{F}^=(R)(m, x) \,\&\, \mathcal{F}^=(R)(x, x)]$$
$$\to \text{Func}(\mathcal{F}^\frown[R\,\pi\,Q; m, a]).$$

That is: If $(R\,\pi\,Q)(\xi, \eta; \zeta, \tau)$ is $2, 2$-functional and no member of the R-series beginning with m follows after itself in the R-series, then $\mathcal{F}^\frown[(R\,\pi\,Q); m, a](\xi, \eta)$ is functional. Theorem 253 is an immediate consequence of these two lemmas.

The proof of Theorem 252 is straightforward.[27]

[27]Viz.: Suppose that $R\xi\eta$ and $Q\xi\eta$ are functional; suppose that $(R\,\pi\,Q)(x, y; z, w)$ and $(R\,\pi\,Q)(x, y; u, v)$. Then, by the definition of coupling, Rxz and Rxu, and Qyw and Qyv. Since $R\xi\eta$ and $Q\xi\eta$ are functions, $z = u$ and $w = v$. QED.

Frege's own proof consists in the application of the definition of 'Func' to yield: $\text{Func}(R) \,\&\, \text{Func}(Q) \,\&\, Rco \,\&\, Rce \,\&\, Qda \,\&\, Qdi \to o = e \,\&\, a = i$. He then applies the ordered pair axiom (Theorem 251) to get: $\text{Func}(R) \,\&\, \text{Func}(Q) \,\&\, Rco \,\&\, Rce \,\&\, Qda \,\&\, Qdi \to (o; a) = (e; i)$. This is Frege's Theorem 251, α. The remainder of the proof consists in the introduction of free variables "A," "D," and "E" for the ordered pairs, which turns out to be exceedingly tedious.

The proof of Theorem 248, in outline, is as follows. By definition, if $\mathcal{F}^\frown[R\,\pi\,Q; m, a](\xi, \eta)$ is not functional, then, for some b, c, and d, we must have $\mathcal{F}^\frown[R\,\pi\,Q; m, a](b, c)$ and $\mathcal{F}^\frown[R\,\pi\,Q; m, a](b, d)$, where $c \neq d$. By the definition of '$\mathcal{F}^\frown$,' then, $\mathcal{F}^=(R\,\pi\,Q)(m, a; b, c)$ and $\mathcal{F}^=(R\,\pi\,Q)(m, a; b, d)$. Now, intuitively, during the inductive 'process' of assigning members of the R-series beginning with m their 'values,' b must have been assigned a value *twice*. For, at each step in the process, exactly one member of the R-series beginning with m is assigned exactly one value (since $(R\,\pi\,Q)(\xi, \eta; \zeta, \tau)$ is $2, 2$-functional); moreover, the process of assigning values proceeds from one member of the R-series beginning with m to a unique next member of the R-series beginning with m (since $R\xi\eta$ is functional). So, if b is to get two values, it must occur twice in this series. But that will imply that b follows after itself in the series, contradicting the supposition that no member of the R-series beginning with m follows after itself.

The central lemma in Frege's proof of Theorem 248 is the famous Theorem 133 from Frege's *Begriffsschrift*, here appearing as Theorem 243:[28]

$$\text{Func}(R) \ \& \ \mathcal{F}^=(R)(m, t) \ \& \ \mathcal{F}^=(R)(m, n)$$
$$\rightarrow [\mathcal{F}(R)(n, t) \vee \mathcal{F}^=(R)(t, n)].$$

That is: If $R\xi\eta$ is functional and both t and n belong to the R-series beginning with m, then either n belongs to the R-series beginning with t or t follows after n in the R-series. It is this theorem which will imply that any object which "occurs twice" in the R-series beginning with m follows after itself in the R-series.

---

[28] *Begriffsschrift, eine der arithmetischen nachgebildete Formelsprache des reinen Denkens*, first published in 1879, reprinted in *Begriffsschrift und andere Aufsätze*, Ignacio Angelelli, ed., New York: Georg Olms, 1988. For an English translation, see *From Frege to Gödel: a sourcebook in mathematical logic*, Jean van Heijenoort, ed., Cambridge: Harvard University Press, 1970, Stefan Bauer-Mengelberg, tr.

One might wonder if the fact that this theorem is *not used* in the proofs of the axioms of arithmetic does not indicate something about how many of the proofs of the theorems of *Grundgesetze* Frege had in 1879. But we are unlikely, at this point, to be able to establish anything very definitive about the date of Frege's discovery of Theorem 263.

What we need is in fact not Theorem 243, but its analogue for the 2-ancestral, Theorem 243_2:

$$\text{Func}_{2,2}(R) \;\&\; \mathcal{F}^=(R)(m,a;n,y) \;\&\; \mathcal{F}^=(R)(m,a;s,t)$$
$$\to [\mathcal{F}(R)(n,y;s,t) \vee \mathcal{F}^=(R)(s,t;n,y)].$$

We need two other lemmas:

(244) $\qquad \mathcal{F}(R\,\pi\,Q)(x,y;z,w) \to \mathcal{F}(R)(x,z)$

(246) $\qquad \mathcal{F}^=(R\,\pi\,Q)(m,a;x,y) \to \mathcal{F}^=(R)(m,x).$

The derivation of Theorem 248 from these lemmas provides a nice illustration of the power of the 2-ancestral. Suppose that $(R\,\pi\,Q)(\xi,\eta; \zeta,\tau)$ is $2,2$-functional, that no member of the R-series beginning with m follows itself, and (as above) that $\mathcal{F}^=(R\,\pi\,Q)(m,a;x,y)$ and $\mathcal{F}^=(R\,\pi\,Q)(m,a;x,z)$. Note that, by Theorem 246, $\mathcal{F}^=(R)(m,x)$. By Theorem 243_2, one of the following three cases must hold:

(i) $\mathcal{F}(R\,\pi\,Q)(x,y;x,z).$

(ii) $\mathcal{F}(R\,\pi\,Q)(x,z;x,y).$

(iii) $x = x \;\&\; y = z.$

But, if either (i) or (ii), then $\mathcal{F}(R)(x,x)$, by Theorem 244, contradicting the assumption that the R-series beginning with m is simple. Hence, (iii) must hold, $y = z$, and $\mathcal{F}^\frown(R\,\pi\,Q;m,a)(\xi,\eta)$ is functional.

The proofs of the lemmas are not difficult and, except for that of Theorem 243_2 (which is in Appendix II), will be omitted.

VII. THEOREM 263 IN THE CONTEXT OF FREGE'S DEVELOPMENT OF ARITHMETIC

We have thus completed our discussion of Frege's proof of Theorem 263. As we have seen, the proof can be reconstructed in Fregean arithmetic, with or without the use of the ordered pair axiom. Moreover, Frege's proofs of the crucial Theorems 254 and 259, of which the Isomorphism theorem is an immediate consequence, can be reconstructed in second-order logic, either with or without the use of the ordered pair axiom. Since Frege was perfectly aware that his uses of Axiom V in such proofs are dispensable,[29] Frege did, in effect, prove Theorem 263 in Fregean arithmetic, augmented by the ordered pair axiom, and the Isomorphism theorem in second-order logic, augmented by the ordered pair axiom. I have suggested, further, that Frege knew his use of ordered pairs to be dispensable and so knew, and effectively showed, these also to be theorems of second-order arithmetic and logic simpliciter.

The significance of Theorem 263, in the context of Frege's project, is far from clear, however; Frege does not say very much about it. In the Introduction, for example, he writes:

> It might be thought that the propositions concerning the number 'Endlos' could have been omitted; to be sure, they are not necessary for the foundation of arithmetic in its traditional compass, but their derivation is for the most part simpler than that of the corresponding propositions for finite numbers and can serve as preparation for it (Gg p. v).

But, surely, the theorem is of greater significance than that: *Grundgesetze* is not just a random collection of results, and it is implausible in the extreme that Frege should have included the proof of Theorem 263 in what was intended to be his *magnum opus* just because it is similar to, and simpler than, the proof of a corresponding result for finite numbers.[30]

Theorem 263, together with Frege's Theorem 207, does yield a characterization of concepts whose number is Endlos. Theorem 207 is just the converse of Theorem 263:

$$Nx : Gx = \infty \rightarrow \exists Q\{\text{Func}(Q) \ \& \ \neg\exists x.\mathcal{F}(Q)(x,x)$$
$$\& \ \forall x[Gx \rightarrow \exists y.Qxy] \ \& \ \exists x\forall y[Fy \equiv \mathcal{F}^=(Q)(x,y)]\}.$$

---

[29] See again "The development of arithmetic."

[30] The result in question is Theorem 327, which states that the number of a concept $F\xi$ is *finite* if: $\exists Q\exists x\exists y\{\text{Func}(Q) \ \& \ \neg\mathcal{F}(Q)(y,y) \ \& \ \forall z[Fz \equiv \mathcal{F}^=(Q)(x,z) \ \& \ \mathcal{F}^=(Q)(z,y)]\}$. Indeed, Frege defines: $\text{Btw}(Q\xi\eta; a, b)(x) \equiv_{\text{df}} \text{Func}(Q) \ \& \ \neg\mathcal{F}(Q)(b,b) \ \& \ \mathcal{F}^=(Q)(a,x) \ \& \ \mathcal{F}^=(Q)(x,b)$. Frege reads "$\text{Btw}(Q; a, b)(x)$" as "$x$ belongs to the Q-series running from a to b," but I write it as I do because it seems more intuitive to read it as "x is between a and b in the Q-series" or just "x is Q-between a and b." Theorem 327 is then: $\exists Q\exists x\exists y\exists z[Fz \equiv \text{Btw}(Q; x, y)(z)] \rightarrow \mathcal{F}^=(\text{Pred})(0, Nx : Fx)$. The converse of Theorem 327 is Theorem 348.

Thus, the number of Gs is Endlos just in case the Gs can be ordered as a simply endless series beginning, say, with an object a. It follows from Frege's Theorem 243 that the Q-series beginning with a is linearly ordered by $\mathcal{F}(Q)(\xi, \eta)$; moreover, Theorem 359, which is a generalization of the least number principle, implies that the Q-series beginning with a is *well-ordered* by $\mathcal{F}(Q)(\xi, \eta)$. And so, Theorems 207, 263, 243, and 359 together imply that the number of Gs is Endlos if and only if the Gs can be ordered as an ω-sequence. Presumably, *part* of Frege's reason for proving Theorem 263 is that he intends so to characterize countably infinite concepts.[31]

Further, Theorem 263 plays an important role in the proof of Theorem 428:

$$Nx : Gx = \infty \ \& \ \forall x(Fx \to Gx)$$
$$\to [Nx : Fx = \infty \lor \mathcal{F}^=(\text{Pred})(0, Nx : Fx)].$$

That is: If the number of Gs is Endlos and every F is a G, then, either the number of Fs is Endlos or the number of Fs is finite; i.e., every concept subordinate to a concept whose number is countably infinite is countable. This result is of some importance, but its significance is not immediately apparent either.[32]

The real point of Theorem 263, it seems to me, is not revealed by examination of its use in *Grundgesetze*. A modern reader would naturally take the theorem—or, more generally, the Isomorphism theorem—to show that any two structures satisfying certain condi-

---

[31] For further discussion of Frege's interest in finitude and infinity, see my "The finite and the infinite in Frege's *Grundgesetze der Arithmetik*," forthcoming in M. Schirn, ed., *Philosophy of mathematics today*, Oxford: Oxford University Press.

[32] Not that this is what Frege had in mind, but this result is the central theorem needed in the proof that an *inner model* for Fregean arithmetic is definable in Fregean arithmetic. The first-order domain consists of the natural numbers, together with Endlos. This result implies that the relativization of Hume's principle to that domain is a theorem of Fregean arithmetic, and it is easy to see that the restriction of each instance of comprehension is also a theorem.

tions are isomorphic. Recall that the Isomorphism theorem is:

$$\text{Func}(Q) \ \& \ \neg\exists y(\mathcal{F}^=(Q)(a,x) \ \& \ \mathcal{F}(Q)(x,x))$$
$$\& \ \forall x(\mathcal{F}^=(Q)(a,x) \to \exists y.Qxy)$$
$$\& \ \text{Func}(R) \ \& \ \neg\exists y(\mathcal{F}^=(R)(m,x) \ \& \ \mathcal{F}(R)(x,x))$$
$$\& \ \forall x(\mathcal{F}^=(R)(m,x) \to \exists y.Rxy)$$
$$\to \text{Map}(\mathcal{F}^\frown(R\,\pi\,Q;m,a))(\mathcal{F}^=(R)(m,\xi),\mathcal{F}^=(Q)(a,\xi))$$
$$\& \ \text{Map}[\text{Conv}(\mathcal{F}^\frown(R\,\pi\,Q;m,a))](\mathcal{F}^=(Q)(a,\xi),\mathcal{F}^=(R)(m,\xi)).$$

The conditions in question are thus:

(1) $\text{Func}(P)$.
(2) $\neg\exists x[\mathcal{F}^=(P)(a,x) \ \& \ \mathcal{F}(P)(x,x)]$.
(3) $\forall x[\mathcal{F}^=(P)(a,x) \to \exists y.Pxy]$.

Earlier, we called a series whose determining relation satisfies these conditions a simply endless series and said that the Isomorphism theorem shows that all simply endless series are isomorphic. Better to understand the significance of this theorem, let us write the conditions slightly differently. We write '0' for 'a' and introduce a concept $N\xi$, as in Theorem 263:

(1) $\text{Func}(P)$
(2) $\neg\exists x[Nx \ \& \ \mathcal{F}(P)(x,x)]$.
(3) $\forall x[Nx \to \exists y.Pxy]$.
(4) $Nx \equiv \mathcal{F}^=(P)(0,x)$.

Conditions (1)-(4) are axioms for arithmetic: The more familiar Dedekind-Peano axioms are easily derived from them (and conversely).[33] What the proof of Theorem 263 shows is thus that any

---

[33]For, if some natural number precedes zero, then $Px0$ and $\mathcal{F}^=(P)(0,x)$, whence $\mathcal{F}^=(P)(0,0)$. And, if Pxz and Pyz, for *natural numbers* x, y, z, then, since $\mathcal{F}^=(P)(0,x)$, $\mathcal{F}^=(P)(0,y)$, and $P\xi\eta$ is functional, we have, by Theorem 243, that either $\mathcal{F}(P)(x,y)$ or $\mathcal{F}(P)(y,x)$, or $x=y$. But, if the former, since Pxz, by Theorem 242, $\mathcal{F}^=(P)(z,y)$; and, since Pyz, $\mathcal{F}(P)(y,y)$. Similarly, if $\mathcal{F}(P)(y,x)$, then $\mathcal{F}(P)(x,x)$. Hence $x=y$.

Conversely, the Dedekind-Peano Axioms imply these. The only one we must prove is (2). We proceed by induction. If $\mathcal{F}(P)(0,0)$, then, by Frege's Theorem 124, for some x, $Px0$. Suppose $\neg\mathcal{F}(P)(x,x)$, x a natural number, and Pxy. Suppose $\mathcal{F}(P)(y,y)$. Then, we have the following strengthening of Theorem 124, Theorem 141: $\mathcal{F}(Q)(a,b) \to \exists x[Qxb \ \& \ \mathcal{F}^=(Q)(a,x)]$. Hence, for some z, Pzy and $\mathcal{F}^=(P)(y,z)$. But $P\xi\eta$ is one-one, so $z=x$; so $\mathcal{F}^=(P)(y,x)$; but Pxy, so $\mathcal{F}(P)(x,x)$. Contradiction.

two structures satisfying *Frege's* axioms for arithmetic are isomorphic.

As was said, that Frege so intended the proof can not be shown by quoting him: He is not very good about explaining the significance of the theorems he proves. There is nevertheless reason to think that he did so intend it: Namely, the hypothesis that the proof is to show that any structures satisfying these axioms for arithmetic are isomorphic resolves an annoying puzzle about the structure of Part II of *Grundgesetze*. What is puzzling is the status accorded to the proof that no natural number follows itself in the Pred-series: This theorem is granted its very own section, advertised as a major result. The modern reader, with the Dedekind-Peano axioms firmly in mind, naturally reads it as but an important lemma in the proof that every natural number has a successor. But Frege accords it pride of place: For it is among *his* axioms for arithmetic.

It is worth noting, too, that Frege's axioms are extremely intuitive. The natural numbers are the members of a series which begins with the number zero (Axiom (4), induction); each number in this series is followed by one (Axiom (3), that every natural number has a successor) and exactly one number (Axiom (1), that $\text{Pred}(\xi, \eta)$ is functional). And, so to speak, each number in the series is followed by a *new* number, by one which has not previously occurred in the series: That is, the successor of a natural number never precedes it in the number-series. Formally:

$$\neg \exists x [Nx \ \& \ P(x,y) \ \& \ \mathcal{F}(P)(y,x)].$$

As is easily seen, however, in the presence of the other axioms, this is equivalent to Frege's Axiom (2), that no natural number follows after itself in the number-series. It is, of course, essential that Frege's axioms should be intuitive and connected with our ordinary applications of arithmetic: For Frege intends to show that *arithmetic* is a branch of logic, not that some formal theory which looks something like arithmetic can be developed within something which looks like logic.

The import of the second axiom is, indeed, explained in detail by Frege, in a passage which is worth quoting in full:

> The proposition ... says that no object which belongs to the number-series beginning with zero follows after itself in the

number-series. Instead of this, we can also say: "No *finite* number follows after itself in the number-series." The importance of this proposition will be made more evident by the following considerations. If we determine the number belonging to a concept $\Phi(\xi)$, or, as one normally says, if we count the objects falling under the concept $\Phi(\xi)$, then we successively associate these objects with the number-words from "one" up to a number-word "N," which will be determined by the associating relation's mapping the concept $\Phi(\xi)$ into the concept "member of the series of number-words from 'one' to 'N'" and the converse relation's mapping the latter concept into the former. "N" then denotes the sought number; i.e., N is this number. This process of counting may be carried out in various ways, since the associating relation is not completely determined.

The question arises whether, by another choice of this relation, one could reach another number-word "M." Then, by our assumptions, M would be the same number as N, but, at the same time, one of these two number-words would follow after the other, e.g., "N" would follow "M." Then N would follow in the number-series after M, which means that it would follow after itself. That is excluded by our proposition concerning finite numbers (*Gg* I § 108).

That is to say: That the result of the process of counting—the number-word one reaches by counting—is well-determined is, according to Frege, a consequence of the fact that no natural number follows after itself. And so, this fact about the natural numbers has a special, and central, role to play in helping us to understand, or in providing a justification for, our ordinary employment of arithmetic.

One might wonder, however, whether the Isomorphism theorem can possibly have the significance I have suggested it has. After all, one might say, Frege does not even write it down.—It should, however, be remembered that *Grundgesetze* is a *mathematical* work and that Frege surely would have assumed that his readers could make certain inferences for themselves: The Isomorphism theorem is, as we have seen, an utterly trivial consequence of Theorems 254 and 259 which no careful reader can miss. Indeed, in Frege's proof of Theorem 263, Theorem 259 is not needed in the general form in which Frege proves it; all he needs is that $\mathcal{F}^\frown(Q\,\pi\,\text{Pred}; a, 0)(\xi, \eta)$ is the converse of $\mathcal{F}^\frown(\text{Pred}\,\pi\,Q; 0, a)(\xi, \eta)$. The fact that he *does* prove Theorem 259 in its more general form indicates that he intends the reader to realize that a proof like that of Theorem 263 could be given of a more general result, namely, the Isomorphism theorem.

Besides, one should not put too much weight on the distinction between what Frege writes down and what he leaves for the reader to infer. Frege never explicitly records *mathematical induction* as a theorem either, and he can hardly be said to have been oblivious to its import.

VIII. CLOSING

If the foregoing account of Frege's proof of Theorem 263 and its significance is correct, then the evaluation of his work in *Grundgesetze* must change. For the most part, *Grundgesetze* has been ignored: When it has not been ignored, it has usually been dismissed, as if that were more charitable than the alternative, which would be ridicule. As we have seen, however, not only does *Grundgesetze* contain a derivation of axioms for arithmetic, in second-order logic, from Hume's principle, it contains a proof, in Fregean arithmetic, that *Frege's own* axioms for arithmetic determine a class of structures isomorphic to the natural numbers. Moreover, it contains a proof, in pure second-order logic, of the more general fact that all structures satisfying these axioms are isomorphic.

Now, I do not want here to compare the relative merits of Frege's and Dedekind's proofs of these theorems, and I am surely not going to suggest that Frege's work in these areas had very much impact on the working mathematicians of his day. However, the recent tendency, among some philosophers, to dismiss Frege's work as largely irrelevant simply can not be sustained, unless one is prepared, as I for one am not, to dismiss *Was sind?* in the same breath.[34]

---

[34]I should like to thank George Boolos, Michael Dummett, Mathieu Marion, Charles Parsons, and Jason Stanley for their advice and criticism.

APPENDIX I: PROOF THAT

$\mathcal{F}^\frown(R\,\pi\,Q; m, a)(\xi, \eta)$ IS ORDER-PRESERVING

We prove that, if the R-series beginning with m and the Q-series beginning with a are simply endless, then $\mathcal{F}^\frown(R\,\pi\,Q; m, a)(\xi, \eta)$ is order-preserving. The ordering relations which $\mathcal{F}^\frown(R\,\pi\,Q; m, a)(\xi, \eta)$ respects are the weak ancestrals of $R\xi\eta$ and $Q\xi\eta$. To show that $\mathcal{F}^\frown(R\,\pi\,Q; m, a)(\xi, \eta)$ and its converse are order-preserving, it is sufficient, in light of Theorem 259, to show that:

$$\mathcal{F}^\frown(R\,\pi\,Q; m, a)(b, x) \ \ \& \ \ \mathcal{F}^\frown(R\,\pi\,Q; m, a)(c, y)$$
$$\rightarrow [\mathcal{F}(R)(b, c) \equiv \mathcal{F}(Q)(x, y)].$$

So suppose that $\mathcal{F}^\frown(R\,\pi\,Q; m, a)(b, x)$ and $\mathcal{F}^\frown(R\,\pi\,Q; m, a)(c, y)$. By definition, $\mathcal{F}^=(R\,\pi\,Q)(m, a; b, x)$ and $\mathcal{F}^=(R\,\pi\,Q)(m, a; c, y)$. Hence, by Theorem 243: (i) $\mathcal{F}(R\,\pi\,Q)(b, x; c, y)$; (ii) $\mathcal{F}(R\,\pi\,Q)(c, y; b, x)$; or, (iii) $b = c$ & $x = y$. Note that $\mathcal{F}^=(R)(m, b)$, by Theorem 232. Now, suppose $\mathcal{F}(R)(b, c)$; if (iii), then $\mathcal{F}(R)(b, b)$, contradicting the fact that the R-series beginning with m is simple. If (ii), then, by Theorem 232, $\mathcal{F}(R)(c, b)$, hence $\mathcal{F}(R)(b, b)$, by the transitivity of the ancestral, again contradicting the simplicity of the R-series beginning with m. Hence, (i) must hold: So, by Theorem 232, again, $\mathcal{F}(Q)(x, y)$.

By a parallel argument, if $\mathcal{F}(Q)(x, y)$, then (i) must hold; so $\mathcal{F}(R)(b, c)$, by Theorem 232. $\square$

APPENDIX II: THEOREMS CONCERNING

THE n-ANCESTRAL, AND PROOFS OF THEM

123_n : $\mathcal{F}_n(R)(\mathbf{a}, \mathbf{b})$ & $\forall \mathbf{x}(R a \mathbf{x} \rightarrow F \mathbf{x})$ & $\forall \mathbf{x} \forall \mathbf{y}(F \mathbf{x}$ & $R \mathbf{xy} \rightarrow F \mathbf{y}) \rightarrow F \mathbf{b}$.

127_n : $\forall F[\forall \mathbf{x}(R a \mathbf{x} \rightarrow F \mathbf{x})$ & $\forall \mathbf{x} \forall \mathbf{y}(F \mathbf{x}$ & $R \mathbf{xy} \rightarrow F \mathbf{y}) \rightarrow F \mathbf{b}] \rightarrow \mathcal{F}_n(R)(\mathbf{a}, \mathbf{b})$.

Proof. Both by definition. $\square$

124_n : $\mathcal{F}_n(R)(\mathbf{a}, \mathbf{b}) \rightarrow \exists \mathbf{x}.R \mathbf{xb}$.

Proof. (123_n), with $\exists \mathbf{x}.R \mathbf{x}\xi$. If $R a \mathbf{y}$, then, plainly, $\exists \mathbf{x}.R \mathbf{xy}$. Suppose $\exists \mathbf{x}.R \mathbf{xy}$ and $R \mathbf{yz}$. Obviously, $\exists \mathbf{x}.R \mathbf{xz}$. $\square$

128_n :$\mathcal{F}_n(R)(\mathbf{a}, \mathbf{b})$ & $F\mathbf{a}$ & $\forall \mathbf{x}\forall \mathbf{y}(F\mathbf{x}$ & $R\mathbf{xy} \to F\mathbf{y}) \to F\mathbf{b}$.

Proof. If $\forall \mathbf{x}\forall \mathbf{y}(F\mathbf{x}$ & $R\mathbf{xy} \to F\mathbf{y})$, then $\forall \mathbf{y}(F\mathbf{a}$ & $R\mathbf{ay} \to F\mathbf{y})$; hence, if $F\mathbf{a}$, then $\forall \mathbf{y}(R\mathbf{ay} \to F\mathbf{y})$. Apply (123_n). □

129_n :$\mathcal{F}_n(R)(\mathbf{a}, \mathbf{b})$ & $R\mathbf{ca} \to \mathcal{F}_n(R)(\mathbf{c}, \mathbf{b})$.

Proof. To apply (127_n), suppose $F\xi$ is hereditary and that $\forall \mathbf{x}(R\mathbf{cx} \to F\mathbf{x})$. Since $R\mathbf{ca}$, $F\mathbf{a}$. By (128_n), $F\mathbf{b}$. □

131_n :$R\mathbf{ab} \to \mathcal{F}_n(R)(\mathbf{a}, \mathbf{b})$.

Proof. Suppose $F\xi$ is hereditary and that $\forall \mathbf{x}(R\mathbf{ax} \to F\mathbf{x})$. Since $R\mathbf{ab}$, $F\mathbf{b}$. Apply (127_n). □

132_n :$\mathcal{F}_n^{=}(R)(\mathbf{a}, \mathbf{b})$ & $R\mathbf{ca} \to \mathcal{F}_n(R)(\mathbf{c}, \mathbf{b})$.

Proof. Suppose $\mathbf{a} = \mathbf{b}$ and $R\mathbf{ca}$. Then $R\mathbf{cb}$, so $\mathcal{F}_n(R)(\mathbf{c}, \mathbf{b})$, by (131_n). Hence, with (129_n):

$$[\mathbf{a} = \mathbf{b} \vee \mathcal{F}_n(R)(\mathbf{a}, \mathbf{b})] \;\&\; R\mathbf{ca} \to \mathcal{F}_n(R)(\mathbf{c}, \mathbf{b}). \quad \square$$

133_n :$\mathcal{F}_n(R)(\mathbf{a}, \mathbf{b})$ & $R\mathbf{bc} \to \mathcal{F}_n(R)(\mathbf{a}, \mathbf{c})$.

Proof. By (127_n). Suppose $F\xi$ hereditary, $\forall \mathbf{x}(R\mathbf{ax} \to F\mathbf{x})$. Since $\mathcal{F}_n(R)(\mathbf{a}, \mathbf{b})$, $F\mathbf{b}$. Since $R\mathbf{bc}$ and $F\xi$ is hereditary, $F\mathbf{c}$. □

134_n :$\mathcal{F}_n^{=}(R)(\mathbf{a}, \mathbf{b})$ & $R\mathbf{bc} \to \mathcal{F}_n(R)(\mathbf{a}, \mathbf{c})$.

Proof. From (133_n), as (132_n) from (129_n). □

136_n :$\mathcal{F}_n(R)(\mathbf{a}, \mathbf{b}) \to \mathcal{F}_n^{=}(R)(\mathbf{a}, \mathbf{b})$.

139_n :$\mathbf{a} = \mathbf{b} \to \mathcal{F}_n^{=}(R)(\mathbf{a}, \mathbf{b})$.

140_n :$\mathcal{F}_n^{=}(R)(\mathbf{a}, \mathbf{a})$.

Proof. All immediate from the definition of "$\mathcal{F}_n^{=}$" (and identity). □

137_n :$\mathcal{F}_n^{=}(R)(\mathbf{a}, \mathbf{b})$ & $R\mathbf{bc} \to \mathcal{F}_n^{=}(R)(\mathbf{a}, \mathbf{c})$.

Proof. From (134_n) and (136_n). □

141_n :$\mathcal{F}_n(R)(\mathbf{a},\mathbf{b}) \rightarrow \exists x[\mathcal{F}_n^=(R)(\mathbf{a},\mathbf{x}) \ \& \ R\mathbf{x}\mathbf{b}]$.

Proof. (123_n), with $\exists x[\mathcal{F}_n^=(R)(\mathbf{a},\mathbf{x}) \ \& \ R\mathbf{x}\xi]$. Suppose $R\mathbf{a}\mathbf{y}$. Then, by (140_n), $R\mathbf{a}\mathbf{y}$ and $\mathcal{F}_n^=(R)(\mathbf{a},\mathbf{a})$; hence, $\exists x[\mathcal{F}_n^=(R)(\mathbf{a},\mathbf{x}) \ \& \ R\mathbf{x}\mathbf{y}]$. Suppose $\exists x[\mathcal{F}_n^=(R)(\mathbf{a},\mathbf{x}) \ \& \ R\mathbf{x}\mathbf{y}]$ and $R\mathbf{y}\mathbf{z}$. Since $\mathcal{F}_n^=(R)(\mathbf{a},\mathbf{x})$ $\&$ $R\mathbf{x}\mathbf{y}$, $\mathcal{F}_n^=(R)(\mathbf{a},\mathbf{y})$, by (137_n). So $\mathcal{F}_n^=(R)(\mathbf{a},\mathbf{y})$ $\&$ $R\mathbf{y}\mathbf{z}$, so $\exists x[\mathcal{F}_n^=(R)(\mathbf{a},\mathbf{x}) \ \& \ R\mathbf{x}\mathbf{z}]$. $\square$

144_n :$\mathcal{F}_n^=(R)(\mathbf{a},\mathbf{b}) \ \& \ F\mathbf{a} \ \& \ \forall x\forall y[Fx \ \& \ Rxy \rightarrow Fy] \rightarrow F\mathbf{b}$.

Proof. Suppose $\mathbf{a} = \mathbf{b}$. Then, if $F\mathbf{a}$, $F\mathbf{b}$. Hence, with (128_n): $[\mathbf{a} = \mathbf{b} \vee \mathcal{F}_n(R)(\mathbf{a},\mathbf{b})] \ \& \ F\mathbf{a} \ \& \ \forall x\forall y[Fx \ \& \ Rxy \rightarrow Fy] \rightarrow F\mathbf{b}$. $\square$

152_n :$\mathcal{F}_n^=(R)(\mathbf{a},\mathbf{b}) \ \& \ F\mathbf{a} \ \& \ \forall x\forall y[\mathcal{F}_n^=(R)(\mathbf{a},\mathbf{x}) \ \& \ Fx \ \& \ Rxy \rightarrow Fy] \rightarrow F\mathbf{b}$.

Proof. Suppose the antecedent, use (144_n), with $\mathcal{F}_n^=(R)(\mathbf{a},\xi) \ \& \ F\xi$. Since $F\mathbf{a}$, by (140_n), $\mathcal{F}_n^=(R)(\mathbf{a},\mathbf{a}) \ \& \ F\mathbf{a}$; i.e., $\mathbf{a}$ falls under $\mathcal{F}_n^=(R)(\mathbf{a},\xi) \ \& \ F\xi$. Suppose, further, $\mathcal{F}_n^=(R)(\mathbf{a},\mathbf{x}) \ \& \ Fx$ and $R\mathbf{x}\mathbf{y}$. Then, $F\mathbf{y}$, by hypothesis, and $\mathcal{F}_n^=(R)(\mathbf{a},\mathbf{y})$, by (137_n); so, $\mathcal{F}_n^=(R)(\mathbf{a},\xi) \ \& \ F\xi$ is hereditary. Hence, by (144_n): $F\mathbf{b} \ \& \ \mathcal{F}_n^=(R)(\mathbf{a},\mathbf{b})$. $\square$

Definition. $\text{Func}_{n,n}(R\xi;\eta) \stackrel{\text{df}}{\equiv} \forall x\forall y\forall z(Rxy \ \& \ Rxz \rightarrow y = z)$.

242_n :$\text{Func}_{n,n}(R\xi;\eta) \ \& \ \mathcal{F}_n(R)(\mathbf{d},\mathbf{n}) \ \& \ R\mathbf{d}\mathbf{a} \rightarrow \mathcal{F}_n^=(R)(\mathbf{a},\mathbf{n})$.

Proof. (123_n), with $\mathcal{F}_n^=(R)(\mathbf{a},\xi)$. We must show that, if $R\xi;\eta$ is n,n-functional and $R\mathbf{d}\mathbf{a}$, then:

    (i) $\forall x[R\mathbf{d}x \rightarrow \mathcal{F}_n^=(R)(\mathbf{a},\mathbf{x})]$.
    (ii) $\forall x\forall y[Rxy \ \& \ \mathcal{F}_n^=(R)(\mathbf{a},\mathbf{x}) \rightarrow \mathcal{F}_n^=(R)(\mathbf{a},\mathbf{y})]$.

(ii) follows immediately from Theorem 137_n. For the proof of (i), suppose that $R\mathbf{d}x$. Since $R(\xi;\eta)$ is n,n-functional and $R\mathbf{d}\mathbf{a}$, we have that $\mathbf{x} = \mathbf{a}$, whence $\mathcal{F}_n^=(R)(\mathbf{a},\mathbf{x})$, by Theorem 139. $\square$

243 :$\text{Func}_{n,n}(R\xi;\eta) \ \& \ \mathcal{F}_n^=(R)(\mathbf{a},\mathbf{b}) \ \& \ \mathcal{F}_n^=(R)(\mathbf{a},\mathbf{c}) \rightarrow [\mathcal{F}_n^=(R)(\mathbf{c},\mathbf{b}) \vee \mathcal{F}_n(R)(\mathbf{b},\mathbf{c})]$.

Proof. (144_n), with: $\mathcal{F}_n^=(R)(\xi,\mathbf{b}) \vee \mathcal{F}_n(R)(\mathbf{b},\xi)$. We must prove, that, if $\text{Func}_{n,n}(R\xi;\eta)$ and $\mathcal{F}_n^=(R)(\mathbf{a},\mathbf{b})$:

    (i) $\mathcal{F}_n^=(R)(\mathbf{a},\mathbf{b}) \vee \mathcal{F}_n(R)(\mathbf{b},\mathbf{a})$.

(ii) $\forall \mathbf{x} \forall \mathbf{y} \{ R\mathbf{x}\mathbf{y} \quad \& \quad [\mathcal{F}_n^=(R)(\mathbf{x}, \mathbf{b}) \lor \mathcal{F}_n(R)(\mathbf{b}, \mathbf{x})] \rightarrow [\mathcal{F}_n^=(R) \\ (\mathbf{y}, \mathbf{b}) \lor \mathcal{F}_n(R)(\mathbf{b}, \mathbf{y})] \}$.

(i) follows immediately, since $\mathcal{F}_n^=(R)(\mathbf{a}, \mathbf{b})$. Suppose, then, that $R\mathbf{x}\mathbf{y}$ and either $\mathcal{F}_n^=(R)(\mathbf{x}, \mathbf{b})$ or $\mathcal{F}_n(R)(\mathbf{b}, \mathbf{x})$. If the latter, then $\mathcal{F}_n(R)(\mathbf{b}, \mathbf{y})$, by Theorem 133_n. Suppose, then, that $\mathcal{F}_n^=(R)(\mathbf{x}, \mathbf{b})$. Then either $\mathbf{x} = \mathbf{b}$ or $\mathcal{F}_n(R)(\mathbf{x}, \mathbf{b})$. If $\mathbf{x} = \mathbf{b}$, then, since $R\mathbf{x}\mathbf{y}$, $R\mathbf{b}\mathbf{y}$, whence $\mathcal{F}_n(R)(\mathbf{b}, \mathbf{y})$, by Theorem 131_n. But, if $\mathcal{F}_n(R)(\mathbf{x}, \mathbf{b})$, then since $R\mathbf{x}\mathbf{y}$ and $R(\xi; \eta)$ is n, n-functional, $\mathcal{F}_n^=(R)(\mathbf{y}, \mathbf{b})$, by Theorem 242. $\square$

TREE OF IMPORTANT THEOREMS
IN THE PROOF OF THEOREM 263

The chart on the following page is a graphical representation of the dependencies among the main theorems used in Frege's proof of Theorem 263. Theorems in italics are general results concerning the ancestral. Uses of definitions or theorems which are immediate consequences thereof have been omitted. The chart and the list of theorems which follows should enable the reader to reconstruct Frege's proof of Theorem 263 without difficulty.

IMPORTANT THEOREMS IN THE PROOF OF THEOREM 263

263 : $\exists Q[\text{Func}(Q) \quad \& \quad \neg \exists x.\mathcal{F}(Q)(x, x) \quad \& \quad \forall x(Gx \rightarrow \exists y.Qxy) \quad \& \\ \exists x \forall y(Gy \equiv \mathcal{F}^=(Q)(x, y))] \rightarrow Nx : Gx = \infty$.

262, η : Func(Q) $\quad \& \quad \neg \exists x.\mathcal{F}(Q)(x, x) \quad \& \quad \forall x(\neg \exists y.Qx, y \rightarrow \\ \neg \mathcal{F}^=(Q)(a, x)) \rightarrow Nx : \mathcal{F}^=(Q)(a, x) = Nx : \mathcal{F}^=(\text{Pred})(0, x)$.

259 : $\forall y \forall z \{ \mathcal{F}^{\wedge}(Q \pi R; a, m)(y, z) \equiv \text{Conv}[\mathcal{F}^{\wedge}(R \pi Q; m, a)](y, z) \}$.

258 : $\mathcal{F}^=(R \pi Q)(m, a; n, y) \rightarrow \mathcal{F}^=(Q \pi R)(a, m; y, n)$.

257 : $\mathcal{F}^=(R \pi Q)(m, a; n, y) \quad \& \quad \forall z \forall w(Fzw \rightarrow \forall u \forall v(Rzu \quad \& \quad Qwv \rightarrow \\ Fuv)) \quad \& \quad Fma \rightarrow Fny$.

256 : Func(Q) $\quad \& \quad \forall y(\neg \exists z.Qy, z \rightarrow \neg \mathcal{F}^=(Q)(a, y)) \rightarrow \\ \text{Map}[\mathcal{F}^{\wedge}(\text{Pred } \pi Q; 0, a)](\mathcal{F}^=(\text{Pred})(0, \xi), \mathcal{F}^=(Q)(a, \xi))$.

254 : Func(R) & Func(Q) & $\neg \exists y(\mathcal{F}^=(R)(m, y) \quad \& \quad \mathcal{F}(R)(y, y)) \quad \& \\ \forall y(\neg \exists z.Qy, z \rightarrow \neg \mathcal{F}^=(Q)(a, y)) \rightarrow \text{Map}[\mathcal{F}^{\wedge}(R \pi Q; m, a)] \\ (\mathcal{F}^=(R)(m, \xi), \mathcal{F}^=(Q)(a, \xi))$.

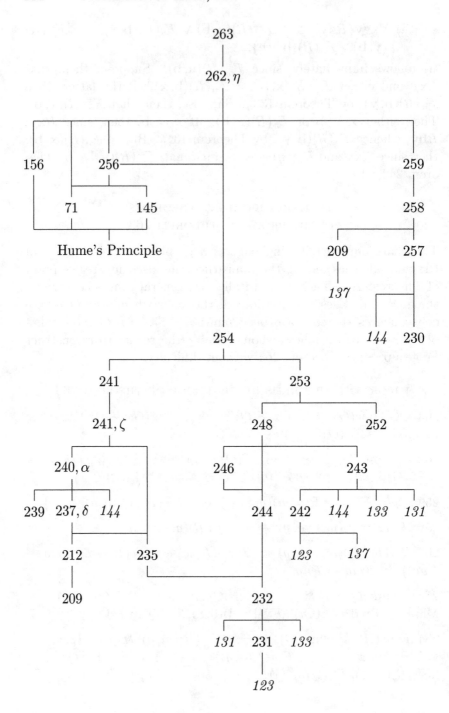

253 :$\text{Func}(R)$ & $\text{Func}(Q)$ & $\neg\exists y[\mathcal{F}^=(R)(m,y)$ & $\mathcal{F}(R)(y,y)]$ $\rightarrow \text{Func}(\mathcal{F}^\frown(R\,\pi\,Q; m, a))$.

252 :$\text{Func}(R)$ & $\text{Func}(Q) \rightarrow \text{Func}_{2,2}(R\,\pi\,Q)$.

248 :$\text{Func}_{2,2}(R\,\pi\,Q)$ & $\neg\exists x[\mathcal{F}^=(R)(m,x)$ & $\mathcal{F}^=(R)(x,x)] \rightarrow$ $\text{Func}(\mathcal{F}^\frown(R\,\pi\,Q; m, a))$.

246 :$\mathcal{F}_2^=(R\,\pi\,Q)(m,a;b,d) \rightarrow \mathcal{F}^=(R)(m,b)$.

244 :$\mathcal{F}_2(R\,\pi\,Q)(m,a;b,d) \rightarrow \mathcal{F}(R)(m,b)$.

243 :$\text{Func}(R)$ & $\mathcal{F}^=(R)(m,t)$ & $\mathcal{F}^=(R)(m,n) \rightarrow [\mathcal{F}(R)(n,t) \vee \mathcal{F}^=(R)(t,n)]$.

242 :$\text{Func}(R)$ & $\mathcal{F}(R)(d,n)$ & $Rda \rightarrow \mathcal{F}^=(R)(a,n)$.

241 :$\text{Func}(\mathcal{F}^\frown(R\,\pi\,Q; m, a))$ & $\forall x[\mathcal{F}^=(Q)(a,x) \rightarrow \exists y. Qxy] \rightarrow$ $\text{Map}(\mathcal{F}^\frown(R\,\pi\,Q; m, a))(\mathcal{F}^=(R)(m,\xi), \mathcal{F}^=(Q)(a,\xi))$.

241, ζ :$\forall x[\mathcal{F}^=(Q)(a,x) \rightarrow \exists y. Qxy] \rightarrow \forall x\{\neg\exists y[\mathcal{F}^\frown(R\,\pi\,Q; m, a)(x,y)$ & $\mathcal{F}^=(Q)(a,y)] \rightarrow \neg\mathcal{F}^=(R)(m,x)\}$.

240, α :$\forall x[\neg\exists y. Qxy \rightarrow \neg\mathcal{F}^=(Q)(a,x)]$ & $\mathcal{F}^=(R)(m,n) \rightarrow$ $\exists y. \mathcal{F}^\frown(R\,\pi\,Q; m, a)(n, y)$.

239 :$\exists x. \mathcal{F}^\frown(R\,\pi\,Q; m, a)(m, x)$.

237, δ :$\forall y[\neg\exists z. Qyz \rightarrow \neg\mathcal{F}^=(Q)(a,y)] \rightarrow \forall y\forall z\{\exists x. \mathcal{F}^\frown(R\,\pi\,Q; m, a)$ (y,x) & $Ryz \rightarrow \exists x. \mathcal{F}^\frown(R\,\pi\,Q; m, a)(z, x)\}$.

235 :$\mathcal{F}^\frown(R\,\pi\,Q; m, a)(c,d) \rightarrow \mathcal{F}^=(Q)(a,d)$.

232 :$\mathcal{F}(R\,\pi\,Q)(m,a;c,d) \rightarrow [\mathcal{F}(R)(m,c)$ & $\mathcal{F}(Q)(a,d)]$.

231 :$\mathcal{F}(R\,\pi\,Q)(m,a;n,b)$ & $\forall x\forall y\forall z\forall w[Fxy$ & Rxz & $Qyw \rightarrow$ $Fzw]$ & $\forall x\forall y[Rmx$ & $Qay \rightarrow Fxy] \rightarrow Fnb$.

230 :$\forall x\forall y(Fxy \rightarrow \forall z\forall w(Rxz$ & $Qyw \rightarrow Fzw)) \rightarrow \forall x\forall y(Fxy \rightarrow$ $\forall z\forall w[(R\,\pi\,Q)(x,y;z,w) \rightarrow Fzw))$.

212 :$\forall y[\neg\exists z. Qyz \rightarrow \neg\mathcal{F}^=(Q)(a,y)]$ & $\mathcal{F}^=(Q)(a,d)$ & $\mathcal{F}^\frown(R\,\pi\,Q;$ $u,v)(y,d)$ & $Ryz \rightarrow \exists x. \mathcal{F}^\frown(R\,\pi\,Q; u, v)(z, x)$.

209 :$\mathcal{F}^=(R\,\pi\,Q)(u,v;c,d)$ & Rco & $Qdb \rightarrow \mathcal{F}^=(R\,\pi\,Q)(u,v;o,b)$.

156 :$\forall x[\mathcal{F}^=(\text{Pred})(0,x) \rightarrow \exists y. \text{Pred}(x,y)]$.

145 :$\neg\exists x.[\mathcal{F}^=(\text{Pred})(0,x)$ & $\mathcal{F}(\text{Pred})(x,x)]$.

71 :$\text{Func}(\text{Pred})$.

12

Eudoxos and Dedekind: On the Ancient Greek Theory of Ratios and its Relation to Modern Mathematics

I. THE PHILOSOPHICAL GRAMMAR
OF THE CATEGORY OF QUANTITY

According to Aristotle, the objects studied by mathematics have no independent existence, but are separated in thought from the "substrate" in which they exist, and treated "as separable"—i.e., are "abstracted"—by the mathematician.[1] In particular, numerical attributives or predicates (which answer the question "how many?") have for "substrate" *multitudes with a designated "unit"* ("how many pairs of socks?" has a different answer from "how many socks?")—cf. *Metaphysics* XIV i 1088^{a}5ff.: "*One* [τὸ ἕν: literally, "the one"] signifies that it is a measure of a multitude, and

From *Synthese*, **84** (1990) 163–82. *Partially* reprinted by kind permission of Kluwer Academic Publishers and the author.

The remainder of the paper (pp. 182–203 of the original, with notes 21–55, pp. 206–210), comprises two sections dealing with matters more especially concerning the history of Greek mathematics. These sections have been omitted here, as not pertinent to the theme of the present volume.

The material presented here is based upon work supported, in part, by the National Science Foundation under Grant No. DIR-8808575.

[1]Aristotle, *Metaphysics* VI i 1026^{a}8–12, 15; XI iii 1061^{a}28–^{b}4, iv 1061^b 22–3, vii 1064^{a}33–4; *Nichomachean Ethics* VI viii 1142^{a}17.

number [ὁ ἀριθμός: literally, "the number"] that it is a measured multitude and a multitude of measures"; cf. also Euclid VII Defs. 1 and 2. It is reasonable to see in this notion of a "measured multitude" or a "multitude of measures" just that of a (finite) *set*: the "measures" or "units" are what we should call the "elements" of the set; the requirement that such units be distinguished is precisely what differentiates a set from a mere "accumulation" or "mass." There is perhaps some ambiguity in the quoted passage: the statement, "Number signifies that it is a measured multitude," might be taken either to *identify* numbers with finite sets, or to imply that the *subjects numbers are predicated of* are finite sets. Euclid's definition—"a number is a multitude composed of units"—points to the former reading (which implies, for example, that there are many *two*s—a particular knife and fork being one of them). Number-words, on this interpretation, would be strictly construed as denoting *infimae species* of numbers. It is clearly in accord with this conception that Aristotle says, for example (in illustrating the "discreteness," as opposed to continuity, of number): "The parts of a number have no common boundary at which they join together. For example, if five is a part of ten the two fives do not join together at any common boundary but are separate; nor do the three and the seven join together at any common boundary" (*Categories* vi 4^{b}28–9). The fundamental operation of the addition of numbers is then just that of the *union of finite sets*—which sets must, however, be supposed disjoint. This requirement, and the corresponding one for subtraction (namely, that the "number"—i.e. set—subtracted must be contained in the one it is subtracted from), would lead to some awkwardness in the formulation of arithmetical relationships, and much awkwardness in the arrangement of proofs. Such requirements are never stated or accommodated in practice. One therefore has to conclude either that the requirements and the procedures for fulfilling them are tacitly understood, or—as seems more likely—that the strict distinction between number as substrate and number as species is ignored by the mathematicians. Perhaps, indeed, this is what Aristotle means when he says that the mathematicians consider their objects "*qua* separable from the substrate."

Corresponding to the question "how many?" which asks about multitude, is the question "how much?" which asks about magnitude (distinguished from number, according to Aristotle, as "continuous," in contrast to "discrete," quantity.)[2] But this question requires fuller specification: that of a "respect," or a *kind* of magnitude—what one now calls, in the physicist's terminology, a "dimension": "how long?"; "how much area?"; "how capacious [or "how much volume"]?"; "how heavy?"; etc. As in the case of number, there is reason to think that the primary reference of magnitude terms is to the substrate—the *bearer* of magnitude. For example, Aristotle remarks[3] that "some quantities consist of parts having position relative to one another, others not of parts having position"; and he instances, as of the former kind, *lines*, *planes*, *solids*, and *places*—which latter, therefore, are all by him taken to be "quantities," and more specifically "magnitudes." It is clear that in speaking of these as constituted of parts having position relative to one another, he must have in mind particular spatial figures in each case. This is strikingly confirmed by another passage (*Metaphysics* V 13 1020^a7–14):

> We call a quantity that which is divisible into constituent parts of which each is by nature a one and a "this." A quantity is a *multitude* if it is numerable, a *magnitude* if it is measurable. We call a multitude that which is divisible potentially into non-continuous parts, a magnitude that which is divisible into continuous parts; in magnitude, that which is continuous in one dimension is length, in two breadth, in three depth. Of these, limited multitude is number, limited length is a line, breadth a surface, depth a solid.

We have here not only the same identification of specific magnitudes with actual spatial configurations, but the striking parallel of "multitude" (or "plurality") with "length, breadth, and depth" as genera, and of "number" with "line, surface, and solid" as species within those respective genera—suggesting once again that, if the particular magnitude of the kind *length* (the "limited length") is a *line* (in the sense, of course, of line-segment), the corresponding particular "numerable quantity" (the "limited multitude") is a set.

---

[2]Aristotle, *Categories* vi 4^b20ff.
[3]Ibid.

One more point seems worth calling attention to in connection with this primarily "concrete" notion of quantities. We are told by Aristotle (*Categories* vi 6^{a}27) that what is most characteristic of quantities is the attribution to them of *equality and inequality*—that these relations are predicated of quantities and of nothing else. And indeed one finds, in Euclid's arithmetic and geometry, that "sameness" is never predicated of numbers, lengths, areas, volumes, or angles: *ratios*, for example, of two areas on the one hand, two lengths on the other, are (in appropriate circumstances) said to be "the same"—but never "equal";[4] on the other hand, the areas of two figures are said to be "equal"—but never "the same" (indeed, most often it is simply said that "the two figures are equal"—that area is the appropriate magnitude-kind is taken to be understood). This difference is quite alien to our present way of thinking about such matters: for us, to say that two distinct triangles are equal in area is to say that they *have "the same area."* But on the suggested reading of the Greek terminology, it would be incorrect to speak of "the area of this triangle": a triangle does not "have" an area, it *is* an area—that is, a finite surface; "this area" means *this figure*, and the two distinct triangles are two different—but equal—areas. On exactly the same principle, then, two different "numbers"—that is, two different finite sets—may be "equal" (cf. Aristotle's reference, above cited, to the "two separate fives" that compose a ten). Thus, we may say that each species of quantity (whether discrete or continuous) is distinguished in Greek mathematics by its own proper equivalence-relation, called in each case just "equality"; and that where our own practice is to proceed to the corresponding equivalence classes, regarding these as particulars (numbers, lengths, etc.), the Greeks did not, in principle, make this abstraction. (On the other hand, as already remarked in connection with numbers, the exigencies of mathematical discourse tended to lead to compromises in practice.)

---

[4]But it is extraordinary in the light of this that in one passage (*Nichomachean ethics* V vi 1131^{a}31) Aristotle himself says that "proportion is equality of ratios [ἡ γὰρ ἀναλογία ἰσότης ἐστὶ λόγων]."

II. QUANTITIES AND THEIR RATIOS: EUDOXOS

The Eudoxean-Euclidean theory of ratio and proportion involves three distinct (interrelated) notions: *number*, *magnitude*, and *ratio*. The notion of magnitude is just presupposed in Euclid's exposition: neither definitions nor explicit assumptions are formulated concerning it; and number, although it is made the subject of a definition, is also in effect simply taken to be understood (for the definition does not provide a basis for arithmetical reasoning); but ratio is defined in a remarkably precise and adequate way.—The phrase "theory of proportion" is used because the notion of *sameness of ratio* is crucial (both to the development of the theory, and to the very definition of ratio); and two pairs of magnitudes that have the same ratio are said to be "proportional" (or "in proportion").

The notion of "kind (or genus) of magnitude" is explicitly invoked by Euclid (Bk. V, Def. 3). No definition and no postulates are given by him for this notion, but he seems to take it for granted that in each magnitude-kind there is an appropriate "combining" operation on the *substrates*—for length, e.g., on *line-segmer*   for area, on *figures*; for volume, on *solids*—analogous to the "j⊾   ɪg" of (disjoint) multitudes, that leads to an "addition" of the magnitudes of that kind (this assumption characterizes the traditional philosophical notion of "extensive magnitude").[5] (It seems quite in accord with this point of view that the word used for the operation of addition, whether of numbers or of magnitudes, is simply καί—that is, the conjunction "and.")

It is not difficult to extract from Euclid's procedures a statement of the properties that must be presupposed, for any given

---

[5]Two points are worth noting, however: (1) Aristotle asserts a relation of proportionality for magnitudes—namely, *speeds* and *densities*—that are not "extensive." (2) Euclid treats *angles* as magnitudes; but since he does not admit any angles greater than or equal to two right angles, he cannot in fact "join" or "compose" arbitrary angles—i.e., arbitrary "substrates" for this genus of magnitude. It was suggested, rather persuasively, by C. L. Dodgson (Lewis Carroll) that Euclid would regard values of this magnitude-kind exceeding (or equal to) two right angles not as themselves "angles," but in effect as what we should now describe as "formal sums" of angles; see Dodgson *Euclid and his modern rivals*, London: Macmillan & Co., 1879, pp. 192–193. To carry out this idea explicitly would go some distance in the direction of the abstract structural notion to be developed in the text immediately below.

magnitude-kind—or, more generally, for any species of quantity (whether discrete or continuous)—in order to apply to it the general Eudoxean theory of proportion. In doing this, it is convenient to take that step in abstraction which, as we have seen, the Greeks evidently did not take in principle, although in some degree they did in practice: to abstract upon the equivalence-relation called "equality" in any species of quantity, so that the "objects" of that species correspond to the equivalence-classes, and equality becomes identity. Accordingly, we postulate a combining operation—to be called "addition"—not upon the "substrates" (a notion that is hard to axiomatize in a manageable way, and is problematic in any case for the theory of magnitude—e.g., how "combine" two bodies whose masses are each equal to that of the galaxy?), but directly upon these more abstract objects. Any species of quantity $\mathbf{Q}$, under its operation of addition, is required to be a *totally ordered strictly positive commutative semigroup*, in which (moreover) *subtraction* (of the lesser from the greater) is always possible; that is, the following conditions must be satisfied:

(1) $a + (b + c) = (a + b) + c$;
(2) $a + b = b + a$;
(3) for any a and b in $\mathbf{Q}$, exactly one of the following alternatives holds:
    (a) for some c in $\mathbf{Q}$, $a = b + c$;
    (b) for some c in $\mathbf{Q}$, $b = a + c$;
    (c) $a = b$.

It easily follows from these conditions that the *cancelation law* (uniqueness of the result of subtraction) holds: if $a + b = a + c$, then $b = c$. As to the *ordering*, we introduce it by defining "$a < b$" (or, equivalently, "$b > a$") to mean that there is a c such that $a + c = b$. It is easily established from our stipulations that if $a < b$ and $b < c$, then $a < c$ (the relation $<$ is *transitive*); that we never have both $a < b$ and $b < a$ (the relation $<$ is *asymmetric*); and that for any elements a, b, of $\mathbf{Q}$, exactly one of the following three conditions holds: $a < b$, $a = b$, or $a > b$ ("*law of trichotomy*"). These properties characterize $<$ as a *strict total ordering* of $\mathbf{Q}$ ("strict" because asymmetric—i.e., analogous to "strictly less than," not to "less than or equal to"; "total" because the relation $<$ holds, in one direction or the other, between any two distinct elements of $\mathbf{Q}$). We also have the important proposition: if $a < b$, then $a + c < b + c$ (the ordering is "compatible with the semigroup structure").—It is worth noting that the procedure of defining the ordering with the help of the relation of addition has a certain correspondence with the last of the "Common Notions" at the beginning of Book I of the received text of Euclid: "The whole is greater than the part." To be sure, the authenticity of this common notion has been questioned (for that matter, Tannery challenged all of them); but even if the passage is an interpolation in the original text, it provides evidence that the traditional concept of the "greater" was just this: that the greater is what is composed of the lesser and something besides.

Addition gives rise in an obvious way to the operation of *multiplying a quantity of kind $\mathbf{Q}$ by a positive integer*: na means "the sum of n terms, each equal to a." This operation is central to the definition of the fundamental notion of ratio.

Euclid's characterization of the concept of ratio is contained in Definitions 3, 4, and 5 of Book V of the *Elements*. The contents of these definitions may be paraphrased as follows—clauses (a), (b), (c), corresponding roughly to Euclid's three "Definitions" (although (b) really contains more than does Def. 4):

(a)   A ratio ρ is a binary relation, of the following general character:

> if ρ is a ratio, then given any species of quantity, and any pair (a, b) of elements of $\mathbf{Q}$, it makes sense to affirm (or deny) that ρ "holds" between a and b (in that order)—which we may symbolize (tentatively) by: "$\rho(a, b, Q)$": "a and b, taken in that order, as elements of $\mathbf{Q}$, have the ratio ρ.

(b)   It is conceivable that a pair (a, b) of quantities of kind $\mathbf{Q}$ "have no ratio at all"—i.e., that all statements of the form $\rho(a, b, Q)$ are *false* for this pair (a, b) and this kind $\mathbf{Q}$. If a and b do "have a ratio," it is unique; and we shall symbolize it by "$(a : b)_\mathbf{Q}$"—the ratio of a to b in $\mathbf{Q}$. (In fact, the context usually makes it clear what $\mathbf{Q}$ is, and we shall actually therefore drop the subscript and just write "$a : b$.") The condition—necessary and sufficient—for a and b to have a ratio in $\mathbf{Q}$ is that for some positive integer m, $ma > b$, and for some positive integer n, $nb > a$. (Note that this condition is symmetric as between a and b; and it guarantees the existence of both ratios, $a : b$ and $b : a$.)[6]

(c)   If a and b are quantities of kind $\mathbf{Q}$ that have a ratio $\rho = (a : b)_{\mathbf{Q}}$, and if $\mathbf{Q}'$ is any species of quantity, and c, d, any elements of $\mathbf{Q}'$, then $\rho(c, d, \mathbf{Q}')$ is true if and only if the following holds:

For each given pair of positive integers m, n:

> either   both $na > mb$ and $nc > md$,
>
> or   both $na = mb$ and $nc = md$,
>
> or   both $na < mb$ and $nc < md$.

Under these conditions, we say that (a, b) and (c, d) are *proportional*, or *have the same ratio*: $(a : b)_{\mathbf{Q}} = (c : d)_{\mathbf{Q}}$.

Note, then, that clause (c) (or Euclid's Definition 5), which is the heart of Eudoxos's construction, characterizes the ratios by introducing the relation of *sameness of ratio*. It is of course crucial for such a characterization that the relation defining "sameness" be an equivalence. Reflexivity and symmetry of the Eudoxean relation are immediately obvious from the definition; as for transitivity, Euclid takes the pains to prove explicitly that it holds (Book V, Proposition 11: "Ratios which are the same with the same ratio are also the same with one another"). It is quite remarkable that this step in abstraction—the analogue of which, as we have seen, appears not to have been made in the essentially simpler case of numbers and of magnitudes—here is taken explicitly and completely: the exigencies of the problem of characterizing ratios and proportions for not necessarily commensurable quantities led to the development of a technique of "mathematical abstraction," whose fully explicit *general* recognition and exploitation (if we make an exception of a

---

[6]The wording of Euclid's Definition V.4 is susceptible of two different literal interpretations. The definition reads: "Magnitudes are said to have a ratio to one another which are capable, when multiplied, of exceeding one another." The defining clause can be taken to mean either, as in the text above, that some multiple of each exceeds the other, or that some multiple of each exceeds some multiple of the other. Since each of these conditions follows trivially from the other, however, the difference is of no mathematical importance.

remark of Leibniz's) was achieved only in the course of the great transformation of mathematics in the nineteenth century.[7]

III. PRELIMINARY COMPARISON WITH DEDEKIND

The relation of Eudoxos's explication of the notion of ratio to Dedekind's well-known construction of the real numbers is easy to see. Let a and b be quantities (of some kind $\mathbf{Q}$) that "have a ratio" in the sense laid down in Euclid's Definition V.4. Consider all pairs (m, n) of positive integers for which $mb \leqslant na$. For each of these pairs (m, n) consider the rational number m/n. It is easy to show that if (m', n') is another pair of positive integers, and if $m/n = m'/n'$ then $mb \leqslant na$ implies $m'b \leqslant n'a$; therefore we may speak of a well-defined *partition* of the set of all positive rational numbers into two subsets, "upper" and "lower," S^* and S_*, characterized by: m/n belongs to S_* just in case $mb \leqslant na$; otherwise—i.e., just in case $mb > na$—m/n belongs to S^*. Note that clause (b), or Euclid's Def. 4, guarantees that if a and b have a ratio, neither S_* nor S^* is empty. It is also easy to see that—still for a given pair, (a, b), of quantities of kind $\mathbf{Q}$ having a ratio—*each rational number in S_* is smaller than each rational number in S^**. Thus the partition into lower and upper sets determined by a given pair of quantities that have a ratio is precisely a "Dedekind cut" in the system of positive rational numbers; and therefore defines in its turn a *positive real number in the sense of Dedekind*. (The fact that Dedekind himself considered cuts in the system of *all* rational numbers—positive, negative, or zero—is obviously of no great importance.)

It is clear that two pairs of quantities, each pair having a ratio in the sense of Euclid's Definition V.4, which moreover have the *same* ratio in the sense of Definition V.5, determine (by the above construction) the same Dedekind cut, and therefore the same positive real number. We have thus a well-defined mapping of the system of all Eudoxean ratios into the system of positive real numbers.

However, it is not the case that the mapping we have constructed is (necessarily) one-to-one. We have, in fact, in proceeding to the partition of the positive rationals, discarded some Eudoxean in-

[7]Cf. Hermann Weyl, *Philosophy of mathematics and natural science*, Princeton: Princeton University Press, 1949, pp. 9–11.

formation. Eudoxos's criterion gives a partition into *three* sets of rationals—one of which (the "middle one") may be empty, and (as one easily sees) contains at most a single element (when the middle set is non-empty, our construction throws its element into the lower set). Let us consider under what conditions this can lead to the assignment of the same real number to more than one Eudoxean ratio.

Suppose that we have quantities a, b, of kind $\mathbf{Q}$ possessing a ratio, and quantities c, d, of kind $\mathbf{Q}'$, possessing a ratio and determining the same upper set as the former pair; thus, for arbitrary positive integers m, n, we have: $mb > na \iff md > nc$. Can it be that, at the same time, there are positive integers j,k, such that jb and ka are unequal but jd and kc equal?—For this to be so, in view of the former condition, we must have $jb < ka$. Let the difference, $ka - jb$, be called o.

Now, since $jd = kc$, for every positive integer N we have $Njd = Nkc$, hence $(Nj + 1)d > Nkc$, hence $(Nj + 1)b > Nka$; and from this it follows that $N(ka - jb) < b$: *every multiple of the quantity o is smaller than b*—we may say that o is "infinitesimal" in relation to b. In particular, o and b do *not have a ratio*.

Conversely, let us now suppose given two quantities, o and a of the same kind $\mathbf{Q}$, with the first infinitesimal in relation to the second; then it is easily seen that the phenomenon under consideration does actually occur within $\mathbf{Q}$. For suppose that m and n are positive integers with $ma > na$—which, of course, simply means that $m > n$. By the infinitesimality of o in relation to a, we shall then have that $(m - n)a > no$, i.e., $ma > n(a + o)$. Since, on the other hand, the last inequality obviously entails $m > n$, we see that the ratios $a : a$ and $a + o : a$ determine the same upper class, and therefore the same real number. But these ratios are not the same in the sense of Eudoxos, because we have, for any positive integer n, $na = na$ but not $n(a + o) = na$.

We have therefore seen that *a necessary and sufficient condition for the mapping we have defined, from the set of Eudoxean ratios into the set of real numbers, to be one-to-one, is that every pair of quantities of any given kind have a ratio.*

This result shows that the Eudoxean theory is in a sense stronger—more refined—than that of real numbers: it allows the discrimination of ratios that are not distinguished by the real numbers they determine. But in fact this refinement is of no use, and even clouds the theory. The trouble is this: It is of central importance to the theory of proportion that the ratio of b to a *determines* b—more precisely, that given a quantity a of kind $\mathbf{Q}$ and a ratio ρ, there is *at most one* quantity b of the same kind such that $b : a = \rho$; and, indeed, Euclid proves a proposition to this effect (Book V, Proposition 9: "Magnitudes which have the same ratio to the same are equal to one another; and magnitudes to which the same has the same ratio are equal.") But Euclid's proposition is *false* in any domain in which infinitesimals exist; for if o is infinitesimal in relation to a—under which circumstance, as we have just seen, $a : a$ and $a + o : a$ are distinct but determine the same real number—it is easy to show that, for any positive integer n, the Eudoxean ratios $a + no : a$ and $a + o : a$ are the same. In fact, the argument given above essentially shows this: it shows that for any o infinitesimal relative to a, the ratio $a + o : a$ determines the same upper set as $a : a$ but differs from the latter in having an empty "middle set"; since the quantities no are obviously all infinitesimal in relation to a, the ratios $a + no : a$ all determine the same upper set and, having empty middle sets, determine also the same lower set; thus, by the Eudoxean criterion, they are all the same.

Conversely, suppose the envisaged situation occurs: that $a : c$ and $b : c$ are the same, but a and b are unequal—say $a > b$. Since a and c have a ratio, there is an integer whose product with a is $> c$; let n be any such integer. Again, there is an integer (necessarily > 1) whose product with c is $\geqslant na$; let the smallest such be $m + 1$. Then $(m+1)c \geqslant na > mc$, and, by the equality of the ratios, it follows that nb satisfies the same double inequality; therefore $n(a - b) < c$. Since n can be as large as one wants, this shows that $a - b$ is infinitesimal in relation to c.

We are thus led to the same necessary and sufficient condition as before: the non-existence of infinitesimals, in a given species of quantity $\mathbf{Q}$, is necessary and sufficient for the non-existence in $\mathbf{Q}$ of different quantities having the same ratio to the same quantity. Clearly, Euclid's proof of Proposition V.9 must have made tacit use of the assumption that all pairs of magnitudes of a given kind have ratios; and indeed we find that his proof of Proposition V.8, upon which the proof of V.9 depends, does do so.[8]

The need for this assumption was pointed out (presumably for the first time) by Archimedes,[9] in connection, not with the theory of ratios, but with the closely related "method of exhaustion"—in effect, the method of limits: another great mathematical creation of Eudoxos. The assumption has come to be known as the Axiom of Archimedes; accordingly, we shall characterize as "Archimedean" any species of quantity each pair of whose elements has a ratio, and shall take the Eudoxean theory of proportion to deal in principle only with Archimedean species; with this stipulation, we are in possession of a well-defined one-to-one mapping of the system of all possible Eudoxean ratios into the system of positive real numbers.

IV. DISCRETE QUANTITY AND MAGNITUDE; OPERATIONS UPON RATIOS

In the foregoing, the theory of proportion has been developed uniformly for all Archimedean species of quantity, whether discrete or continuous; in particular, therefore, for *numbers*.—Of course, it is clear that all our conditions, including the Archimedean condition, are satisfied by the positive integers (provided that we, unlike the Greeks, count 1 as a number—for otherwise condition (3) of section II above will not be met).

Euclid does not in fact proceed in this way: after developing the Eudoxean theory for *magnitudes* in Book V, he gives an independent treatment for *numbers*, in terms of "multiples" and "parts"

---

[8]T. L. Heath, in his edition of Euclid, cites Definition V.4 as justifying the step in question—see Heath, *The thirteen books of Euclid's* Elements, New York: 2nd ed. Dover, 1925, Vol. III, p. 14. But this is inconsequent: the definition merely states the condition for two magnitudes of the same kind to have a ratio; what is required is assurance that that condition is always satisfied.

[9]See Heath, *The works of Archimedes* (1912), New York, Dover reprint n.d., pp. 233–234) (the introduction to *Quadrature of the parabola*, which contains what is presumably the first explicit occurrence of the assumption); p. 155 (in the introduction to *On spirals*); and pp. 1–2 (the introduction to *On the sphere and cylinder*, Book I: this passage does not mention the assumption—or "lemma," as Archimedes calls it—in question; but it is here that Archimedes attributes to Eudoxos the first rigorous proofs of those propositions which, in the first cited passage, he says have been established by his predecessors with the help of that lemma).

(i.e., submultiples), in Book VII (see Definition VII.20). It has often been noted, however, that this leads to an incoherence in Euclid's exposition, when he speaks of the *identity* of some ratio of magnitudes with a ratio of numbers. And it is quite remarkable (this too is a well-known circumstance) that Aristotle already speaks of a unified theory of proportion for all species of quantity; he says (*Posterior analytics* I v 74^{a}18–25):

> That proportionals alternate [i.e., that $a : b = c : d$ implies $a : c = b : d$—assuming that all the quantities involved are of the same kind] might seem to hold for the terms *qua* numbers and *qua* lines and *qua* solids and *qua* times; since it used to be proved separately, although it is possible to prove it of all by a single demonstration. But because there was no one name for all these—numbers, lengths, times, solids—and they differ in species from one another, they were taken up separately. But now it is proved universally; for the property indeed did not belong to them *qua* lines or *qua* numbers, but *qua* this [unnamed attribute] which is supposed to belong [to them universally].

If we consider an arbitrary Archimedean species of quantity **Q**, a number of interesting questions arise concerning the ratios of quantities of the kind **Q**. (Let us call these, for short, simply "ratios on **Q**.") We may, for instance, ask: *Onto what subset of the system of the real numbers are these ratios mapped by the correspondence we have defined?* Again, we know that each ratio on **Q** is a one-to-one relation from some subset of **Q** to some subset of **Q**; we may ask, Is each of these subsets, for each such ratio, identical with **Q**?—in other words: *Does each ratio on **Q** define a function on **Q**?* (For it is easy to see that if each ratio on **Q** does define a function on **Q**, these functions must map **Q** *onto* itself; indeed, each ratio on **Q** has an inverse that is also a ratio on **Q**.) A third question: Given quantities a, b, c, d, all in **Q**, do there exist quantities e, f, g in **Q**, such that $a : b = e : g$ and $c : d = f : g$—i.e.: *Do the ratios $a : b$ and $c : d$ admit a common denominator in **Q**?*

Of these three questions, the first and last may be asked "absolutely" as well—that is to say, may be asked of the system of *all ratios that ever occur between quantities of any domain whatsoever*.

In other words, we may ask *which real numbers correspond to Eudoxean ratios*; and *whether every pair of Eudoxean ratios can be expressed* (in *some* species of quantity) *with a common denominator*.

Now, questions of this "absolute" kind were never raised by the Greeks: the first obviously not, since the concept of real number was lacking; but the last also not: the question can—as we have just seen—be posed in the terminology of Greek mathematics, but considerations of "all possible species of quantity" were alien to the subject as far as the Greeks developed it. Aristotle offers an enumeration of species of quantity—discrete: *number* and *language* (the latter with reference to prosody); continuous: *line*, *surface*, *solid*, *time*, and *place*—and proceeds to remark that these are, taken strictly, the *only* quantities—that whatever else is quantitative is so in some way "derivatively."[10] However, one cannot assume that the exhaustiveness of this list was generally accepted—even by Aristotle himself. In the first place, there certainly are "derivative" magnitudes that Aristotle refers to elsewhere, namely velocity and density (cf. n. 5). And in the second place, of these, although one— velocity—may be taken to be "derivative" from length and time (not, to be sure, in the sense of the term "derivative" that Aristotle indicates when discussing his enumeration), the other—density— can only be referred back to volume and weight; which leaves weight itself as a presumed addendum to the list. But however this may be, the clear fact is that Greek mathematics always presupposes, quite in the spirit of the Aristotelian view of the objects of mathematics as "existing in" *things*, that such kinds of quantity as there are are simply *given* in the natural world; that it is, one may say, the business of the mathematician to study those he finds, not to speculate about what others are possible. More precisely: since certain conditions are assumed to hold of every kind of magnitude (or, more generally, of quantity), simple *universal* assertions about magnitude-kinds can be warranted; but no more complex sorts of generality than this—that is, than the generality expressible by free variables ranging over magnitude-kinds—are accessible.

The issue of the "functionality" of the ratio-relations takes the

---

[10] Aristotle, *Categories* vi 4^{b}23–26, 5^{a}39–^{b}10.

form, in Greek mathematics, of the question of the existence of a
"fourth proportional" to three given quantities: Given three quantities a, b, c, of which the first two are of the same kind, does there
exist a quantity d of the same kind as c, such that $a : b = c : d$? In
Book IX, Proposition 19, Euclid discusses the question when this is
and when it is not true in case the quantities involved are all numbers.[11] In Book VI, Proposition 12, he shows how to *construct* a
fourth proportional to three given line-segments. But in the general
theory of proportion in Book V (proof of Proposition 18), he makes
tacit use of the assumption that a fourth proportional to three given
magnitudes *always* exists. It seems not unreasonable, in the light of
this, to suppose that the existence of a fourth proportional was taken
to be a property of magnitudes in general—that is, of "continuous,"
as opposed to "discrete," quantity.

It is easy to see that the functionality of ratios for a given species
of quantity guarantees the existence of common denominators for
that species. (The converse, of course, is false, as the example of
the natural numbers shows.) This property is of particular interest
in connection with what we may call (in the modern sense) the
"algebra" of ratios: that is, ratios as a domain upon which *operations*
are defined.

Here what one finds in Euclid is quite interesting. He does indeed
define six such operations: *inversion*, *composition*, *separation*, and
conversion, which take $a : b$ to $b : a$, $(a + b) : b$, $(a - b) : b$, and
$a : (a-b)$, respectively (the last two, of course, under the assumption
that $a > b$) (Defs. V. 13–16); and *duplication* and *triplication*,
which we should describe as the "square" and the "cube" of a ratio
(Defs. V. 9–10). In each of these cases, there is a theorem to be
proved: namely, that if one substitutes for the quantities a, b, two
others, c, d, having the same ratio, then the result of the operation
expressed in terms of the second pair will be the same as the result
expressed in terms of the first pair: In the case of inversion, this
is immediately obvious from the definition of sameness of ratios;
for composition and separation, Euclid proves what is required in
Propositions V. 17 and 18. For conversion, the required theorem is

[11] The discussion itself is, in fact, defective; but (whether this is a fault in the
original, or—as T. L. Heath assumes in his comment on the proposition—results
from a corruption of the text) the defect is of no importance here.

not given by Euclid; but it is clear that this case is essentially like that of separation.

For duplication and triplication, the situation is rather more complicated. In the first place, the theorem that these operations do indeed depend upon the given *ratios*, not upon the particular magnitudes by which they are represented, is not itself stated by Euclid. In the second place, however, he does state a much more general theorem, which concerns in effect a much more general operation— namely, the operation called "composition" in more modern terminology; but composition of an *arbitrary number* of ratios (in a given order).

First a point of terminological clarification; Euclid actually uses the same verb, $\sigma\upsilon\nu\tau\acute{\iota}\theta\eta\mu\iota$, for the operation he *defines* as "composition," and for a second operation—the one we now call by that name; that is, if we think of ratios as functions, the operation that takes two (or more) functions (in given order) to the function that results from the successive application of those functions (in that order). In deference to what has become—at least since Heath—the custom among English-writing commentators on Euclid, I shall call the latter operation "compounding" rather than composition.

Now, although Euclid does not formally define this second use of his *term* $\sigma\acute{\upsilon}\nu\theta\epsilon\sigma\iota\varsigma$, he does define the *operation* of compounding an arbitrary sequence of ratios: he calls the result $\delta\iota'$ $\acute{\iota}\sigma\upsilon$ $\lambda\acute{o}\gamma\varsigma$— "a ratio *ex aequali*," in the standard rendering; and defines it as follows (Def. V. 17): "A ratio *ex aequali* arises when, there being several magnitudes, and others equal to them in multitude, which taken two and two are in the same proportion, as the first is to the last among the first magnitudes, so is the first to the last among the second magnitudes; or, in other words, it means taking the extreme terms by virtue of the removal of the intermediate terms."

This, it should be noted, is a bit odd—it seems, almost, to confuse the notions of *ratio* and of *proportion*; for no "ratio *ex aequali*" is actually defined. The definition in fact introduces a term that serves Euclid essentially as an abbreviated reference to a theorem— namely, to Proposition V. 22: "If there be any number of magnitudes whatever, and others equal to them in multitude, which taken two and two together are in the same ratio, they will also be in the same ratio *ex aequali*." (The term "*ex aequali*" in the enunciation of the theorem itself plainly serves no purpose, except as an abbreviation or tag to describe the situation considered.) But the *ratio* that plays the central role here—that of the first to the last of the magnitudes— is just the ratio compounded of the ratios of the successive terms (in each series); and the asserted result, "sameness of the ratio *ex aequali*," is nothing but the required independence of the operation of compounding from the particular magnitudes in terms of which the compounded ratios are presented.—Finally, it should be noted that for the existence, in general, within a given magnitude-kind, of *compounds of arbitrary ratios* in that kind, what is required is the existence of common denominators;[12] thus, if we proceed on the view that discrete quantity is represented just by *number*, and continuous quantity by *magnitude-kinds for which fourth proportionals always exist*, compounding of arbitrary ratios will always be possible *within a given species of quantity*. (The possibility has not so far been ruled out that there are pairs of ratios that never hold respectively for two pairs of magnitudes all of the same kind; the result of compounding two such ratios would then be undefined.)—It should be noted, in connection with the operation of compounding, that Proposition V. 23 establishes the *commutativity* of this operation (a result that is by no means trivial; it is of course not true in general for the composition of relations or functions).

The compounding of ratios corresponds, of course, to what we call *multiplication* of real numbers. The question arises, did the Greeks have an operation upon ratios that corresponds to our "addition"?

The answer to this is a little ambiguous. The standard answer would be a simple negative. Now, it is certainly true that the Greek mathematicians never speak of adding ratios: *quantities* are added; and ratios are not, for them, quantities (see, however, the remark just below, on Archimedes). But it is equally true that these mathematicians never speak of "multiplying" ratios; yet they do have the operation *we* call by that name, and they have a name for it. (The terminology of Archimedes here deviates from Euclid's; he avoids the ambiguity of using the same verb and its derivatives for two operations. Archimedes has more than one term for "compound-

---

[12]More precisely: to compose the ratios $a : b$ and $c : d$, taken in that order, one needs to represent them respectively in the forms $e : f$ and $f : g$; but this is just to represent the ratios $a : b$ and $d : c$ with a common denominator f.

ing," and it is somewhat curious to note that one of his expressions for "the ratio compounded of that of a to b and that of b to c" is simply "the ratio of a to b *and* [in Greek: καί] that of b to c"— which tempts one to think of compounding as a kind of "addition" of ratios: a notion that fits with the terms "duplicate, triplicate, [etc.]), ratio," and also with that, still current for us, of the *"mean proportional."*)[13]

For the addition of ratios in our sense, the Greeks have no name at all; and this justifies what I have called the standard negative answer to our question. Nevertheless, the operation of taking the ratio, to a given magnitude as "denominator," of a sum of magnitudes taken as "numerator," is of frequent occurrence; and one finds in Euclid just the theorem required to establish that *the result depends only upon the respective ratios, to the common denominator, of the magnitudes summed to form the numerator*—Proposition V. 24: "If a first magnitude have to a second the same ratio that a third has to a fourth, and also a fifth have to the second the same ratio as a sixth to the fourth, the first and fifth added together will have to the second the same ratio that the third and sixth have to the fourth." In this not unimportant sense, then, the Greeks did have available the operation on ratios that we know as addition.

V. DEDEKIND ONCE AGAIN; THE
CONTINUITY OF THE STRAIGHT LINE

To be able to answer the remaining question raised in the preceding section, we need something new.

It is possible to argue—I think convincingly—that one way to arrive at a definite answer is already implicit in Greek mathematics. I have suggested that "continuous quantity" or "magnitude" may be characterized, for the Greeks, by the existence of the fourth proportional; but although it is plausible to regard this as a distinguishing mark, for them, of magnitude in contrast to number, it cannot be considered as expressing what they meant by the "continuity" of

---

[13]See Heath, *The works of Archimedes*, p. clxxix. (Heath actually renders Archimedes's terms καί ['o] and προσλαβὼν [τὸν] as "multiplied by"; but this has no linguistic justification, as the first means simply "and," the second "taken besides."

magnitude (as opposed to the "discreteness" of number). It will be convenient before considering the Greek conception further, to digress briefly to the modern conception.

In the year 1872 there appeared both the well-known little monograph of Dedekind on continuity and irrational numbers,[14] and a work, perhaps less well known to a philosophical audience, in which Georg Cantor sketched a treatment of essentially the same subject, from a different point of view, but with equivalent results.[15] In particular, both Cantor and Dedekind pointed out the necessity, for a complete theory of classical geometry, of introducing an axiom to assert what Dedekind called "the continuity of the straight line":[16] namely, taking the line as ordered by the relation "to the left of," the principle that if the line is in any way divided into two parts, of which each point of one is to the left of each point of the other, then either the "lefthand" part has a rightmost point, or the "righthand" part has a leftmost point.

Now, such an axiom is by no means necessary for the geometry contained in Euclid's *Elements*—a fact already emphasized by Dedekind himself in the preface to the first edition of his monograph on the natural numbers.[17] Indeed, if E is the smallest subfield of

[14]Richard Dedekind, *Stetigkeit und irrationale Zahlen* (1872) reprinted in his *Gesammelte mathematische Werke*, ed. R. Fricke, E. Neather, and O. Ore, vol. III, Braunschweig: F. Vieweg & Sohn, 1932, pp. 315–344; "Continuity and irrational numbers" (English translation of the foregoing), in Richard Dedekind, *Essays on the theory of numbers*, Wooster Woodruff Beman, trans., New York: reprint, Dover Publications, 1963, pp. 1–27.

[15]Cantor's considerations bearing on our subject appeared in one of his papers on the theory of trigonometric series—see Cantor, "Über die Ausdehnung eines Satzes aus der Theorie der trigonometrischen Reihen" (1872); reprinted in his *Abhandlungen mathematischen und philosophischen Inhalts*, ed. Ernst Zermelo, reprint, Hildesheim: Georg Olms Verlagsbuchhandlung, 1966.

[16]See section III of Dedekind's monograph *Stetigkeit und irrationale Zahlen*. —The property that Dedekind called "continuity" is formally identical with what, in current topological terminology, is called the connectedness of the line. Cantor's construction, which relies upon the metric structure of the line (whereas Dedekind's makes use only of its ordering), leads formally to the property now called (metric, or, more generally, "uniform") *completeness*; but for a uniform space with a dense subset isomorphic to the rational numbers, the two concepts are equivalent.

[17]Dedekind, *Was sind und was sollen die Zahlen?* (1888), reprinted in his *Gesammelte mathematische Werke*, vol. III, pp. 335–390; "The nature

the field of real numbers that is closed under the operation of taking the positive square root of a positive quantity, then analytic geometry over the field E admits all Euclidean constructions, and in it all the theorems of the *Elements* are true; but (of course) this geometry does not satisfy Dedekind's axiom of continuity. E itself, one should note, cannot be the subset of the reals corresponding to the full domain of ratios for Euclid's geometry (even if it be taken to correspond to the full domain of ratios of straight line-segments)— for the ratio of the area of a circle to that of the square on its radius can be proved to exist (strictly on Euclid's principles); and this ratio does not belong to E (as Lindemann demonstrated in 1882). If, furthermore, we admit the existence of the straight line-segments aimed at (or implied by) the famous construction problems of classical geometry—implied, that is, by the existence of a cube double in volume to a given cube; an angle one-third of a given angle, or, more generally, an angle equal to an arbitrary "part" of a given angle; a square equal in area to a given circle (or, equivalently, a line whose length is equal to the circumference of a given circle)—one gets, in the first two cases, new ratios altogether, and in the third, new ratios of straight line-segments. It is unclear where this process should end.

On the other hand, there are reasons—suggestive, and, I think, plausible, even if not conclusive—for believing that the Greek geometers would have accepted Dedekind's axiom, just as they did that of Archimedes (once it had been stated). For instance, in the treatise *Measure of the circle*, Archimedes admits the existence of a straight line-segment equal to the circumference of a given circle; and in the treatise *On spirals*, he *proves* the existence of such a line: namely, he proves the beautiful theorem (Proposition 18) that the tangent to his spiral at the endpoint A of its first turn meets the line through the origin O of the spiral and perpendicular to OA in a point whose distance from O is equal to the circumference of the circle of radius OA. But such a line does not exist in the geometry based upon the field E; so it is clear that Archimedes must make use, in his proof, of some mode of argument that transgresses the

and meaning of numbers" (English translation of the foregoing), in Richard Dedekind, *Essays on the theory of numbers*, 31–115. See esp. pp. 39–40.

framework of the *Elements*. Now, the only point in the proof that can possibly be challenged is the assumption that the tangent in question *exists*. And it is hard to think of a natural principle upon which the proof of its existence can be based that is not equivalent to Dedekind's principle of continuity.

This geometrical consideration can be supplemented by a philosophical one (if we are willing to take Aristotle as authoritative here). Aristotle offers the following definition of the "continuous":[18] First, he defines "contiguous" as "next in succession, and touching." Then he declares continuity to be a kind of contiguity: "I call [contiguous things] *continuous* when the extremes that touch and hold them together become one and the same." This is undeniably a little vague; but it does suggest a warrant for arguments of the following type: "Let a half-line with origin O be divided into two parts, so that all the points of the first have a distance from O no greater than the circumference of a given circle, and all the points of the second have a distance from O no smaller than that circumference. Clearly, nothing stands between these parts; therefore they are contiguous. But the line is a continuous magnitude; therefore these parts have a common extremity."

To be sure, there is a weak point in this argument if we base it upon Aristotle's very words: one might object that the fact that "nothing stands between" the two parts does not establish their contiguity, since it does not establish that the two "touch." But it is hard to believe that Aristotle would countenance this objection. (On another count, Aristotle does reject the argument: in *Physics* VII iv 248^{a}18–^{b}7, he contends that a circular arc cannot be greater or smaller than a line segment—on the grounds that if it could be greater or smaller, it could also be equal. This, however, denies, not the validity of arguments of the type suggested above, but the truth of the premises in the example; moreover, Aristotle's intimation that the possibility of "greater" and "less" implies that of "equal" can even be taken to support the view that the reasoning in the example is sound: that if the premises *were* correct, the conclusion would follow.)

It is worth putting on record here an ingenious argument, sug-

[18] *Physics* V iii 227^{a}9–13; *Metaphysics* XI xii 1069^{a}2–8.

gested to me orally by W. Tait, for Dedekind's principle as implicit in Greek geometry—an argument drawn after all from the *Elements* itself, namely from Definition 3 of Book I: "The extremities of a line are points." Clearly, this in no sense defines anything; but on Tait's reading, the "definition" is in effect a formulation of Dedekind's principle. In the argument given above, for instance, one considers the part of the line that is "closer" to O, and reasons thus: "That part is a *finite line*" (this is Euclid's only term for a "line-segment"; indeed, he ordinarily refers to such simply as a "line"); "therefore *its extremities are points*: namely, O, and a second point which can only be at a distance from O equal to the circumference of the circle."

I do not offer any of the foregoing considerations as more than plausible; but if the conclusion from any of them is accepted, it does follow that every real number corresponds to a Eudoxean ratio, indeed to a ratio of straight line-segments.

VI. MAGNITUDE RECONSIDERED: FROM
THE ANCIENT TO THE MODERN MODE

Dedekind, of course, aimed to make the theory of the real numbers, and of continuity and limits, fully independent of geometric intuition (or "geometric evidence").[19] It seems of interest to examine how far the Eudoxean theory itself can be made to serve this end.[20]

---

[19] See the introductory remarks to *Stetigkeit und irrationale Zahlen*; and, for some reflections on the motivation of Dedekind's concern, see my paper, "Logos, logic, and logistiké," in William Aspray and Philip Kitcher, eds., *History and philosophy of modern mathematics*, *Minnesota studies in the philosophy of science*, Vol. XI, Minneapolis: University of Minnesota Press, 238–59, esp. pp. 242–49.

[20] The discussion that follows is to a certain extent adumbrated by Frege, in the second volume of his *Grundgesetze der Arithmetik*. Just as Frege based his theory of the natural numbers on the demand that the intrinsic structure of such a number reflect its use in the representation of the sizes of sets, he wanted to construct a logical concept of the real numbers to reflect their use in representing the measures of magnitudes; and, in his characteristically ponderous and thorough way, he moves towards such a construction through most of the course of that volume (without actually attaining its formal completion—this was to have been given in the third volume, which was never written). See *Grundgesetze* II, §§ 55–164 in which Frege criticizes numerous attempts at a foundation of the

If it is to do so, it is absolutely necessary that the concept of magnitude be freed from dependence upon anything empirically "given." Thus, in the spirit of modern mathematics—with explicit acknowledgment that we here take a step that is foreign to the Greek view— we make of the conditions earlier laid down for a species of quantity a *definition* of the notion of such a species. By a species of quantity, then, we henceforth understand any domain and binary operation on that domain satisfying the conditions (1)–(3) of section II above together with the Axiom of Archimedes.

It has already been pointed out that the Eudoxean ratios instantiated in any species of quantity for which common denominators exist allow the mode of composition we now call "addition"; and even that this mode of composition was known (in effect) by the Greek mathematicians. It is trivial to verify that this operation on any such ratio-domain satisfies all our conditions for a species of quantity. (Once again it should be emphasised: the Greeks did not regard ratios as quantities. Nevertheless, *their* ratios formed, in the indicated sense, a domain—or domains, when we consider various classes of ratios—of quantities, in *our* present sense of the term.)

In particular, then, the ratios of numbers form (with the indicated operation) a species of quantity; and in accordance with modern usage, we call this the domain of "positive rational numbers." The ratios of such rational numbers are of course themselves already ratios of (whole) numbers, hence are themselves rational numbers.

Now we may, following Dedekind (lightly modified), consider the domain of all *cuts* in the system of positive rational numbers. This domain can be given the structure of a species of quantity (indeed, of magnitude in the strongest possible sense of the term), by introducing a suitable operation of "addition of cuts": the "upper set" of the sum of two cuts is, by definition, the set of sums of elements belonging respectively to the upper sets of the summands. It is easy to see that this procedure does indeed define a new Dedekind cut (in the sense of the discussion in section III above—including the provision that the upper set contain no smallest element); and, further, that the system of all these cuts, under this operation, constitutes

theory of the real numbers, and finds a clue to the way he proposes to follow); and II, §§ 165–245 (in which he begins his own construction).

an Archimedean species of quantity or magnitude-kind. Let us call this magnitude-domain D^+ (the "positive Dedekind domain"). If, now, we consider a particular magnitude d in D^+, and we let $\mathbf{1}$ be the Dedekind cut whose lower set consists of all the positive rational numbers $\leqslant 1$, then it turns out that the Dedekind cut that corresponds (in the sense of section III above) to the ratio $d : \mathbf{1}$ is just d itself; so indeed every cut in the system of positive rational numbers "is" (that is, corresponds to, or "determines") a ratio—and beyond this, corresponds to a ratio of elements of one particular magnitude-kind, fixed once for all: the domain D^+. On the other hand, because the elements of D^+, by virtue of what we have now established, correspond one-to-one to the system of all Eudoxean ratios, and by a correspondence that is easily seen to take sums to sums, we are led to the conclusion that *the domain of all Eudoxean ratios, with its "natural" operation of addition, itself constitutes an Archimedean magnitude-kind possessing Dedekindian continuity; all Eudoxean ratios of quantities of any Archimedean species occur as ratios of magnitudes of this one kind.*

In effect, therefore, with the help both of our more abstract and general notion of a species of quantity, and of Dedekind's construction applied to the rational numbers, we have established the *identity* of the system of Eudoxean ratios with the system R^+ of the positive real numbers; for since the operation of "compounding" is available for the ratios, alongside the operation of addition, the structure obtained is not only that of a positive ordered semigroup, as required for a species of quantity, but that of a positive ordered "semifield"— i.e., the structure of the positive elements of an ordered field; and moreover of a "complete" ordered field, in the modern algebraic sense.[21]

---

[21] *Editor's Note*: An extended endnote to this section—note 21 on pp. 205–6 of the original printing of the paper—has been omitted.

13

Frege's Theory of Real Numbers

I. INTRODUCTION

Among the many hundreds of works, large and small, that have been written about Frege, very few concern themselves at all with his theory of real numbers.[1] In view of the importance of such a theory for

From *History and philosophy of logic*, **8** (1987) 25–44. Reprinted by kind permission of Taylor & Francis Ltd. and the author.

I should like to thank Gottfried Gabriel, Michael Resnik and Barry Smith for their advice, although I did not always take it.

[1]F. von Kutschera, "Frege's Begründung der Analysis," *Archiv für mathematische Logik und Grandlagenforschung*, **9** (1966) 102–111; reprinted in M. Schirn, ed., *Studien zu Frege I. Loqik und Philosophie der Mathematik*, Stuttgart-Bad Cannstatt: Fromman-Holzboog, 1976, pp. 301–312. Kutschera's set-theoretic reconstruction is in several ways unfaithful to Frege's approach; for instance it is wrongly claimed (110; p. 311 of the reprint) that Frege *defines* the real numbers as ordered pairs (cf. section IV below) . Kutschera also distinguishes between the real numbers (so 'taken') and the 'Fregean real numbers' (ibid.) in a way which Frege would have been unable to accept. See also footnote 22 at the end of this chapter. All references to Kutschera are to this paper.

There is also a short section on the theory of real numbers in G. Currie, *Frege: an introduction to his philosophy*, Brighton: Harvester Press, 1982, pp. 57–59. See also Currie's "Continuity and change in Frege's philosophy of mathematics," in L. Haaparanta and J. Hintikka, eds., *Frege synthesized: essays on the philosophy and foundational work of Gottlob Frege*, Dordrecht: Reidel, 1986, pp. 345–374 which appeared after I first drafted this paper. This new paper also deals with the real numbers, and gives a most useful account of the development of the programme of logicism, this is surprising. But there are a number of reasons for the neglect. Frege's published work on the reals is incomplete: although it takes up most of the second volume of *Grundgesetze der Arithmetik* the treatment is broken off while still in the early stages, and the second volume is in any case completely overshadowed by Russell's paradox. It is clear that Frege had much more material than he published, which was no doubt being saved for a third volume of *Grundgesetze* but the antimony put a stop to the project, and nothing more of it has come down to us in his *Nachlass*. Another reason is that while there is an admirable informal account of the natural numbers in *Grundlagen*, which makes their treatment in *Grundgesetze* I much easier to follow, there is no such account of the reals: we have only a few brief remarks in *Grundgesetze* II. Finally, as Kutschera points out, [p. 102 (p. 301 of the reprint)] when Frege was writing, the work by Cantor and Dedekind on the definition of real numbers in terms of natural numbers was already well known, so the impression could easily arise in informed circles that Frege would have nothing essentially new to say.

That the fragmentary account of real numbers has been given almost no attention is a great pity, since it throws unexpected new light on the nature of Frege's logicism, which in turn enables us to raise a number of philosophically interesting questions about logicism and mathematics in general. For another thing, Frege brings perceptive criticisms of then current theories of reals, among others those of Cantor, Dedekind and Weierstrass, which are not without contemporary relevance. His theory is no less insightful and persuasive than his theory of natural numbers, and what is more, it is very unlike the standard contemporary approaches deriving from Cantor and Dedekind. For these reasons I want to give a short description of his theory and consider how Frege would probably have

Frege's views from his *Habilitation* to the end of his life. However, I think Currie somewhat misrepresents Frege's views on magnitudes in *Grundgesetze*. As the proposed reconstruction below makes clear, the relationship between magnitudes and real numbers is not left obscure (cf. p. 360), the applicability of real numbers is not essentially different from that of natural numbers (cf. p. 361), in the sense that actual magnitudes play an analogous role in the definition of reals to that of concepts in the definition of natural numbers.

completed it, as a modest attempt to try to do for real numbers what the later sections of *Grundlagen* do for natural numbers. Of course just as his theory of natural numbers is no longer tenable, so the theory of reals here tentatively reconstructed can no longer be maintained, but we may still learn from it about both the real numbers and about Frege.

II. FREGE'S CRITICISMS OF OTHER THEORIES

Grundgesetze is unsystematically divided: the theory of cardinal numbers up to and including aleph-0 straggles across from Volume I to take up the first 68 pages of Volume II. Only then does Frege begin Part III, 'The real numbers,' with a long polemic (85 pages) against his contemporaries, followed by a short sketch (9 pages) of his own theory. The remaining part, called *Die Grössenlehre* ('The theory of magnitude') is formal and takes up 90 pages. It breaks off abruptly before reaching completion, and we are left to face Russell's paradox.

The polemical preliminaries are a counterpart to the similar sections concerning natural numbers in *Grundlagen*. Frege prefaces his criticisms with a recapitulation of the requirements he places on definitions, especially definitions executed in a natural language (II, §§ 55ff.). In particular, he is against all forms of piecemeal, step-by-step definition, which he contends rob concepts of fixed, sharp boundaries. He ridicules mathematicians who are reluctant to take the momentous decision to introduce a new sign and prefer to soldier on with one familiar sign which must then be given several different meanings. Piecemeal extension of definitions is easily avoidable in practice and its use brings theoretical difficulties, since one must always show that when old signs get new meanings no difficulties arise through there being sentences containing the defined sign which are true when it has its old meaning and false when it has its new meaning. In particular, theorems previously proved should not lose their validity.

Frege draws the familiar conclusion that functions should be completely defined, so that they have values for all arguments. He admits that this leads to artificialities, but gives arguments to show why it is difficult to set limits to the natural domain of application of functions. We do not have to decide here whether the requirement that functions be everywhere defined is tenable, since one could adopt a less stringent requirement and still agree with the criticisms Frege gives of the then current views.

Frege attacks two different sorts of theory of real numbers (II, §§ 68ff.). The first sort is formalism. In the then current sense, formalism held that numbers are signs manipulated according to arbitrarily stipulated game-like rules, and nothing more. The theory of Frege's Jena contemporary Johannes Thomae comes in for particularly extensive drubbing. Even the most ardent admirer of Frege cannot fail to succumb to boredom here: he had already successfully killed formalism several times over with argument and ridicule, but he carries on flogging the dead horse for section after section (II, §§ 86–137). From the point of view of a realist like Frege, the cardinal sin of a formalist is that because he takes signs to be the subject-matter of mathematics, he fails to distinguish signs from what they actually signify, and even incomparably more able mathematicians such as Cantor—although Cantor is no formalist in the sense of Thomae—often blur over the distinction. Frege notes that the 'twilight' engendered by word/object confusions is precisely what is required for formalists and others to pull off the sleight of hand required to conjure up the real numbers (II, § 67).

In his discussion, Frege incidentally makes a clear distinction between a subject and theories about it; he notes that the game of chess (unlike 'contentual' mathematics) is no theory, says nothing, and contains no theorems, whereas the theory of chess has a content and may indeed contain theorems (II, § 93). What is it then that raises arithmetic above the level of a mere game to that of a science? Frege replies that it is the *applicability* of arithmetic (II, § 91). A game, lacking any content or subject-matter, is thereby incapable of being applied. The requirement of applicability is at work in Frege's own choice of theory.

Frege is much happier with Dedekind's theory, and praises its sharp distinction between sign and signification, its acceptance of = as signifying identity, and its taking numbers to be what numerical signs signify rather than the signs themselves (II, § 138). He takes issue with Dedekind over a completely different matter, namely his constructivism, the idea that in performing certain arithmetical operations the mathematician creates or constructs new objects. Dedekind speaks quite frankly of creating new irrational numbers corresponding to certain Dedekind cuts of rationals.[2] Frege criticizes the notion of creation. Can we create numbers at all? Are there limits to our creative capacity? Creative definitions, he remarks sarcastically, are a first-rate invention and bring the irremunerable reward that they make it unnecessary to prove many results which would otherwise require proof (II, § 143).[3] We cannot be so free in our creation that we lapse into contradiction, yet the surest way to show a theory is consistent is to exhibit a model of it which is not created. As if prescient of the ironic ring these words were to attain on Russell's discovery of the paradox, Frege immediately goes on to defend the introduction of his own courses-of-values and Basic Law V (II, §§ 146f.). Basic Law V is not a definition, and it is furthermore implicit in all usages concerning 'sets,' 'classes' and the like. It is the only means available for grasping logical objects, without which there can be no scientific foundation for arithmetic. If Basic Law V involves creation, it is at least neither arbitrary nor unregulated, and it is highly profitable. It is easy to see from this impassioned defence how shattering Russell's discovery was to be.

Weierstrass also comes in for plenty of criticism (II, §§ 148ff.), chiefly because he embraces the 'ginger biscuit' theory of elementary arithmetic according to which numbers are concrete aggregates, against which Frege had polemicized in *Grundlagen*. In fact Weierstrass wavers over whether numbers are such aggregates, or a property (the 'value' or 'validity') of aggregates, or aggregates of abstract objects, or perhaps one abstract object occurring repeatedly. The followers of Weierstrass are thus in the pleasant position of being able to choose among these alternatives whichever one suits

---

[2]§ IV of Dedekind's *Continuity and irrational numbers* is called "Creation of irrational numbers."

[3]It should be recalled that Frege does not mean by 'creative definitions' what we normally mean by this today, a sense which was coined by Leśniewski. Creative definitions in the old sense seem to be either acts which bring the objects required into existence, or else existence postulates. It was Dedekind's use of such postulates which evoked Russell's famous remark, 'The method of "postulating" what we want has many advantages; they are the same as the advantages of theft over honest toil.' *Introduction to mathematical philosophy*, London: Unwin, 1919, p. 71.

the needs of the moment (II, § 153). When it comes to the move to higher arithmetic, Weierstrass chronically fails to distinguish between concepts and the objects (if any) failing under such concepts, and his examples from commerce and geometry have only illustrative, not definitory value (II, § 155).

III. A MODERN COMPARISON

Frege's unerring instinct of attacking the weakest point of opposing theories makes it hard to give his remarks contemporary relevance, since it is often possible to preserve the mathematical content of a theory which is defectively presented while discarding its undesirable features. Such is the case with Frege's august contemporaries Cantor and Dedekind, whose substantial positive contributions Frege somewhat irritatingly passes over in silence.[4] The originality of Frege's own theory and the force of his criticisms may nevertheless be usefully appreciated by contrasting it with a standard modern account which retains many features of the step-by-step approaches of Cantor and Dedekind while avoiding the more obvious pitfalls.

Suppes describes the sixth chapter, "Rational numbers and real numbers," of his well-known textbook *Axiomatic set theory*, as follows: 'In Chapter 6 the standard construction of the rational and real numbers is given in some detail. Cauchy sequences of rational numbers rather than Dedekind cuts are used to define the real numbers.'[5] The set-theoretic framework is ZF, and the natural numbers have been previously defined as the von Neumann ordinals. Suppes defines (non-negative) fractions as ordered pairs $\langle m, n \rangle$ of natural numbers such that $n \neq 0$; rewrites '$\langle m, n \rangle$' suggestively as 'm/n'

---

[4]In II, § 164 Frege mentions Cantor' s result that there are more sets of the natural numbers than natural numbers, and in II, § 175 he uses a definition of continuity which in all probability derives from Dedekind; but in neither case does he mention the other mathematician. This rather miserly attitude to his contemporaries did not go unnoticed: in a letter of 7 November 1903 Hilbert writes to Frege of *Grundgesetze* II: 'I agree in general with your criticisms; except that you do not do full justice to Dedekind and especially Cantor.' See Brian McGuinness, ed., Gottlob Frege: *Philosophical and mathematical correspondence*, Oxford: Blackwell, 1980, p. 52.

[5]P. Suppes, *Axiomatic set theory*, New York: Van Nostrand, 1960, p. vii.

and defines an equivalence $\simeq_f$ on fractions by

$$m_1/n_1 \simeq_f m_2/n_2 \text{ iff } m_1 n_2 = m_2 n_1;$$

and defines the less-than relation for fractions, their addition and multiplication, showing these to have important invariance properties under $\simeq_f$. On this basis the non-negative rationals are defined as equivalence classes of fractions and derived concepts of less than, addition and multiplication, zero and one are defined and shown to have the right properties. Next ordered pairs of non-negative rationals are taken with the intended interpretation of $\langle M, N \rangle$ as $M - N$, another equivalence is defined, less than, addition, multiplication, zero and one defined for the resulting equivalence classes, which are the rationals. Finally, Cauchy sequences of rationals are introduced, relations of less than, addition, multiplication, zero and one defined for them and shown to have the right properties, a suitable equivalence on them is defined, the real numbers are taken to be equivalence classes of rationals, and concepts of less than, addition, multiplication, zero and one defined for the reals and shown to have the required properties. In particular, the reals, unlike the rationals, are continuous, in that every subset of reals bounded above has a least upper bounds in the reals.

To illustrate the necessary tortuousness of the development, we may follow the fate of 'the' number 1, or rather the several objects which successively do the job of 1. The natural number 1 is simply the set $\{\emptyset\}$. The fractional 1, 1_f, is $\langle 1, 1 \rangle$ which, since Suppes uses the Wiener-Kuratowski construction for ordered pairs, turns out to be the set $\{\{\{\emptyset\}\}\}$. The non-negative rational 1, 1_v, is the equivalence class of 1_f, that is the set of all ordered pairs $\langle n, n \rangle$ for $n \neq 0$, that is, the set of all sets of the form $\{\{n\}\}$, where n is any natural number other than zero. The rational 1, 1_s, is the set of all pairs of non-negative rationals $\langle M, N \rangle$ such that $M = N +_v 1$, where $+_v$ is addition of non-negative rationals. The unit Cauchy sequence is the object $\langle 1_s, 1_s, \ldots, 1_s, \ldots \rangle$, and finally the real 1, 1_r, is its equivalence class, i.e., roughly speaking, the set of all Cauchy sequences which converge to 1_s. This set-theoretical object is far too complicated for its description to be written down here.

No criticism is intended of Suppes's development. There is no twilight and no sleight of hand. He carefully distinguishes objects, predicates and functions at each stage by means of subscripts, while retaining a suggestive symbolic analogy. He does not try to pretend that the various sets successively picked out are 'somehow the same,' but speaks of objects at one level *corresponding* to objects at another. He also proves the requisite structural properties at each stage or leaves the proof as an exercise. There is no suspicious 'creation' of numbers. Although Suppes speaks of 'construction' of the rationals and reals, this is loose talk: the sets which are taken to be numbers are not constructed by the definitions, but picked out by them; it is really the definitions that are constructed. Each object thus picked out is a well-defined antecedently given member of the cumulative set-theoretic hierarchy ZF.

It is precisely because Suppes's development is so clean and above-board that we can easily see why Frege would not have liked it. One can imagine his sarcastic query as to which of the many objects which 'play the part' of 1 really *is* 1.[6] The variety is not confined to the actual construction. Suppes notes that the various kinds of object could be introduced in a different order, and he opts for Cauchy sequences rather than Dedekind cuts for a practical reason only.[7] There is thus an infinite variety of possible versions of the rational and real numbers for the mathematician to choose from. Surely, Frege might go on, the mathematician with such riches at his fingertips will look down on previous theories as incomparably miserly, and praise the wonders of modern set-theoretical technology. He would no doubt be prepared to accept a multiplicity of *routes* from the natural numbers to the reals, but not a multiplicity of systems of real numbers. The multiplicity of possible set-theoretical interpretations of mathematical objects has already been noted for the case of natural numbers.[8] The problem here is not different in

---

[6] In a letter to Russell of 21 May 1903 (*Philosophical and mathematical correspondence*, p. 156), Frege admits that Russell's own construction of the irrational numbers as classes of rationals is 'logically unassailable,' but objects that whereas Russell needs two steps, 'I want to go at once from the [natural] numbers to the real numbers as ratios of magnitudes.'

[7] *Axiomatic set theory*, p. 161. The reason is that the methods used in the treatment of Cauchy sequences are more like those used in analysis in general.

[8] See P. Benacerraf, "What numbers could not be," *The philosophical review*, **74** (1965) 47–73; and M. Resnik, *Frege and the philosophy of mathematics*, Ithaca: Cornell University Press, 1980, pp. 228–233.

principle, only in magnitude and complexity. Frege disapproved of the set theories of Schöder and Dedekind, and appears simply to have ignored Zermelo's cumulative version.[9] Perhaps he would have regarded the absolute and conditional existence axioms of ZF as being at least as mysteriously creative as the more modest moves of Dedekind and Cantor, so that the 'mystery' surrounding numbers is not completely dissolved, but only referred on to the greater mystery surrounding sets.

On the other hand, while Frege would have been unhappy with the amount of convention going into the definitions, it is clear that the need to make some kind of selection of objects in the ZF hierarchy is deeply rooted, as the untenability of the Frege-Russell definition of cardinal numbers showed. Furthermore (*ad hominem*), Frege himself was not above making stipulations, in particular when considering courses-of-values and which objects are to be taken as the True and the False (*Grundgesetze* I, § 10). While the step-by-step approach of Suppes would have won Frege's disapproval, it could easily be seen as a heuristic device for getting us all acquainted with the objects which end up as the reals: the important thing, after all, is that these have the right structural properties. This they do, but had we simply been presented with the appropriate objects and relations without any preparation we should have had no idea how to go about showing they have the right structure. We can, so to speak, throw the set-theoretical ladder away once we have climbed up it, and carry on using the usual notation without needing to concern ourselves any more as to what further set-theoretical properties the real numbers have.

Frege's strongest objection, however, would be that the set-theo-

---

[9]Frege's disapproval of Schröder's set theory is expressed in his review of Schröder's *Vorlesungen über die Algebra der Loqik*, reprinted in Brian McGuinness, ed., Gottlob Frege: *Collected papers on mathematics, logic and philosophy*, Oxford: Blackwell, 1984, pp. 210–228, Peter Geach, tr. His remarks allow one to infer that he would not accept any form of set theory in which sets are 'built on' their elements. Although Frege must have heard of Zermelo's axiomatic set theory, he appears not to have taken it seriously: even when asked by Hönigswald in 1925 to comment on later developments, including Zermelo (*Philosophical and mathematical correspondence*, p. 53), Frege merely states that the contradictions show set theory to be impossible (*Philosophical and mathematical correspondence*, p. 55).

retic construction gives us no information at all about the application of the real numbers in the physical sciences. On the contrary,
the construction is a demonstration of structural parallels wholly internal to pure mathematics. If we did not already know that the real
numbers *are* used in measurement of magnitudes, we should never
divine this however long we stared at their set-theoretical counterparts or replacements. It was just this that Frege held against the
arithmetical accounts of his day: they either completely ignored
measurement or else patched a theory of measurement on from the
outside, whereas the applicability of real numbers in measurement
must be built into their essence (II, § 159). One of the features of
Frege's theory of natural numbers in *Grundlagen* was precisely that
he attempted to build the application of natural numbers in counting into their definition. We shall have occasion below, however, to
question the purity of Frege's deference to applications.

IV. NUMBERS AND MAGNITUDES: THE INFORMAL ACCOUNT

From his discussion of other theories, Frege has already ruled out a
formalistic treatment. Of the 'contentual' approaches hitherto prac-
tised, he sees two alternatives, neither of which completely satisfies
him, and between which he attempts to steer a middle course (II,
§ 159). The first approach is the arithmetical construction approach
of Dedekind and Cantor, which we have already discussed. The second has recourse to geometry, and defines real numbers as ratios
between the lengths of straight line segments. Frege's criticism of
this second approach is that it is too specific, since there are many
magnitudes, such as angle, mass, temperature, time-span, and light-
intensity, whose magnitude ratios are also given by real numbers,
but such ratios are not in themselves geometrical (II, § 158). This
argument is analogous to one found in *Grundlagen* against taking
natural numbers to be properties of physical objects: we can count
abstract objects as well as concrete ones. It is part and parcel of
Frege's logicism, which is, it should be noted, propounded not as a
dogma but as an hypothesis to be tested. He is prepared to let the
decision rest on the success or failure of the attempt (II, § 159), and
he proved to be true to his word, since in the end he gave up the
attempt and opted for a geometric foundation for real numbers.

It is a mistake to confuse real numbers with magnitudes themselves (II, § 160), once again because there are many different kinds of magnitude, but only one system of real numbers. A real number is rather a ratio of magnitudes, which is to say, the extension of a relation between magnitudes: the same ratio holds between the mass 2 g and the mass 1 g as holds between the mass 4 g and the mass 2 g or the length 10 cm and the length 5 cm. That real numbers are closely associated with ratios of magnitudes has been known for a very long time: the fifth book of Euclid's *Elements* attempts to establish a theory of ratios of arbitrary magnitudes which does not presuppose number, and both Dedekind and Weierstrass appear to have been indebted to Euclid. Frege was an admirer of Euclid and in *Grundlagen* (§ 19) quotes Euclid together with Newton, whom he also mentions as a precursor of his views in *Grundgesetze* (II, § 157n.).

Frege then turns naturally to the question what a magnitude is (II, § 160). He notes that various attempts at a definition have failed, probably because there are so many different kinds of magnitude that it is difficult to conceive what property distinguishes magnitudes from non-magnitudes. In any case such a criterion would be powerless to tell us when two magnitudes are of like kind. He gets round the problem by noting that magnitudes do not occur in isolation, but together with other magnitudes of the same kind. The length 2 cm is one member of a system of lengths comparable to it, and this system must have a structure. So Frege turns the question round (II, § 161):

> Instead of asking, what properties must an object have in order to be a magnitude? we must ask, how must a concept be constituted for its extension to be a domain of magnitudes [*Grössengebiet*]? Now instead of saying 'extension of a concept' we shall say 'class' for short. Then we can put the question as follows: what properties must a class have in order to be a domain of magnitudes? Something is not a magnitude just by itself, but only in so far as it belongs with other objects to a class which is a domain of magnitudes.

Currie comments on this passage that it shows Frege to be still faithful to a kind of context principle.[10] In my view this understands

[10] *Frege: an introduction to his philosophy*, p. 160.

'context principle' too loosely: the principle that magnitudes do not occur in isolation but only in domains is rather different from the *Grundlagen* principle that the meaning of a word should be asked after not in isolation but only in the context of a sentence. However, Currie is right that Frege did not abandon the context principle in the form stated in *Grundlagen*: our reconstruction of his theory of real numbers will provide an instance of its application.

There is no internal necessity that all magnitude domains should have the same structure, and Frege is aware of this. His goal is a theory of the real numbers, negative as well as positive, which means that the magnitude domains must embody something corresponding to the reversal of sign. He notes that this rules out absolute magnitudes, where there is nothing corresponding to negative numbers (II, § 162). Presumably he had magnitudes like mass, speed and distance in mind, which have a zero as lower limit. The absolute magnitude domain corresponding to the term 'mass' will consist of concepts (i.e. functions from objects to truth-values). One such concept is that signified by the expression-form 'ξ has a mass of 1 g,' though the concept itself is independent of the system of measurement employed—the same concept is signified by 'ξ has a mass of 0.0352739 oz.' Ratios of absolute magnitudes yield as a structure at best that of the non-negative reals.

Quoting Gauss, Frege argues that the opposition between positive and negative rests on the notion of the converse of a relation (II, § 162), and he is surely right. So the magnitudes whose ratio is to yield the reals must be dyadic relations. For example, if we take as our magnitude domain directed distances on the line of the earth's axis, the domain will consist of relations such as that signified by the expression-form 'ξ is 10 km north of η.' Once again, the system of measurement is irrelevant: if a is 10 km north of b, it is that far north of b no matter what units we use to express it. And a is 10 km north of b iff b is 10 km south of a: the two relations are converses, and we have the basis for the introduction of negative numbers: to be -10 km north of a is just to be 10 km south of a.

We must now fix terminology. Frege normally uses the term '*Beziehung*' for that which dyadic predicates signify: a dyadic function from objects to truth-values, just as a concept is a monadic function from objects to truth-values. Just as he uses 'class' to abbreviate 'extension of a concept,' so he uses the word '*Relation*' for the extension of a *Beziehung*. Since both words are naturally translated by 'relation,' I distinguish them by writing 'relation' for '*Beziehung*' and 'Relation' for '*Relation*.' A Relation is thus the extension (course-of-values) of a relation. Frege summarizes his characterization as follows: 'The ratios between magnitudes or real numbers will then be regarded as Relations of Relations. Our domains of magnitudes are classes of Relations, namely extensions of concepts subordinate to the concept *Relation*' (II, § 162). Reversal of sign will then correspond to replacing a Relation by its converse, addition will correspond to the relative product, difference will correspond to the product of one Relation with the converse of another, and zero will correspond to the product of a Relation with its converse.

While Frege does not go into details, it is clear that his general approach is very similar to that involved in defining the natural numbers as classes (extensions of concepts). It is crucial that he does not regard the natural numbers as a subset of the reals, but rather as a different system altogether (II, § 157). He reflects this in his notation, writing '$a \cap (b \cap \mathbf{f})$' for the natural numbers rather than '$a + 1 = b$,' and '$\emptyset$,' 'Λ' for the natural numbers instead of '0,' '1,' the real numbers. He calls cardinal or counting numbers '*Anzahlen*,' real numbers simply '*Zahlen*.' The rationale is that cardinal numbers are not ratios but rather answers to 'How many?' questions, whereas real numbers measure how large a certain magnitude is in comparison with a unit magnitude. Even when he later rejected logicism, and attempted to give a geometrical foundation for real and complex numbers, he distinguished real and complex numbers from counting numbers, which by then he had come to disparage as 'kindergarten numbers.'[11]

Frege now (II, § 164) faces the problem of where to obtain the Relations whose ratios are to form the real numbers, especially as there are not countably but continuum-many reals, the result of Cantor which Frege cites in passing. If he gets anywhere a magnitude relation which is empty, the system breaks down, because all empty relations have the same extension. For logicism to be upheld, the ob-

---

[11]See the manuscript "Numbers and arithmetic," in H. Hermes et al., eds., Gottlob Frege: *Posthumous writings*, Oxford: Blackwell, 1979, pp. 275–277.

jects cannot be borrowed from the empirical world, since they would not then exist of necessity. The related problem for natural numbers was solved by defining zero using the empty class, the extension of a necessarily empty concept, and using this necessarily existing object to start us off on the trip to the first infinite number. The natural numbers then all exist necessarily, independent of empirical fact. Frege's trick here is even more subtle: he makes use of the availability of the natural numbers as already defined to show the existence of suitable Relations on which to base the reals. Hence the separateness of the naturals from the reals is of strategic importance for his plan, since without it he would be moving in a circle.

Every positive real number may be expressed as the sum of a non-negative whole number and an infinite sum of negative powers of two (a 'bicimal'), as follows (II, § 164):

$$a = r + \sum_{k=1}^{\infty} \left[\frac{1}{2^{n_k}} \right],$$

where the function $k \to n_k$ from the positive whole numbers to themselves is monotone increasing. So we can associate with each positive real number a pair whose first member is a natural number and whose second member is an infinite class of natural numbers, corresponding to those places in the infinite bicimal expansion which have non-zero numerator. This class clearly cannot contain 0. So for instance all whole numbers correspond to pairs whose second member is the class of all positive natural numbers, since e.g. $4 = 3 + 0.111\ldots$ (the latter in bicimal notation). The number $1/3$ corresponds to the pair whose first member is 0 and whose second member is the class of all positive even real numbers, and so on. Frege has here used antecedent knowledge of the properties of real numbers, but this is psychological or epistemological, not logical antecedence. Now consider any three positive reals a, b, c such that $a + b = c$. The representations of a, b and c are then related as follows: the infinite set of natural numbers corresponding to the bicimal part of c is obtained from those for a and b by a recursive method corresponding to the addition of infinite bicimals. The first member of the pair representing c is the sum of the first members of the pairs representing a and b, plus 1 if 1 is to be 'carried' from the bicimal sum. For example, consider the sum

$$2/3 + 1/2 = 7/6,$$

in bicimal notation

$$0.1010101 \cdots + 0.0111111 \cdots = 1.0010101\ldots,$$

which corresponds to the operation '+' on pairs:

$$\langle 0, \{1,3,5,7,\ldots\}\rangle \; '+' \; \langle 0, \{2,3,4,5,6,7,\ldots\}\rangle = \langle 1, \{3,5,7,\ldots\}\rangle.$$

If we take any particular fixed b, the Relation between pairs corresponding to reals a and c such that $a + b = c$, is completely determined by the pair belonging to b. We thus generate a family of Relations on pairs with continuum-many members (as many as there are infinite subsets of the natural numbers). To these we add their converses, and get a one-to-one correspondence between these Relations and non-zero reals. Addition of numbers $b+b'$ corresponds to the relative product of the Relations associated with b and b', and 0 corresponds to the product of a Relation with its converse. The class of these Relations is a domain of magnitudes, showing that the Relations required for defining the reals are not empty.

The reals themselves are not the classes of natural numbers, nor the pairs, nor the relations between the pairs, just as the natural number $\emptyset$ is not the necessarily empty concept *not identical with itself*, nor its extension, but rather the extension of the concept 'equinumerous with the concept *not identical with itself*' (*Grundlagen* § 74). The contrast with other theories is marked. Frege, like all those advocating a non-geometrical foundation for the real numbers, makes use of the natural numbers, but he plans to obtain the reals in one go rather than step by step. It is also clear that the reals are understood to be not constructed or created but to be preexisting courses-of-values picked out by the definitions, closely related to but distinct from the various objects used in showing they exist.

Having outlined his strategy, Frege tackles the problem from the other end, without logically presupposing the reals at all: 'In this way we shall succeed in defining the real numbers purely arithmetically or logically as ratios of magnitudes which can be shown to be available, so that no doubt can remain that there are irrational numbers' (II, § 164).

V. THE FORMAL DEVELOPMENT

Since addition is commutative and associative, the relative product of Relations in domains of magnitudes must also be commutative and associative. The associativity law always holds, not only among Relations but among all objects (Thm. 489). The commutative law does not hold in general, but Frege first proves it holds among finite powers of a Relation (Thm. 501). It would be possible to consider a class of finite powers of a Relation as a magnitude domain and define the positive rational numbers as ratios between members of the class (II, § 173). The negative rationals could be obtained by adding the converses of the Relations, but we should still be missing the irrational numbers. To be able to define the *greater than* Relation on a domain of magnitudes this must fulfil certain conditions which have yet to be determined. Once this Relation is defined, we can define what it is for two magnitudes to belong to the same domain of magnitudes: both are greater than some third magnitude[12] (II, § 173).

However, rather than work directly with *greater than* Frege prefers to define it in terms of the notion of the positive: a is greater than b iff the difference $a - b$ is positive. He thus considers what conditions a class must fulfil to be a *positive class*. Once we have this, we can determine a domain of magnitudes as consisting of all Relations in the positive class, together with those whose converses are in the positive class, and what he calls the Null Relation, which is the product of a positive Relation and its converse (II, § 173, Definition X). It remains to be seen what positive classes must be like. For the reals to be obtained the resulting magnitude domain must be continuous. Frege uses the following definition of continuity, which he probably got from Dedekind (II, § 175): if a magnitude has a property F and all smaller magnitudes have F, but some magnitude in the domain does not have F (i.e. F delimits a *lower segment* of the domain), then there is in the domain a least upper bound to the magnitudes with F. The connection with Dedekind cuts is obvious. To get to the point where these concepts can be defined, Frege introduces the broader concept of a *positival class*, with which he will de-

---

[12]We find something similar in Euclid. Definition 4 of Book V of the *Elements* reads: 'Magnitudes are said to have a ratio to one another when the less can be multiplied so as to exceed the other.'

fine the concept of least upper bound. A positival class S is a class of Relations fulfilling the following conditions (II, § 175, Definition Ψ):

(1) each relation in S is one-one;

(2) the relative product of a Relation in S and its converse does not belong to S;

(3) S is closed under taking relative products; and

(4) for any two members s and t of S, the relative product of s with the converse of t and the relative product of the converse of s with t must both belong to the domain of magnitudes associated with S.

Frege proves that if a relation belongs to a positival class S, its converse does not (Thm. 538); that if p and q belong to the same positival class, the product of p with its converse is identical with the product of q with its converse (Thm. 556); several more theorems about the zero of a domain of magnitudes, about greater and less than, and finally that Relations in a positival class are identical iff neither is greater than the other (Thm. 587).

The definition of positival class does not constrain the ordering structure defined in terms of *greater than* to be continuous, or indeed even dense. A positival class could have the ordering structure of the positive integers, the positive rationals, or, to take an example we shall meet again, those positive rationals which are expressible as multiples of a negative power of 2 (e.g. 27/64, 675/512).

Frege now defines the least upper bound of a class of Relations in a positival class. The Relation d is a least upper bound of a class P in the positival class S (he says 'd is an S-limit of P') iff the following conditions are fulfilled (II, § 193, Definition AA):

(1) S is a positival class;

(2) d is in S;

(3) every member of S which is less than d belongs to P; and

(4) every member of S which is larger than d is larger than at least one Relation in S which is not in P.

He proves that if there is a least upper bound to a class in a positival class then it is unique (Thm. 602).

Frege now returns to define 'positive class.' A positive class fulfils the following conditions (II, § 197, Definition AB):

(1) it is a positival class;

(2) it is not bounded below (for each Relation in it there is a smaller Relation in it); and

(3) every lower segment in it has a least upper bound in it.

The second condition now rules out discrete orderings like the positive integers, and the third ensures continuity. He now proves for positive classes the Archimedean principle that for any two Relations in a positive class there is a multiple (i.e. power) of one which is not less than the other (Thm. 635). The proof takes several sections. Finally he proves, over the remarkable span of 38 pages, that Relations belonging to the magnitude domain of a positive class obey the commutative law (Thm. 689). This is as far as the formal development goes.

VI. THE LIKELY CONTINUATION

The final pre-Russellian section of *Grundgesetze* II, § 245, sketches very briefly the steps Frege was planning to take next: firstly he will show that there is indeed a positive class, as indicated above, and then he will show that the real numbers themselves are magnitudes in the domain of magnitudes of a positive class.

What would the continuation have looked like? We have an outline, but not much detail. We also know when we should be satisfied. Just as the foundation of natural numbers may be considered complete once the second-order Peano axioms have been derived, so the foundation of the real numbers could be considered complete once one has demonstrated the existence of logical objects satisfying second-order principles corresponding to the Hilbert or Tarski axiom systems for the theory of real numbers:[13] However, our interest here is merely to conjecture the most important steps which Frege is likely to have taken along this path.

---

[13] See Hilbert's *Foundations of geometry*, La Salle: Open Court, 1971, Chapter III; and Tarski, *Introduction to logic and to the methodology of deductive sciences*, 3rd edn., New York: Oxford University Press, 1965, Ch. X. Frege himself certainly knew Hilbert's book, which first appeared in 1899 (see his correspondence with Hilbert in Gottlob Frege: *Philosophical and mathematical correspondence*) but it is unlikely, in view of his opinion of axiom systems of Hilbert's kind, that he would have cared to make use of Hilbert's theory of real numbers.

To get the objects required to show a positive class exists, Frege needs several things: the natural numbers, infinite subsets of the natural numbers, ordered pairs, addition of natural numbers, and an operation on infinite subsets of the natural numbers corresponding to the addition of infinite bicimals. It is this last which promises to be most complicated. The concept of a natural number is that of an object standing in the proper or improper ancestral of the successor relation to $\emptyset$, and while Frege does not actually define a course-of-values corresponding to this concept in *Grundgesetze* I, he could have done so. having defined 'endless,' the number of natural numbers, or aleph-0, as Cantor called it, Frege can easily define what it is to be an infinite subset of the natural numbers excluding $\emptyset$: it is a class of natural numbers in the series beginning with 1 whose number is endless. He defined ordered pairs in *Grundgesetze* I (Definition $\varXi$). Addition of natural numbers is, rather surprisingly, not defined in *Grundgesetze*. One way to define it would be to define first the nth power of a Relation, so that for any natural numbers m and n their sum is any number p (there will be only one) such that p stands in the nth power of the successor Relation to m. Frege could well have defined the nth power of a Relation R like this: it is the course-of-values corresponding to the relation that x has to y when there is a function f taking the numbers $\emptyset$ to n into objects $f(\emptyset), \ldots, f(n)$ such that $f(\emptyset) = x$, $f(n) = y$ and for all i and j such that i is greater than or identical with $\emptyset$, j is less than or identical with n, and j is the successor of i, $f(i)$ stands in R to $f(j)$. A corresponding definition in Frege's notation can easily be written down.

How can we get from the infinite sets corresponding to bicimals to the infinite set corresponding to the bicimal remainder (less than 1) of their sum? We have a recursive procedure for adding binary numbers: adding digitwise we use the table $0 + 1 = 1 + 0 = 1$, $0 + 0 = 0$, and $1 + 1 = 0$ with 1 'carried' to the next left-most digit and added on there. But in this case the bicimal is infinite, and Frege would not in any case have been content with a merely recursive account.

Of course I cannot be sure how Frege would have tackled the problem, but the following line of thought is plausible in view of his previous remarks. Each infinite bicimal is represented by an infinite set of natural numbers: we have a one-to-one correspondence between reals in the interval $(0, 1]$ and infinite subsets of the positive natural numbers. But of course one cannot use the reals here: the interpretation has suggestive force only. A number n is in such a set if the nth bicimal digit is 1 and it is not in the set if the nth bicimal digit is 0. Let a be a proper infinite bicimal and A its corresponding set of positive natural numbers. For each positive natural number m let $T_m(a)$ be the mth *truncate* of a, that is, the finite bicimal consisting of just the first m binary digits of a. Let $T_m(A)$ be the corresponding finite set of positive natural numbers, which is the intersection of A with $\{1, \ldots, m\}$. All the truncates of a are multiples of finite negative powers of 2, and for all m, $a > T_m(a)$.

But the larger m becomes, the nearer $T_m(a)$ gets to a: we have the inequality

$$a - T_m(a) \leqslant 1/2^m,$$

since the worst that could happen is that all the truncated digits are 1s, corresponding to 1 in the mth place. It is possible to say, for any three truncated bicimals $T_m(a)$, $T_m(b)$ and $T_m(c)$, when $T_m(c)$ is the bicimal part of the sum of $T_m(a)$ and $T_m(b)$. For any positive natural number k let the kth digit of a be a_k, likewise for b and c. Each such digit will be 0 or 1. The required relationship holds among the truncates iff there exist 'carry' digits $z_0, z_1, \ldots, z_m$ with $z_m = 0$, the remaining z_i all being 1 or 0, such that for all k, $1 \leqslant k \leqslant m$,

$$c_k \text{ is } 0 \text{ if } (a_k + b_k + z_k) \text{ is } 0 \text{ or } 2,$$
$$c_k \text{ is } 1 \text{ if } (a_k + b_k + z_k) \text{ is } 1 \text{ or } 3,$$
$$\text{and } z_{k-1} = 1 \text{ iff } (a_k + b_k + z_k) \geqslant 2.$$

This condition may be expressed in Fregean terms using suitable functions. We may then say that three infinite bicimals a, b and c are such that c is the bicimal part of the sum of a and b iff the above defined relation holds among their respective truncates $T_m(a)$, $T_m(b)$ and $T_m(c)$ for all positive natural numbers m. The only exceptional case to note is if the truncate sums always end in a series of 0s after a certain m. In that case the equivalent c is obtained by changing the last 1 to 0 and all the succeeding 0s to 1s. Note that if $a + b > 1$, then after some m the first 'carry' digit z_0 will always be 1, which is then to be added to the sum of the natural numbers in the first members of the pairs. Translating this back into the terms of sets of natural numbers gives us a condition on infinite subsets A, B, C of the positive natural numbers, so we get a Relation between such subsets corresponding to B. The Unit Relation, that corresponding to 1, is the Relation determined by the pair $\langle \emptyset, \{ 1, 2, \dots \} \rangle$. It then has to be shown that the Relations so defined form a positive class, so that the associated domain of magnitudes has all the desired properties.[14]

The antecedent knowledge of real numbers here made use of is a heuristic device, not a logical ground. Frege's logicism is in effect a giant existence proof: he has to show there are logical objects satisfying the formal and intuitive requirements of the natural and real numbers. That in picking out suitable logical objects and Relations we make use of antecedent knowledge of that for which we are trying to provide a logical foundation is not a logical circularity, since (he thinks) the right objects exist and have these properties independently of our knowledge.

Assume then we have a domain of magnitudes. How do we get from there to the definition of 'real number'? We note four constraints which the definition must fulfil to be acceptable to Frege:

(1) It must apply equally to all domains of magnitudes;
(2) It must represent reals as ratios of magnitudes;
(3) It must yield a natural explanation of the term 'ratio'; and
(4) It must allow us to show that the reals themselves form a domain of magnitudes.

The first constraint suggests a definition in the same spirit as the definitions of the natural numbers in *Grundgesetze* I. This in turn suggests we look informally at the four-place relationship 'the ratio of a to b is equal to the ratio of c to d.' Since by (1) we cannot rule out comparisons across different domains of magnitudes, a and b may belong to a different domain of magnitudes from c and d. Frege's strategy with natural numbers was to define the whole proposition 'the number of Xs is equal to the number of Ys' or 'Xs are equinumerous with Ys' in terms of 'there are as many Xs as Ys,' this understood in terms of one-one correspondence, and by taking

---

[14]Kutschera shows that Relations in a domain of magnitudes model Tarski's axioms.

'equal' to mean 'identical' and the functor expressions to signify extensions he got what he was looking for: the number of Xs is the extension of the concept *equinumerous with Xs*. Something similar ought to work here: once we have defined 'the ratio of a to b is equal to the ratio of c to d,' we can then define the ratio of a to b as the extension of the relation (N.B. not concept—there are two gaps) equal to the ratio of a to b. This will be a Relation of Relations, as Frege said the reals ought to be. It is in such definitions that Frege's much-vaunted principle in fact does its work: we are to define 'ratio' in terms of 'same ratio.' The reason why the context principle needs to be invoked here, as in *Grundlagen*, but not in *Grundgesetze* itself, is that here and in *Grundlagen* we are using natural language to explain what is meant, whereas in *Grundgesetze* it is the formal development which is uppermost, and there the choice of signs obliterates any appearance of circularity.

The question is how we define the required equivalence. Here we may reflect on both earlier remarks of Frege about the rational numbers as definable in terms of powers of Relations and also on his admiration for Euclid and Leibniz. What can we do with magnitudes in a single domain? First of all we can compare them for greater, identical and less, but that is not enough to give us ratios. We recall that Frege proved a form of Archimedes' principle about multiples of a magnitude. So we can form multiples and compare these too. As we showed above, we can easily define the nth power of a Relation in Frege's system, and for any positive natural m and n, and any magnitudes a and b in one domain, we can compare a^n and b^m (using the usual notation for powers of a relation).[15] We can say for example that a is more than one-and-a-half times bigger than b iff $a^2 > b^3$.

In this way we may in effect compare fractional multiples of any two magnitudes in a single domain, which leads us to the following definition, due to Euclid:[16]

---

[15] Notice the notational infelicity: multiples of numbers correspond to powers of Relations, just as sums of numbers correspond to products of Relations.

[16] This is Definition 5 of Book V of Euclid's *Elements*. In his fifth reply to Clarke, Leibniz praises Euclid's definition. Both Euclid and Leibniz can therefore be seen to have anticipated the context principle in its strict (i.e. mathematical) application. It is possible that Frege was acquainted with Leibniz's

The ratio of a to b is equal to the ratio of c to d iff for all natural numbers n, m, a^n is greater than, identical with or less than b^m according as c^n is greater than, identical with or less than d^m respectively.

This definition uses only concepts and objects previously available to Frege in the system of *Grundgesetze*. The ratio of a to b is then defined as suggested above. Something r is then a real number iff there are Relations a and b in a domain of magnitudes such that r is the ratio of a to b. Because a domain of magnitudes had been previously shown to exist, all the real numbers exist and are as we should wish them. Furthermore, they are purely logical objects. To prove they exist and have the properties they do have one need make no empirical assumption. The most important reals are 0, 1 and -1. Recalling the bicimal construction, 1 can be defined as the ratio of the Unit Relation to itself, -1 as the ratio of the converse of the Unit Relation to the Unit Relation and 0 as the ratio of the product of the Unit Relation with its converse to the Unit Relation. If m and n are any natural numbers, the ratio of the Relations corresponding to the pairs

$$\langle m, \{\mathit{1}, \mathit{2}, \mathit{3}, \ldots\}\rangle \text{ and } \langle n, \{\mathit{1}, \mathit{2}, \mathit{3}, \ldots\}\rangle$$

will be the rational number $(m + 1)/(n + 1)$. And so on.

Such a definition would fulfil all that Frege required. Interestingly, it cannot manage without an intermediate step analogous in strength to the construction of the rational numbers. But the rationals (including the whole numbers) are all *defined* together with the irrational numbers at one stroke, and what is more in a way which makes it clear both that they are magnitude-neutral ratios and that ratios are internally connected to multiples and comparisons. The basic (brilliant) idea is to be found in Euclid: what Frege essentially was doing was to give a general framework for this idea which freed it from any parochiality regarding the magnitudes considered. In his theory of real numbers, Frege seems in general to have been more indebted to other mathematicians than in his theory of counting numbers. Not only does he positively mention and use the ideas of

remark.

Euclid, Newton and Gauss, but he also uses some ideas of his contemporaries Cantor and Dedekind, though we have seen he is rather sparing in his acknowledgements of debt to the latter two.

Frege would then need to go on and fulfil his promise to show that the reals, thus defined (and I have rather little doubt that this is essentially what his definition was) themselves form a domain of magnitudes. Let us leave him at his exercise.

VII. DIFFICULTIES

Interesting though it may be to reconstruct Frege's likely subsequent moves, we should be under no illusions that such a construction can work. The logic on which Frege hoped to build his logicism was inconsistent, and if he could not see his way clear to rework the system enough to recapture the much simpler theory of natural numbers, there is no doubt that the more involved theory of real numbers too would not have been reworked. Late papers show him attempting a geometrical foundation for arithmetic.[17] Despite the intellectual resilience which such an attempt shows, I cannot find it right-minded: he had already argued with great cogency in *Grundgesetze* II that there should be the same kind of internal connection between the real numbers and measurement of magnitudes of very many kinds, not just geometrical magnitudes, and the late attempt flies in the face of this argument. It has probably been the majority opinion that lengths are a somehow privileged kind of magnitude. Certainly we learn to cope with length very early among magnitudes, and our practice of using instruments where a magnitude is shown by an analogue in length (like an ordinary thermometer) also lends support to this view. But at the time he wrote *Grundgesetze* Frege

---

[17]See "A new attempt at a foundation for arithmetic," (1924/25) in *Posthumous writings*, pp. 278–81. Note here that Frege goes straight for the complex numbers, whereas in *Grundgesetze* he defines only the reals. It is interesting that in a letter to Peano of 1896 (*Philosophical and mathematical correspondence*, pp. 125–126) Frege recommends defining mathematical functions like addition and multiplication straight away for the complex numbers, which would mean that he would strictly have to wait until producing a logistically acceptable account of the complex numbers before defining addition for the reals or else replace an interim definition for the reals (with 'don't care' values outside them) with a new definition taking account of the complex numbers. Otherwise he would offend against his own prohibition on piecemeal definition.

would no doubt have regarded the priority of length, if there is one, as merely psychological or epistemological, but certainly not logical in nature.

The more intriguing question remains as to how much of Frege's ideas, perhaps reconstructed as suggested here, can be salvaged in an acceptable framework. We know that a set-theoretical interpretation of the reals can be given, and some aspects of his work can be transposed into those terms.[18] However that would not solve the application problem which he so gallantly strove to overcome. For mathematical purposes it may be assuring to be told there are set-theoretical objects having all the right properties which we may call the natural numbers, the reals, the complex numbers and so on, but the philosophical motives behind the search for foundations of arithmetic and analysis are left unsatisfied, since we are simply handed new mathematical objects for old, and the connection with application has been lost. Similarly the semantical status of arithmetical truths with which Frege started, the problem as to whether they are analytic or synthetic, is not clarified by the move. Further, we have the problem of multiple representation, which suggest we have not "really" captured the right creatures at all, but only a myriad of reflections of them.

Perhaps we need to go back and question the deeper assumptions that numbers are well-defined objects satisfying the law of excluded middle, or that number words are names. Frege anticipated both of these possibilities, though at first he rejected them (*Grundgesetze* II, *Nachwort*, 254f.) and only at the end of his life toyed again with the second.[19] Both of these alternatives are far-reaching in their implications, and we cannot discuss them here.

I conclude by considering what Frege's unfinished theory of real numbers tells us about his logicism, since it gives us a second look at his way of going about supporting that view. The outcome is surprising. Let us assemble a number of relevant facts. He uses the natural numbers—already logically defined to his satisfaction—in showing the existence of the real numbers, but the reals are quite different objects. The reals are defined together as a structure, and

[18] See Kutschera.

[19] See the diary entries on number in *Posthumous writings*.

their definition makes essential use of the structural properties they are required to have, essentially those captured in the notion of a domain of magnitudes corresponding to a positive class. The application of real numbers in measuring magnitudes is kept in view at all times and it permeates their definition. Frege hints that different kinds of domain of magnitudes (he mentions absolute magnitudes) might come with a different kind of structure.

It is this last fact which is especially intriguing. Consider again mass as an example. Its domain of magnitudes would be plausibly assumed to have the structure of a positive class. Two possibilities emerge. One is to regard so-called absolute magnitudes as derivative from relative magnitudes, a move made by Russell and Whitehead.[20] Alternatively (and this is more like Frege's procedure) we need to define logical objects which are Relations of classes and yet which have the same structure as the non-negative reals. Do we then get a third kind of number? Consider also his hint that the rationals may be considered as ratios of powers of a Relation. Suppose we find suitable measurement practices and define the corresponding logical objects. Do we have another set of numbers distinct from those found among the reals? We could also obtain something with the structure of the non-negative rationals by taking ratios of classes or concepts according to their cardinality. What about the complex numbers? They find manifold applications in physics. If we succeed in finding logical objects in Frege's fashion to be the complex numbers, do these include the reals as hitherto defined, or are they separate as the reals are separate from the naturals? We need no longer stop at number. All kinds of pure mathematical object, such as topological spaces, graphs and lattices find application today, and theories from pure mathematics which have yet to find extra-mathematical application could do so in future. This suggests a recipe for logicism: take a theory, such as the theory of finite directed graphs, which has applications. Construct in terms of other mathematical objects (which are already given and known to be logistically accept-

---

[20]See *Principia mathematica*, Part VI, 'Quantity.' Whitehead and Russell call directed magnitudes 'vectors.' So the mass of 1 g is conceived as the mass *difference* between two objects, the first of which is 1 g more massive than the second. A project for another time is the comparison of Frege's views on reals with those of Whitehead and Russell.

able) objects which exemplify the known structural properties of the targeted objects, and define suitable courses-of-values to be the required objects, taking the applications into account in the manner of Frege.

This caricature has several points. Firstly, rather than speak of Frege's logicism it is more appropriate to speak of his logicisms. The logicistic thesis is applied locally, not globally. It is propounded and tested not—as was Russell's—for mathematics as a whole (as we knew anyway from Frege's rejection of logicism for geometry), and not even for very large theories or for several theories at once, but rather for one limited theory at a time. Success in one quarter is no guarantee of success in another. Secondly, the sticking point in trying to prove logicism for a given theory will not lie in a shortage of suitable objects: it will lie in finding, specifying and tying in the applications. This however suggests a further point. Suppose, as seems likely, that some applications differ very markedly from others, such that for instance in one case we look for a structure of classes, and in another case for the same structure among Relations. Are we forced to conclude there are two sets of logical objects which do the appropriate mathematical jobs? Such a case is more likely when one structure is embedded in another, as the natural numbers or the rationals or positive reals in the reals. It will be seen that once again the problem of multiple representation has begun to rear its head in a more general form.

I suggest that Frege and every other realist philosopher of mathematics faces a choice. On the one hand he may closely tie the objects he selects as the natural numbers, the reals, the complex numbers etc. to certain applications, as was Frege's practice. He is then faced with a proliferation of systems corresponding to logically distinct codifications of the various applications, where however there are numerous structural homomorphisms among the systems. He will then need to argue that some applications are preferable to others, or else accept the multiplicity and give a satisfactory explanation of it, or some combination of the two.

On the other hand he may loosen the tie between objects and applications, so that it becomes plain that settling upon these or those objects as numbers etc. is a matter of convention, or convenience, or indifference. Here the motto will be that the structure alone matters, while how and where it is realized does not, and that provided the structure is there, the application will follow.[21]

Frege was strongly committed to the first way of thinking, whose difficulties however he perhaps hardly began to see. The second way of thinking probably represents majority informed opinion today. But despite the difficulties raised against Frege's attitude, I do not think the case for the second way has been made conclusively; it is simply that mathematicians and logicians have voted with their feet.

Finally, can Frege's local logicisms yield adequate accounts of the relations between pure mathematics and the applications on which he (rightly) lays so much store? In both the logicisms he propounds there are disturbing indications that an adequate account may not, despite his efforts, have been attained: in the case of both natural and real numbers he in fact tailors the applications to fit the theory. For the natural numbers this is done in the antechamber to logic, by legislating that concepts have sharp boundaries, which means that 'How many ... ?' questions always have determinate answers. Real-life concepts of course very rarely have sharp boundaries, and can still be of use for practical arithmetic. In the case of real numbers real-life measurement is similarly not adequately reflected, because no actual measurements can ever be fine enough to distinguish the rationals from the reals. What Frege regards as applications turn out to be idealizations which differ from real measurement. His way of supporting logicism thus places a barrier in the way of understanding how the pure theories are applied. If this barrier is insuperable, it follows that even in his own terms (cf. II, § 92, § 159) his attempt was bound to fail. What is more, it would have failed even if the accommodating logical theory had not been inconsistent. The discovery of the contradiction in Frege's logical system might have diverted everyone's attention from a different kind of difficulty confronting his logicism.

---

[21] This view is held by Quine: see *Set theory and its logic*, Cambridge: Harvard University Press, 1969, pp. 81, 135. Thanks are due to Michael Resnik for drawing this to my attention. Quine's view probably falls under what Frege called 'patching applications on from outside' (*Grundgesetze* II, § 159).

14

Frege's Theory of Real Numbers

I. THE CONCEPT OF QUANTITY

By the end of section (*f*) of Part III.1 of *Grundgesetze* it has been fully established that Frege is proposing to define the real numbers, positive and negative, as ratios of quantifies. The last section (*g*), comprising II, §§ 160–4, sketches in outline how he intends to explain the notion of a ratio of quantities. The first question is naturally what a quantity is. This, he claims, has never yet been satisfactorily stated. 'When we scrutinise the attempted definitions, we frequently come upon the phrase "of the same type" or the like. In these definitions, it is required of quantities that those of the same type should be able to be compared, added and subtracted, and even that a quantity be decomposable into parts of the same type.'[1] To this Frege objects that the phrase 'of the same type' says nothing at all: 'for things can be of the same type in one respect, which are of different types in another. Hence the question whether an object is of the same type as another cannot be answered "Yes" or "No": the first demand of logic, that of a sharp boundary, is unsatisfied.'

'Others,' Frege continues, 'define "quantity" by means of the words "greater" and "smaller," or "increase" and "diminish"; but

Frege: philosophy of mathematics, Cambridge: Harvard University Press, 1991, Chapter 22. Reprinted by kind permission of the author.

[1] Frege here refers to Otto Stolz as an example.

nothing is thereby achieved, for it remains unexplained in what the relation of being greater, or the activity of increasing, consists.' The same goes for words like 'addition,' 'sum,' 'reduplicate' and 'synthesis';[2] 'when one has explained words in a particular context, one ought not to fancy that one has associated a sense with them in other contexts. One here simply goes round in a circle, as it seems, by always defining one word by means of another which is equally in need of definition, without thereby coming any closer to the heart of the matter.'

The mistake underlying all these attempts consists, Frege says in II, § 161, in posing the question wrongly. The essential concept is not that of a quantity, but of a *type* of quantity, or, as he prefers to say, a quantitative domain (*Grössengebiet*):[3] distances form one such domain, volumes form another, and so on. 'Instead of asking, "What properties must an object have in order to be a quantity?"' Frege says, 'we must ask, "What must be the characteristics of a concept for its extension to be a quantitative domain?"'; something is a quantity, not in itself, but in virtue of belonging, with other objects, to a class constituting a quantitative domain.

II. QUANTITATIVE DOMAINS

II, § 162 opens with the abrupt declaration that, to simplify the construction, 'we shall leave absolute quantities out of account, and concentrate exclusively on those quantitative domains in which there is an inverse,' that is, which contain positive and negative quantities. Temporal distances provide a natural example of the latter, in that they have a direction; temperatures provide a good instance of the former, since, while they have a natural zero, there can be no temperature lower than absolute zero. Given a domain of absolute quantities, we can indeed always associate with it a domain of signed ones, 'by considering e.g. one gramme as + one gramme, i.e. as the relation of a mass m to a mass m' when m exceeds m' by one gramme,' as Russell and Whitehead put it; and, as they continue,

---

[2] The last of these Frege quotes from Hankel.

[3] The term 'quantitative domain' appears very early in Frege's writings, with essentially the same meaning, namely in his Habilitationsschrift of 1874, *Rechnungsmethoden, die sich auf eine Erweiterung des Grössenbegriffes gründen.*

given a zero, we can get back to the absolute domain, since 'what is commonly called simply one gramme will ... be the mass which has the relation + one gramme to the zero of mass.'[4] Frege, however, does not even trouble to offer this much of an explanation. The restriction impairs his claim to give a comprehensive analysis of the concept of quantity, as also does his neglect of cyclic domains such as the domain of angles;[5] the magnitudes of all these, relative to a unit, are after all also given by real numbers.

Frege immediately quotes an extensive passage from Gauss.[6] This discusses the conditions under which positive and negative integers may be assigned to elements of some totality. Gauss says that the integers must be assigned, not to objects, but to relations on an underlying set of objects with a discrete linear ordering, unbounded in both directions. The relations are those any one of which any object in the set has to another separated from it in a specific direction in the ordering by a specific number of intervening objects; thus these relations are closed under composition and inverse, and include the identity relation as a zero, and form, in fact, a group of permutations.

Frege seizes upon these suggestions as supplying the main features of his characterisation of a quantitative domain. All the persuasive skill he showed in *Grundlagen* and elsewhere in convincing readers that he had given the correct analysis of intuitive concepts here deserts him. He was of course entirely right in insisting that the concept to be explained is that of a quantitative domain, not that of an individual quantity; but those at whom he jeered in II, § 160 were quite right to seize on the addition and comparability of quantities of a given type as central features, whether those quantities are absolute or distinguished as positive and negative. It is essential to a quantitative domain of any kind that there should be an operation of adding its elements; that this is more fundamental than that they should be linearly ordered by magnitude is apparent from the existence of cyclic domains like that of angles. The point

---

[4]A. N. Whitehead and B. Russell, *Principia mathematica*, Vol. III, 1913, part VI, "Quantity," p. 233.

[5]Dealt with by Russell and Whitehead in section D of their part VI.

[6]C. F. Gauss, review of his own "Theoria residuorum biquardraticorum: Commentatio secunda," *Werke*, vol. II, Göttingen, 1863, pp. 175–76.

was put very forcefully in Frege's Habilitationsschrift of 1874. He first remarks that 'one will not give a beginner a correct idea of an angle by placing a drawing of one before him ... One shows [him] how angles are added, and then he knows what they are.'[7] He subsequently generalises the point, saying that 'there is so intimate a connection between the concepts of addition and of quantity that one cannot begin to grasp the latter without the former.'[8]

We know, then, that there must be defined on any quantitative domain, in the general sense that includes absolute and cyclic ones, an operation playing the role of addition, and, on most such domains, a linear ordering playing the role of an ordering by magnitude; but we do not yet know which operation and which relation these will be, nor which objects can be elements of a quantitative domain. Frege, however, proceeds immediately to offer answers to the first and third of these questions; the second, concerning the ordering relation, receives a corresponding answer in Part III.2. Because he has decided to confine himself to quantitative domains containing negative quantities, he follows Gauss in requiring such a domain to consist of permutations of some underlying set and in taking the addition operation to be composition, under which the domain is closed; since it will also be closed under inverse, it will be a group of permutations, and, when the ordering is suitably defined, an ordered group. (Frege nowhere uses the term 'group' in *Grundgesetze*, although he must have been familiar with it.)[9]

This falls very far below Frege's usual standards of conceptual analysis. It could be argued that 'quantitative domain' should be understood as a purely structural term, on the ground that any group that has the right group structure, as subsequently analysed by Frege, will admit application of the notion of ratio as a relation between its elements, and an assignment of real numbers to those ratios, whatever those elements may be, and whatever the group operation is. But this is not Frege's position: he requires the elements to be permutations and the group operation to be composition, although he leaves the underlying set uncharacterised; but he offers

---

[7] *Rechnungsmethoden*, p. 1. (See note 3.)

[8] Ibid., p. 2.

[9] For example, from the second volume of Heinrich Weber's *Lehrbuch der Algebra*, which appeared in 1896.

no good argument for the requirement. Group-theoretically, there is no loss of generality, since every group is isomorphic to a group of permutations; but since these are, in general, permutations on the elements of the original group, this is not explanatory. The question is precisely on what underlying set the permutations Frege identifies as elements of a quantitative domain operate. In view of the generality required, this cannot be specified in the formal definition; but we need to have an idea what that set will be, in representative cases, before we can accept or even understand Frege's analysis of the notion of quantity. When the domain consists of spatial or temporal distances, there is no problem: the underlying set is naturally taken to comprise points or instants. What, however, when the domain consists of masses? The suggestion of Whitehead and Russell, as it stands, represents signed masses as permutations on absolute masses; if we follow it, we need to know what a domain of absolute quantities is before we can know what a domain with positive and negative quantities is. It might be proposed that the underlying set should be taken to consist of the physical objects to which absolute masses are assigned. We could not then assume, however, that the group of permutations with which Frege identifies the quantitative domain contained all the elements it was required to have to be a quantitative domain on his definition: it is not true a priori that, for every conceivable mass, there is an object that has that mass.

Frege has thus not achieved a convincing analysis of the concept of a quantitative domain. His illustration, in II, § 163, does not greatly help: it is the usual one, used by Veronese, Hölder and Cantor, of distances along a straight line; the underlying set comprises its points, and the permutations forming the quantities of the domain are displacements along it. The example shows, indeed, that some quantitative domains conform to Frege's model; it is powerless to show that all can be so characterised. An adequate general characterisation of the notion of quantity would pay much more attention to how it is applied in practice; it would also embrace absolute domains, cyclic domains, and domains of vectors of more than one dimension. Frege is so anxious to press on to his definition of real numbers that he ignores all quantitative domains save those that have the structure of the real line; as a result he offers a highly defective analysis of the concept on which he fastens so much attention. Possibly this deficiency would have been corrected in a Part IV which never saw the light of day.

What would not have been corrected was the philosophical naivety of taking it for granted that every quantity has a precise value representable by the assignment to it of a real number relatively to a unit but discoverable by us only to within an approximation. We are led to adopt this picture by devising ever more accurate methods of measurement; but with what right do we assume that its limit is a point, and not an interval, or at least that it is an interval with precise end-points, rather than with fuzzy ones? It would be absurd to say that we impose the system of natural numbers upon reality; but it is not at all absurd to say the same about the mathematical continuum. We are not given physical reality as a set of instantaneous states arrayed in a dense, complete ordering: we apprehend it only over temporal intervals. The idea of discontinuous change is not, of itself, conceptually abhorrent; we commonly think of ourselves as experiencing it, as when darkness succeeds illumination when the light is switched off. More exact examination shows that such changes, at the macroscopic level, are in fact continuous; but that does not make the idea of such simple discontinuities absurd. We could, for instance, understand the idea that the colour of a surface might abruptly change from, say, red to green. What is conceptually absurd is to apply to such a change the distinction that can be made with Dedekind cuts, asking what colour the surface was at the instant of change: there are not two distinct possibilities, according as it was then red or then green. Yet more absurd would be the idea of the surface's being red through an interval, save at one particular moment, when it was green. These are not *physical* absurdities, violating well known laws of physics: they are much deeper absurdities, *conceptual* absurdities. And they suggest that the mathematical continuum fits physical reality somewhat imperfectly, yielding apparent logical possibilities that are no possibilities at all. We are familiar with the thought that quantities obtained by differentiation, like velocity and acceleration, do not possess their values at any particular moment in logical independence of what their values are at all other moments; but the foregoing examples suggest that the same is true of *all* quantities, even the fundamental ones, so that these are not 'loose and separate,' as Hume absurdly said. But, if so, the mathematical continuum is not the correct model for physical reality, but only one we use because we do not have a better. In regarding real numbers as 'measurement-numbers,' Frege was treating of a wholly idealised conception of their application, instead of giving an analysis of our actual procedures of measurement and their underlying assumptions. By doing so, he skimped the task he had set himself.

III. HÖLDER

Frege was not as out of step with other mathematicians as he imagined. Only two years before the second volume of *Grundgesetze* appeared, Otto Hölder published an article treating of much the same topic as Part III of that work.[10] A comparison between them is extremely instructive. Hölder is aiming at a general theory of measurable quantity. He is as explicit as Frege about the need for generality, and criticises earlier work by Veronese for failing to separate the general axioms of quantity from the geometrical axioms governing segments of a straight line.[11] Hölder characterises absolute quantitative domains, without a zero quantity; he does so axiomatically in terms of an operation of addition, assumed associative, and a linear ordering relation, assumed dense, complete and left- and right-invariant, both taken as primitive. Such a domain is then an ordered upper semigroup, although, like Frege, Hölder does not use explicit group-theoretic terminology. He appears to have been the first to give a correct proof of the archimedean law from the completeness of the ordering, and also to prove the commutativity of addition from the archimedean law. As we shall see, Frege

[10]O. Hölder, "Die Axiome der Quantität und die Lehre vom Mass," *Berichte über die Verhandlungen der Königlich Sächsischen Gesellschaft der Wissenschaften zu Leipzig: mathematische und physikalische Klasse*, **53** (1901) 1–64. It was this Otto Hölder after whom the Jordan-Hölder theorem is (in part) named. In his article, Hölder does not mention Frege, but expresses himself as of the same opinion as he in regarding arithmetic as purely logical. He is, however, quite unaware of the advances in logic that Frege had pioneered, and remarks in a footnote (p. 2, fn. 1) that arithmetical proofs cannot be rendered in any existing logical calculus.

[11]Hölder, p. 37, fn. 1; see G. Veronese, "Il continuo rettilineo e l'assioma V d'Archimede," *Atti della Reale Academia dei Lincei*, series 4, memorie della classe delle scienze fisiche, matematiche e naturali, **6** (1889) 603–24.

obtained similar theorems in his Part III.2; but Frege's theorems are more powerful than those of Hölder, because his assumptions are considerably weaker.[12]

For n a positive integer, and a a quantity, Hölder easily defines the multiple na in terms of addition. He proceeds to characterise the notion of a ratio between two quantities, and associates a real number with every such ratio. Unlike Frege, however, he does not *construct* the real numbers by this means. Rather, he first defines the positive rational numbers, in effect as equivalence classes of pairs of positive integers.[13] He then takes the real numbers to be defined by Dedekind's method, which he sets out without Dedekind's own appeal to mathematical creation, identifying the real numbers with the corresponding cut in the rational line in which the lower class has no greatest element.[14]

The correct definition of ratio, given addition and therefore multiples, was well-known, having been framed by Euclid,[15] and Hölder appeals expressly to it; it allows the comparison of ratios between pairs of elements of different domains, provided each has an operation of addition, but Hölder confines himself to comparisons within a single domain. Intuitively, we shall want to associate the rational number n/m with the ratio of a to b when $ma = nb$. Euclid defines a as having the same ratio to b as c has to d when, for all positive integers n and m, ma is smaller than (equal to, greater than) nb if and only if mc is smaller than (equal to, greater than) nd.[16] Hölder's contribution is to notice the close connection between the

---

[12]If the conjecture that Vol. II of *Grundgesetze* was already written when Vol. I was published is correct, Frege could have had the priority if he had published sooner; but the mathematical community would not have accorded it to him, because nobody troubled to read Vol. II.

[13]He actually says, rather vaguely, that all equivalent pairs 'represent, in accordance with our (arbitrary) interpretation, an object which we designate a *rational number*' (Hölder, p. 20).

[14]The phrase used is again slightly vague: a cut 'can be regarded as representing' a rational or irrational number, and, in the first case, 'identified with it straight out,' and, in the second, 'called an irrational number straight out'; Hölder, p. 22.

[15]Euclid, *Elements*, book 5, definition 5.

[16]The bracketed expressions occur in Euclid's definition, but are here superfluous.

ideas of Euclid and of Dedekind.[17] For, in view of the archimedean law, every ratio between quantities determines a Dedekind cut in the rational line, and hence has the real number corresponding to that cut associated with it.

In the second part of his paper, Hölder applies his theory to everybody's favourite example, of directed segments of a straight line. The interest of the example, for him, lay in its indicating how to handle dual domains of opposite quantities, together forming a domain of positive and negative quantities of the kind Frege concerned himself with; but we need not follow the details of Hölder's treatment.

It is a matter for the deepest regret that neither Frege nor Hölder ever became aware of the other's work. Had he had to comment on Hölder's theory in his Part III.1, he could not have dismissed it so lightly as he in fact dismissed Dedekind's theory: it shows very clearly how that theory can be applied to ratios of quantities. In doing so, it also brings out more sharply than before the exact nature of Frege's objection to such a theory as Dedekind's. Hölder, like everyone else except Frege, first defines the rationals, essentially as ratios between positive integers, and then defines the real numbers in terms of them. For that reason, although the principles underlying the use both of rationals and of irrationals to give the magnitude of a ratio between quantities are very direct, they are still external to the definitions of the numbers themselves. Frege, by insisting that rationals and irrationals should be defined together, made it necessary that that application of them be internal to their definition. Put in that way, the difference between Frege and Dedekind, once we set aside the matter of free creation by the human mind, becomes much narrower than one might suppose from Part III.1 of *Grundgesetze*. There is a significant methodological difference: for Frege, the theory of quantity is an integral part of the foundations of analysis, not a mere addendum of interest primarily to applied mathematicians. But the mathematical difference becomes more slender. In particular, if he had reached the point in Part III.2 of defining ratios, Frege would have had to use the Euclidean defini-

---

[17]In the note to p. 29, Hölder very properly points out that Dedekind himself acknowledged the affinity in the Preface to *Was sind und was sollen die Zahlen?*

tion, or something very like it, and would thus have come quite close
to Dedekind's conception of the real numbers.

IV. THE EXISTENCE OF A QUANTITATIVE DOMAIN

In II, § 164, which concludes Part III.1, Frege resolves the doubt
expressed in II, § 159. In order to ensure the existence of the real
numbers, at least one quantitative domain must be proved to exist,
containing quantities bearing irrational ratios to one another; for, if
it did not, the real numbers, defined as ratios of elements of such a
quantitative domain, would all be equal to one another and to the
null relation. Furthermore, the proof must use only logical resources.
As in all cases, the domain will consist of permutations on an underlying set. Frege observes that the set underlying such a domain
must have a cardinality higher than the class of natural numbers;
he mentions the fact (not proved in Part II) that the number of
classes of natural numbers is greater than the number of natural
numbers, but fails to make any acknowledgement to Cantor.[18] He
therefore proposes to use classes of natural numbers in specifying
the underlying set.

If we temporarily assume the irrational numbers known, Frege
continues, we can regard every positive real number a as representable in the form

$$r + \sum_{k=1}^{k=\infty} 2^{-n_k}$$

where r is a non-negative integer, and n_1, n_2, ... form an infinite
monotone increasing sequence of positive integers. This amounts
to giving the binary expansion of a (in descending powers of 2, as
a decimal expansion is in descending powers of 10); the expansion
is chosen to be non-terminating, so that $1/2$ is represented by the
infinite series $1/4 + 1/8 + 1/16 \ldots$. Thus to every positive number a,
rational or irrational, is associated an ordered pair, whose first term
is a non-negative integer r and whose second term is an infinite class
of positive integers (which suffices to determine the sequence); these
may be replaced respectively by a natural number and an infinite
class of natural numbers not containing 0.

---

[18]This omission is truly scandalous; Frege would never have displayed such
ill manners at the time of writing *Grundlagen*.

This, then, is the underlying set; the permutations on it are to be defined in some such way as the following. For each positive real number b there is a relation holding between other positive real numbers a and c just in case $a + b = c$. This relation can be defined, Frege says, without invoking the real numbers a, b and c, and thus without presupposing the real numbers. He does not here give the definition; the following should serve the purpose. Suppose given an ordered pair $\langle s, B \rangle$, where s is a natural number and B an infinite class of natural numbers not containing 0: we want to define a relation between similar such pairs $\langle r, A \rangle$ and $\langle t, C \rangle$. Let us first say that a natural number n is *free* if, for every $m > n$ such that m belongs both to A and to B, there is a number k such that $n < k < m$ belonging neither to A nor to B. We may then say that our relation holds if the following two conditions are fulfilled:

(i) for each n, n belongs to C if and only if n is positive and either is free and belongs to one of A and B but not the other, or is not free and belongs either to both A and B or neither;

(ii) $t = r + s$ if 0 is free, and $t = r + s + 1$ if 0 is not free.

This definition is intended to determine the relation as holding between $\langle r, A \rangle$ and $\langle t, C \rangle$ just in case $a + b = c$, where $\langle r, A \rangle$, $\langle s, B \rangle$ and $\langle t, C \rangle$ intuitively represent the real numbers a, b and c respectively. As Frege observes, we now have such relations corresponding to every pair $\langle s, B \rangle$; taken together with their inverses, these correspond one to one to the positive and negative real numbers; and to the addition of the numbers b and b' corresponds the composition of the corresponding relations. 'The class of these relations (*Relationen*),' he says, 'is now a domain that suffices for our plan,' but adds that 'it is not thereby said that we shall hold precisely to this route.'

He could not hold precisely to it, because, in the coming series of formal definitions, he requires a quantitative domain to consist of permutations on an underlying set; that is to say, he requires the relations it comprises to be one-one, all to be defined on the same domain and to have a converse domain identical with their domain. The relations mentioned in II, § 164, and formally defined above, are not, however, permutations: the operation of adding the positive real number b carries the positive real numbers into the real numbers greater than b. In Volume II, he does not reach the formal proof of the existence of a quantitative domain. If, when he did, he had wanted to use additive transformations, he would have had to take the underlying set to be isomorphic to all the real numbers, positive, negative and 0, which would have been somewhat more complicated; if he had wanted the underlying set to be isomorphic just to the positive reals, he would have had to use multiplicative transformations, which would have been very much more complicated to define with the resources available. There is, of course, no actual doubt that either could be done.

V. THE FORMAL TREATMENT

When the reader comes to the formal development in Part III.2, much has been settled. The first problem is to characterise a quantitative domain; and he knows that it must be an ordered group of permutations satisfying a number of conditions. The mathematical interest of the work is considerable; it is a thoroughgoing exploration of groups with orderings, yielding, as already noted, theorems more powerful than those proved by Otto Hölder in the paper discussed above. The interest is not due to Frege's ultimate purpose: he could simply have laid down all the conditions he wanted a quantitative domain to satisfy and incorporated them in a single definition. The interest is due, rather, to Frege's concern for what we should call axiomatics, that is, for intellectual economy: as he explains in II, § 175, he wants to achieve his aim by making the fewest assumptions adequate for the purpose, ensuring that those he does make are independent of one another. Hence, although a quantitative domain will prove in the end to be a linearly ordered group in the standard sense, in which the ordering is both left- and right-invariant, many theorems are proved concerning groups with orderings not assumed to be linear or to be more than right-invariant.

Before we proceed further, a word is in place concerning Frege's formal apparatus. A reader unfamiliar with it may have felt uncertain whether his quantitative domains contain objects, relations or functions. The answer is that they contain objects, but objects which are extensions of relations. The formal system of *Grundgesetze* contains expressions for functions both of one and of two arguments; these include both one-place and two-place predicates, that is, expressions both for concepts and for binary relations (*Beziehung-en*). There is, however, no special operator for forming terms for value-ranges of functions of two arguments: this is accomplished by reiterated use of the abstraction operator (symbolised by the smooth breathing on a Greek vowel)[19] for forming terms for value-ranges of functions of a single argument. Thus '$\dot{\varepsilon}(\varepsilon+3)$' denotes the value-range of the function that maps a number x on to $x+3$; and so '$\dot{\alpha}\dot{\varepsilon}(\varepsilon+\alpha)$' denotes the value-range of the function that maps a number y on to $\dot{\varepsilon}(\varepsilon+y)$. This 'double value-range' is then taken by Frege as the extension of the binary function of addition. In the same way, '$\dot{\varepsilon}(\varepsilon<3)$' denotes the class of numbers less than 3, while '$\dot{\alpha}\dot{\varepsilon}(\varepsilon<\alpha)$' denotes the value-range of the function that maps a number y on to the class of numbers less than y. This, being the double value-range of a relation (*Beziehung*), in this case the 'less-than' relation, is identified by Frege with its extension, standing to it as a class to a concept; the extension of a relation, being an object is called a *Relation*, to distinguish it from a relation proper. This is just an example of how, throughout *Grundgesetze*, Frege is able to work with value-ranges in place of concepts, relations and functions. A quantitative domain contains *Relationen*—extensions of relations— rather than relations in the true sense: specifically, extensions of one-one relations on an underlying set. We may, for brevity, call these 'permutations'; throughout Part III.2, Frege works exclusively with value-ranges of various kinds, concepts, relations and functions hardly ever making an appearance. For this reason, the word 'relation' itself will henceforth be understood in the sense of '*Relation*,' namely as applying to the extension of a relation (*Beziehung*) in the proper sense.

Frege begins by announcing that addition—that is, composition of permutations—in a quantitative domain must satisfy the commutative and associative laws. He then proves that composition of relations is always associative.[20] It is by no means always commu-

---

[19]Some commentators on Frege write the smooth breathing over Greek consonants, which looks extremely odd. Of course, there is no logical mistake; but Frege never used Greek consonants as bound individual variables, and it would surely have offended his sense of propriety to write a breathing over them if he had.

[20]Composition of relations was defined in I, § 54.

tative, as he remarks. A special case in which it is is first singled out by Frege, namely the class consisting of a relation p together with all its iterations $p|p$, $p|(p|p)$, ... (Here the symbol $|$ is used for composition, in place of Frege's own; no attempt will be made to reproduce his symbolism.) Frege uses his definition of the ancestral to express membership of this class without reference to natural numbers (and hence to multiples of the form np as used by Hölder). Even when p is a permutation, the class of its positive multiples will not always be a group. In this connection, Frege defines an important notion, that of the *domain* of a class P of relations. This consists of P together with the identity and the inverses of all members of P. If P is the class of all multiples of a permutation p, its domain will of course be the cyclic group generated by p; but it should be noted that the domain of a class of permutations will not always be the group generated by it, since it is not required to be closed under composition.

The next problem is how to introduce the notion of order. Frege chooses to do it by defining the conditions for a class to consist of the positive elements of a group of permutations on which there is an ordering, and defining the ordering in terms of that class. His first approach is to introduce the notion of what he calls a *positival class*. A positival class is a class of permutations on some underlying set satisfying the following four conditions:

(1) if p and q are in P, so is $p|q$;

(2) the identity e is not in P;

(3) if p and q are in P, then $p|q^-$ is in the domain of P;

(4) if p and q are in P, then $p^-|q$ is in the domain of P.

Here 'p^-' denotes the inverse of p. If P is a positival class according to the foregoing definition, the domain of P will be the group generated by P. Frege goes on to introduce an order relation on the group by setting p less than q if and only if $q|p^-$ is in P. It follows immediately that the order relation $<$ thus defined is right-invariant,[21] that is, that if $p < q$, then $p|r < q|r$ for any element r of the group,

---

[21] Frege's permutations are one-one relations, not functions, and his symbol for composition is defined like Russell's relative product: if x stands to y in the p-relation iff x is the father of y, and in the q-relation iff x is the mother of y, then x stands to y in the $p|q$-relation iff x is the maternal grandfather of y. In standard group-theoretical notation, this would be written qp, the symbol for

and, further, that P is the set of elements of the group greater than the identity e (the set of positive elements). Furthermore, it follows easily from (1) and (2) that $<$ is a strict partial ordering of the group (i.e. is transitive and asymmetrical).

Frege is, however, extremely worried that he is unable to establish whether or not condition (4) is independent of the other three. In fact, it is;[22] Frege, uncertain of the point, proceeds to prove as much as he can, from II, §§ 175–216, without invoking clause (4), and calls attention, in § 217, to the fact that at that stage he finds himself compelled to do so.

If clause (4) does not hold, the domain of P will not constitute the whole group generated by it, which will in fact be the domain together with the elements $p^-|q$ for p and q in P. We may nevertheless still consider the order relation as defined over the whole group. Clause (3) in effect says that $<$ is a strict linear ordering of P, and is equivalent to the proposition that it is a strict upper semilinear ordering of the group. This means that it is a strict partial ordering such that the elements greater than any given one are comparable, and that, for any two incomparable elements, there is a third greater than both of them: pictorially, it may branch downwards, but cannot branch upwards. Clause (4) says that $<$ is a strict linear ordering of the negative elements (those less than e), and is equivalent to the proposition that $<$ is a strict lower semilinear ordering of the group (where this has the obvious meaning). (3) and (4) together are therefore tantamount to the proposition that $<$ is a strict linear ordering of the group. If the ordering is left-invariant, clause (4) must hold, since, if $p < q$, by left-invariance $e < p^-|q$, i.e. $p^-|q$ is in P. (The converse, however, does not hold: a group may have a right-invariant linear ordering that is not left-invariant.) Frege's independence problem thus amounts to asking whether there is a group with a right- but not left-invariant upper semilinear ordering that is not linear. Since in fact there is, the theorems that he takes

---

the operation to be applied first being written first. Using that notation, one would say that Frege defined his order relation to be *left*-invariant; but it seems less confusing to stick to a notation that accords with Frege's in respect of the order in which the variables are written.

[22] See Peter M. Neumann, S. A. Adeleke, and Michael Dummett, "On a question of Frege's about Right-Ordered Groups," Chapter 15, below, Theorem 2.1.

care to prove without invoking clause (4) hold for a genuine class of groups.

The notion of a positival class was only a preliminary approach to that which Frege wants, namely that of a *positive* class. This is a positival class P such that the ordering $<$ is dense and complete. To characterise the notion of completeness, Frege has of course to define the notion of the least upper bound of a subclass A of P. His definition does not agree with what appears to us the obvious way of defining the notion. He uses as an auxiliary notion what we might call that of an 'upper rim' of the class A: r is an upper rim of A in P if and only if A contains every member of P less than r (Frege gives no verbal rendering of this notion, but only a symbol). What he calls an 'upper bound' (*obere Grenze*) or simply 'bound' of A in P is now defined to be an element r of P which is an upper rim of A in P and is not less than any other upper rim of A in P which belongs to P. Since $<$ linearly orders P, there can be at most one upper bound, in this sense, of a class A: it is the greatest lower bound, in our sense, of the complement of A. If A is such as to contain every element of P smaller than any element it contains, Frege's upper bound of A will be its least upper bound in the usual sense. Frege's formulation of the condition for $<$ to be complete in P is that, if some member of a class A is an upper rim of A in P, but there is an element of P not in B, then some member of P is an upper bound of A in P.

Frege continues his policy of avoiding appeal to clause (4) even after introducing the notion of a positive class. Oddly, he does not raise the question whether clause (4), if independent of clauses (1), (2) and (3), remains independent after the addition of the assumptions of completeness and density; as we shall see, it does not. Frege is concerned with the archimedean law, that, for any positive elements p and q, there is a multiple of p which is not less than q; he formulates it with the help of the class of multiples of an element mentioned above. The most important theorems that he proves are as follows:

Theorem 635 (II, § 213). *If $<$ is a complete upper semilinear ordering, then the archimedean law holds.*

Hölder had derived the archimedean law from the completeness of the ordering in his paper of two years earlier, but he was using considerably stronger assumptions than Frege's, namely that the ordering is dense, left-invariant and linear. The completeness of the ordering is needed to obtain the real numbers; but it is the archimedean law that is important in the subsequent theorems.

Frege employed, though did not name, an interesting and fruitful concept, namely that of a restricted kind of left-invariance which we may express as the ordering's being 'limp' ('*left-invariant* under *m*ultiplication by *p*ositive elements'). The ordering has this property if, whenever $q < r$, and p is positive, then $p|q < p|r$. The next theorem uses this notion.

Theorem 637 (II, § 216). *If $<$ is an upper semilinear, archimedean ordering, then $<$ is limp.*

These two theorems have been so expressed in virtue of Frege's avoidance, in their proofs, of appeal to clause (4). The next two theorems do appeal to it.

Theorem 641 (II, § 218). *If $<$ is a linear, archimedean ordering, then $<$ is left-invariant.*

Theorem 689 (II, § 244). *If $<$ is a dense, linear, archimedean ordering, then the group is abelian: that is, the commutative law holds for composition.*

Hölder also derived commutativity from the archimedean law, but he had to assume left-invariance, whereas, for Frege, left-invariance was automatic by Theorem 641. The assumption of density is unnecessary; but Frege's appeal to it in his proof is not a fault, since different proofs are needed for the two cases.

With the help of Frege's Theorem 637, a further improvement can be obtained, namely the

Theorem. *If $<$ is an archimedean, upper semilinear ordering, $<$ is linear and the group is abelian.*[23]

Thus clause (4) is no longer independent in the presence of the assumption of completeness, or even just of the archimedean law, which then suffices to prove commutativity.

[23]See Theorem 3.1 of Chapter 15.

With Theorem 689, Frege reached the end of the quest for a proof of the commutative law announced at the very beginning of Part III.2, and therewith the end of Volume II (save for the Appendix on Russell's paradox). A quantitative domain, in the narrow sense, could now with assurance be identified with the domain of a positive class.

In his brief concluding II, § 245, Frege announces as the next task to prove the existence of a positive class, along the lines indicated in II, § 164. That, he says, will open up the possibility of defining real numbers as ratios of quantities belonging to the domain of the same positive class. 'And we shall then also be able to prove that the real numbers themselves belong as quantities to the domain of a positive class.'

The missing conclusion of Part III.2 would have been laborious, but would have presented no essential difficulties. The device of II, § 164 would have had to be amended a little; but this would have required nothing but work. Frege would have had essentially to use Euclid's definition of the condition for the ratio of a quantity p to another quantity q of some domain D to coincide with that of a quantity r to a quantity s, both belonging to a domain E, whether the same as D or distinct from it. He would not have defined a phrase containing 'the same' or 'coincides with,' but would have defined an equivalence relation between ordered pairs of quantities. (He had defined an ordered pair in I, § 144, as the class of relations in which the first term stood to the second.) Nor, when he had hitherto refrained from appealing to the natural numbers in characterising multiples of quantities, would he have been likely to start doing so at this point; but his definition would of necessity have been essentially Euclid's, all the same. This definition would give the criterion of identity for ratios; we might therefore naturally expect Frege then to define the real numbers by logical abstraction, i.e. as equivalence classes of ordered pairs of quantities. This, however, would not yield the result he demands in his last sentence, that the real numbers should themselves form a quantitative domain, because they would then have to be extensions of relations, which are not, for Frege, classes of ordered pairs, but double value-ranges. He would therefore have had to use a variation of the method. A real number would have to be the relation between a quantity r and a quantity s of the same domain which obtained when r stood to s in the same ratio as some fixed quantity p stood to another fixed quantity q of the same domain, i.e. when the pair $\langle r, s \rangle$ stood in the relevant equivalence relation to the pair $\langle p, q \rangle$.

If we imagine the axioms governing value-ranges to be quite different, yielding a consistent theory analogous to ZF set theory, and Frege's notion of an ordered pair replaced by the modern one, there would be no trouble about any of the work in Volume II, Part III, and none about the proof of the existence of a positive class. The definition of the real numbers as ratios would, however, be blocked, because their domain, as relations, would be the union of all domains of positive classes, and the class of such domains would certainly be a proper class. This, of course, was precisely the fate of Frege's definition of cardinal numbers, including the natural numbers. The paradoxes of set theory imposed limits quite unexpected by him upon definition by logical abstraction.

PETER M. NEUMANN
S. A. ADELEKE
MICHAEL DUMMETT

15

On a Question of Frege's about Right-ordered Groups

Dedicated with respect and good wishes to Graham Higman
to mark his seventieth birthday, January 1987

I. THE PROBLEM AND ITS BACKGROUND

Let G be a group and P a subset of G. We shall be interested in
the following assertions:

(1) $p, q \in P \implies pq \in P$;
(2) $1 \notin P$;
(3) $p, q \in P \implies pq^{-1} \in P \vee p = q \vee qp^{-1} \in P$;
(4) $p, q \in P \implies p^{-1}q \in P \vee p = q \vee q^{-1}p \in P$.

Frege's question[1] is whether (4) follows from (1), (2) and (3). We
shall amplify the question and answer it negatively below.

A binary relation $<$ on G may be defined in terms of P by the
rule

$$a < b \text{ if and only if } ba^{-1} \in P.$$

It is easy to see that $<$ is right-invariant (that is, $a < b \implies ag < bg$
for all $g \in G$) and $P = \{ p \in G \mid 1 < p \}$. Properties of $<$ may be

From *Bulletin of the London Mathematical Society*, **19** (1987) 513–21. Reprinted by kind permission of the London Mathematical Society and the authors.
[1]Posed in *Die Grundgesetze der Arithmetik*, Band II, Jena, 1903; reprinted
with Band I, Hildesheim: Georg Olms, 1966.

characterised either directly or in terms of P. For example, as is well known and easy to prove, $<$ is a strict partial order if and only if (1) and (2) hold.

A binary relation ρ on a set A will be said to be *weakly connected* if for any $a, b \in A$ there exists a sequence $a_0, a_1, \ldots, a_n$ connecting a to b, that is, a sequence of members a_i of A such that $a_0 = a$, $a_n = b$ and $a_i \rho a_{i+1}$ or $a_{i+1} \rho a_i$ for $0 \leqslant i \leqslant n - 1$.

Lemma 1.1. *The relation $<$ is weakly connected if and only if P generates G.*

Proof. Suppose first that P generates G. It is sufficient to show that for each $g \in G$ there is a sequence connecting g to 1. Our supposition means that g may be written as a product $g_0 g_1 \ldots g_m$, where $g_i \in P$ or $g_i^{-1} \in P$ for $0 \leqslant i \leqslant m$. We set $a_i := g_i g_{i+1} \ldots g_m$ for $0 \leqslant i \leqslant m$, and $a_{m+1} := 1$. Then $a_i a_{i+1}^{-1} = g_i$ for each i, so that $a_{i+1} < a_i$ if $g_i \in P$ and $a_i < a_{i+1}$ if $g_i^{-1} \in P$. Thus $a_0, a_1, \ldots, a_{m+1}$ is a sequence connecting g to 1.

Conversely, suppose that $<$ is weakly connected and, for $g \in G$, suppose that $b_0, b_1, \ldots, b_n$ is a sequence connecting 1 to g. We show by induction that $b_i \in \langle P \rangle$ for each i. Certainly $b_0 = 1 \in \langle P \rangle$. If $b_i < b_{i+1}$ then $b_{i+1} b_i^{-1} \in P$ and if $b_{i+1} < b_i$ then $(b_{i+1} b_i^{-1})^{-1} \in P$. In either case, if $b_i \in \langle P \rangle$ then $b_{i+1} = (b_{i+1} b_i^{-1}) b_i \in \langle P \rangle$. By induction, $g = b_n \in \langle P \rangle$. Thus $\langle P \rangle = G$. $\square$

A binary relation ρ will be said to be a (strict) *upper semilinear ordering*[2] if

(i) ρ is a (strict) partial ordering;
(ii) $a\rho b$ & $a\rho c \implies b\rho c \vee b = c \vee c\rho b$;
(iii) $a\rho b \vee a = b \vee b\rho a \vee (\exists c)(a\rho c$ & $b\rho c)$.

Lemma 1.2. *Suppose that P satisfies (1) and (2), and that P generates G. Then (3) holds if and only if $<$ is a (strict) upper semilinear order on G.*

[2]Following S. A. Adeleke and Peter M. Neumann, "Semilinearly ordered sets, betweeness relations and their automorphism groups" (in preparation); see also Manfred Droste, *Structure of partially ordered sets with transitive automorphism groups*, Memoirs of the American Mathematical Society **334**, Providence, R. I.: American Mathematical Society, 1985.

Proof. Suppose first that (3) holds. Certainly $<$ is a strict partial ordering of G (because (1) and (2) hold), and (3) says that $<$ is a strict linear ordering of P. Suppose that $a, b, c \in G$, $a < b$ and $a < c$. Then $ba^{-1}, ca^{-1} \in P$, whence $ba^{-1} < ca^{-1}$ or $b = c$ or $ca^{-1} < ba^{-1}$. By right-invariance of $<$ we have $b < c$ or $b = c$ or $c < b$, so that $<$ satisfies (ii). To prove (iii) we use the fact (Lemma 1.1) that $<$ is weakly connected. For $a, b \in G$, let $a_0, a_1, \ldots, a_n$ be a sequence connecting a to b and chosen so that n is as small as possible. Then certainly $a_i \neq a_{i+1}$, if $a_i < a_{i+1}$ then $a_{i+2} < a_{i+1}$, and if $a_{i+1} < a_i$ then $a_{i+1} < a_{i+2}$, otherwise in each case we could have deleted a_{i+1} and obtained a shorter sequence connecting a to b. If $a_{i+1} < a_i$ and $a_{i+1} < a_{i+2}$, however, then (by (ii)) a_i and a_{i+2} would be comparable and so a_{i+1} again could be deleted to give a shorter sequence. Thus the only possibilities are that $n = 0$ (in which case $a = b$), $n = 1$ (in which case $a < b$ or $b < a$), or $n = 2$ and $a = a_0 < a_1$, $b = a_2 < a_1$. Therefore (iii) holds and $<$ is a strict upper semilinear ordering of G.

Now suppose conversely that $<$ is a strict upper semilinear ordering. If $p, q \in P$ then by (ii) we have that $p < q$ or $p = q$ or $q < p$, that is, $qp^{-1} \in P$ or $p = q$ or $pq^{-1} \in P$, which is (3). $\square$

Lemma 1.3. *Suppose that P satisfies (1) and (2), and that P generates G. Then (3) and (4) both hold if and only if $<$ is a strict linear ordering of G.*

Proof. By an obvious duality, (4) holds if and only if $<$ is a strict lower semilinear ordering of G. Thus if (3) and (4) both hold then $<$ is both an upper semilinear order and a lower semilinear order, hence is a linear order. Conversely, if $<$ is a linear order then it is both an upper and lower semilinear order and so (3) and (4) both hold. $\square$

In the light of Lemmas 1.2 and 1.3 we may transform Frege's problem to the question *is there a group with a right-invariant, upper semilinear ordering that is not linear?* It is to this form that we shall give a positive answer in § 2 below.

Frege's question comes from his uncompleted logical construction of the real numbers in *Grundgesetze*. Having constructed (in Volume I) the natural numbers, but not the integers or rational numbers, Frege wished to represent the real numbers as ratios between quantities belonging to the same domain (for example, that of lengths or masses): the problem was then how to characterise a domain of quantities. Unknown to Frege, Hölder had tackled the same problem.[3] He had considered a domain of absolute quantities, not containing a zero quantity, but forming an ordered semigroup, which he described axiomatically in terms of an ordering $<$ and a binary operation $+$ of addition. Frege, on the other hand, considers domains of positive and negative quantities, which he takes to consist of permutations (and which ultimately turn out to be ordered permutation groups—although neither Frege nor Hölder uses explicit group-theoretic ideas). Thus for Frege a domain of quantities is to be, for some set R, a subset, ultimately a subgroup G of the symmetric group $\mathrm{Sym}(R)$. He proceeds in two stages, first defining what he calls a 'positival' class, later adding conditions for it to be a 'positive' class.

A *positival class* is a set P of permutations of the set R satisfying conditions (1), (2), (3), (4) above (and Frege's domain of quantities is $P \cup \{1\} \cup P^{-1}$, which is then the group generated by P). However, concerned as he is to ensure the independence of the clauses of his definitions, Frege is perturbed by his inability to establish whether (4) is independent of (1), (2), (3) (see *Grundgesetze* II, §§ 175, 245). He therefore proves as much as he can without appeal to clause (4), and he calls attention to the fact that he has not so far invoked it at the point (II, § 217) where he first feels compelled to do so.

For the purpose of resolving Frege's independence problem, the fact that P is a subset of $\mathrm{Sym}(R)$ may be ignored. For if H is any group with a right-invariant, weakly connected, partial ordering $\prec$, we may take P to be the subset of $\mathrm{Sym}(H)$ consisting of the right translations $x \mapsto xp$ by elements p of H with $1 \prec p$. Then P will satisfy (1) and (2), and the subgroup G of $\mathrm{Sym}(H)$ generated by P will consist of all right translations of H, hence will be isomorphic to H; moreover, if $a, b \in G$, $a : x \mapsto xg$ and $b : x \mapsto xh$ for all $x \in H$, then $ab^{-1} \in P$ if and only if $h \prec g$. This justifies our treatment of Frege's question in the general context of abstract right-ordered

[3] In his paper, "Die Axiome der Quantität und die Lehre vom Mass," *Berichte über die Verhandlung der königlichen sächsichen Gesellschaft der Wissenschaften zu Leipzig, mathematische-physicalische Klasse*, **53** (1901) 1–64.

groups.

Suppose now that G is a group and that $<$ is a right-invariant, weakly connected, strict partial ordering of it (so that $<$ is determined by a subset P that generates G and satisfies (1) and (2)). We shall be concerned with various further properties that $<$ and G may have:

(a) the relation $<$ is a strict upper semilinear ordering of G;
(b) the relation $<$ is a strict linear ordering of G;
(c) the relation $<$ is limp;
(d) the relation $<$ is left-invariant;
(e) the relation $<$ is dense;
(f) the relation $<$ is complete;
(g) the relation $<$ is archimedean;
(h) the group G is abelian.

Property (c) requires explanation: we say that $<$ is *limp* (*l*eft-*i*nvariant under *m*ultiplication by *p*ositive elements) if for all $p, q, r \in G$ we have that

$$1 < p \ \& \ q < r \implies pq < pr.$$

We have seen in Lemma 1.2 that (a) is equivalent to the assertion that P satisfies (3), and in Lemma 1.3 that (b) is equivalent to the assertion that P satisfies both (3) and (4). Condition (c) is equivalent to

$$(5) \qquad\qquad p, q \in P \implies pqp^{-1} \in P;$$

condition (d) is equivalent to

$$(6) \qquad\qquad p \in P \ \& \ g \in G \implies gpg^{-1} \in P;$$

and condition (e) is equivalent to

$$(7) \qquad\qquad (\forall p \in P)(\exists q \in P)(pq^{-1} \in P).$$

The proofs are straightforward and we omit them.

Although the language and notation he uses are very different from ours (in particular, he uses the phrase 'p *ist grösser als q*' to describe the relation $pq^{-1} \in P$ rather than using a symbol to denote what we have called $<$), Frege is concerned with these properties (as is Hölder also). He proves the following facts.

Theorem 1.4 (*Grundgesetze*, Theorem 635).   *If $<$ is a complete, upper semilinear ordering then $<$ is archimedean: that is, (a) & (f) $\implies$ (g).*

Theorem 1.5 (*Grundgesetze*, Theorem 637).   *If $<$ is an archimedean, upper semilinear ordering then $<$ is limp: that is, (a) & (g) $\implies$ (c).*

We shall rehearse Frege's proof of this in § 3 below.

Theorem 1.6 (*Grundgesetze*, Theorem 641).   *If $<$ is an archimedean, linear ordering then $<$ is left-invariant: that is, (b) & (g) $\implies$ (d).*

Theorem 1.7 (*Grundgesetze*, Theorem 689).   . *If $<$ is a dense, archimedean, linear ordering then G is abelian: that is, (b) & (e) & (g) $\implies$ (h).*

The first correct proof of the archimedean law from the completeness of the ordering appears to have been Hölder's,[4] although he used stronger assumptions than Frege's, namely, that $<$ is dense, left-invariant and linear. Hölder was also the first to prove commutativity from the archimedean law, but he had to assume left-invariance of $<$, whereas Frege derived (h) from (b), (e) and (g) alone (because, by Theorem 1.6, left-invariance is automatic). It is well known that the assumption of denseness is not needed: in fact (b) & (g) $\implies$ (h).[5]

Frege's original problem was whether (a) implies (b). In § 2 we shall exhibit an example of a group with a right-invariant, upper semilinear order relation that is not linear to show that (a) does not imply (b). In our example it is in fact the case that the ordering is limp and dense: thus we shall show that (a), (c), (e) together do not imply (b); equivalently, that (1), (2), (3), (5) and (7) together do not imply (4).

Our example suggests a further question, namely, whether there exists a group with a right-invariant, upper semilinear ordering that

[4] "Die Axiome der Quantität und die Lehre vom Mass," § 4.

[5] See E. V. Huntington, "A complete set of postulates for the theory of absolutely continuous magnitude," *Transactions of the American Mathematical Society*, **3** (1902) 264–279, 1966, 22, Case I, or L. Fuchs, *Partially ordered algebraic systems*, Oxford: Pergamon Press, 1963, pp. 45, 164.

is not linear but is archimedean, or even complete. This question is closely related to Frege's work because he defines a *positive* class to be a positival class in which $<$ is dense and complete (that is, (1), (2), (3), (4), (e), (f) are satisfied). We shall devote § 3 to a proof that (4), although independent of (1), (2) and (3), may be deduced from (1), (2), (3), (e) and (f): in fact, we shall show that if $<$ is an archimedean, upper semilinear ordering then G is abelian; that is, (a) & (g) $\implies$ (h); hence of course (a) & (g) $\implies$ (d); therefore (1), (2), (3) and (g) imply (4). Since it is readily shown that linearity and denseness of $<$ do not imply that it is left-invariant ((b) & (e) $\not\implies$ (d)), whereas linearity and limpness of $<$ do imply left-invariance ((b) & (c) $\implies$ (d)), all independence questions relating to conditions (a), ... , (h), on our assumption that $<$ is a right-invariant, weakly connected, strict partial ordering of G, are thus resolved.

II. THE CONSTRUCTION

Theorem 2.1. *If G is a free group of rank 2 then there is a right-invariant, limp, dense, semilinear order relation on G that is not linear.*

The task of finding a right-invariant, semilinear ordering of a group G is equivalent to finding a semilinearly ordered set $(\Lambda_0, <)$ that admits G acting regularly as a group of automorphisms. For, if $<$ is a right-invariant, semilinear order on G then the right translations $x \mapsto xg$ form a group of order-automorphisms of $(G, <)$ that acts regularly; conversely, if $(\Lambda_0, <)$ is semilinearly ordered and G acts regularly as a group of automorphisms, then the prescription $a < b$ if and only if $\lambda_0 a < \lambda_0 b$, where λ_0 is some fixed member of Λ_0, describes a right-invariant, semilinear order on G. We shall therefore construct such a set $(\Lambda_0, <)$ on which the free group of rank 2 acts regularly. The main ingredient in our construction is supplied by the following lemma.

Lemma 2.2. *Let Γ be a free group of rank 2, freely generated by γ_1, γ_2. There is a linear order $\prec$ on Γ such that*

(i) *$\prec$ is both right- and left-invariant;*
(ii) *$\prec$ is dense;*

(iii) $\{\gamma_1^n \mid n \in \mathbb{N}\}$ is cofinal (that is, unbounded above) in Γ;

(iv) if $\omega \in \Gamma$ and $1 \prec \omega$ then there exists $n \in \mathbb{N}$ such that $1 \prec \delta_n \prec \omega$, where $\delta_0 := \gamma_2$, and $\delta_{m+1} := \delta_m \gamma_1 \delta_m^{-1} \gamma_1^{-1}$ for $m > 0$.

Proof. The fact that a free group carries a linear order that is both right- and left-invariant, that is, that a free group is orderable in the usual sense, has been known for many years.[6] Furthermore, any ordering in this sense of a group with trivial centre (such as the free group of rank 2) will be dense because if $1 \prec x$ then there exists y such that $xy \neq yx$ and we have either $1 \prec y^{-1}xy \prec x$ or $1 \prec yxy^{-1} \prec x$. The construction described by Fuchs (p. 48) can be adjusted to ensure that (iii) and (iv) hold. To do that we start with the lower central series $\Gamma > [\Gamma, \Gamma] > [\Gamma, \Gamma, \Gamma] > \dots$ of the free group Γ generated by γ_1, γ_2. This series has the property that its intersection is $\{1\}$ and that its factors are free abelian groups of finite rank, freely generated (as abelian groups) by the basic commutators that can be formed from γ_1, and γ_2.[7] We refine the lower central series to give a descending sequence $\Gamma = \Gamma_1 > \Gamma_2 > \Gamma_3 > \dots$ of normal subgroups of Γ such that $\bigcap \Gamma_i = \{1\}$ and, for all i, Γ_i/Γ_{i+1} is infinite cyclic, generated modulo Γ_{i+1} by some basic commutator γ_i. In particular, $\Gamma = \langle \gamma_1, \Gamma_2 \rangle$, $\Gamma_2 = \langle \gamma_2, \Gamma_3 \rangle$, and, since the elements δ_n of Γ are basic commutators (of a special kind), for each $n \in \mathbb{N}$ there exists m such that $\delta_n = \gamma_m$. We now specify $\prec$ as follows: if $\omega_1, \omega_2 \in \Gamma$ and $\omega_1 \neq \omega_2$, then there is a unique integer m such that $\omega_1 \omega_2^{-1} \in \Gamma_m - \Gamma_{m+1}$; there is then a unique integer $n \neq 0$ such that $\omega_1 \omega_2^{-1} \equiv \gamma_m^n \pmod{\Gamma_{m+1}}$; we define $\omega_2 \prec \omega_1$ if $n > 0$ and $\omega_1 \prec \omega_2$ if $n < 0$. It is easy to see that this relation $\prec$ has all the properties specified in the statement of the lemma. $\square$

Proof of Theorem 2.1. Let Γ be as in Lemma 2.2, and let Π be the set of all finite subsets of Γ. Given $\alpha \in \Gamma$ we define an equivalence relation ρ_α on Π by the rule that $a \equiv b \pmod{\rho_\alpha}$ if and only if $a \cap (\alpha, \infty) = b \cap (\alpha, \infty)$, where $(\alpha, \infty) := \{\omega \in \Gamma \mid \alpha < \omega\}$. Define

$$\Lambda := \{A \subset \Pi \mid (\exists \alpha \in \Gamma)(\exists a \in \Pi)(A = \rho_\alpha(a))\},$$

[6] See L. Fuchs, *Partially ordered algebraic systems*, Oxford: Pergamon Press, 1963, pp. 47–50.

[7] See, for example, Wilhelm Magnus, Abraham Karrass and Donald Solitar, *Combinatorial group theory*, New York: John Wiley & Sons, 1966, § 5.7.

where $\rho_\alpha(a)$ denotes the ρ_α-class containing a. For later use we observe that if $A \in \Lambda$ then there is a unique element α of Γ such that A is a ρ_α-class. We shall call α the *level* of A and write it $l(A)$. Thus $l : \Lambda \to \Gamma$ and the fibres $l^{-1}(\alpha)$ of this 'level map' consist of the equivalence classes modulo ρ_α.

Now $\subset$ (strict inclusion) is a strict semilinear order relation on Λ.[8] For, certainly $\subset$ is a strict partial order on Λ. Suppose that $A, B, C \in \Lambda$ and $A \subset B$, $A \subset C$. We have $A = \rho_\alpha(a)$, $B = \rho_\beta(b)$, $C = \rho_\gamma(c)$ for some $\alpha, \beta, \gamma \in \Gamma$, $a, b, c \in \Pi$. Now $\beta \prec \gamma$ or $\beta = \gamma$ or $\gamma \prec \beta$. If $\beta \prec \gamma$ then $\rho_\beta < \rho_\gamma$ (in this context the symbol $<$ should be read 'is a proper refinement of'); since $a \equiv b \pmod{\rho_\beta}$ and $a \equiv c \pmod{\rho_\gamma}$, we have $b \equiv c \pmod{\rho_\gamma}$ and therefore $\rho_\beta(b) \subset \rho_\gamma(b) = \rho_\gamma(c)$; that is, $B \subset C$. Similarly, if $\beta = \gamma$ then $B = C$ and if $\gamma \prec \beta$ then $C \subset B$. Thus clause (ii) in our description of strict semilinear orderings holds. To prove clause (iii) suppose that $A, B \in \Lambda$ and that it is not the case that $A \subset B$, $A = B$ or $B \subset A$. Choose $a \in A$, $b \in B$ (so that $A = \rho_\alpha(a)$, $B = \rho_\beta(b)$ for some $\alpha, \beta \in \Gamma$), let $c := a \cup b$ and $y := \max c$. Then $a \equiv b \equiv c \pmod{\rho_\gamma}$ and so if $C := \rho_\gamma(c)$ then $A \subset C$ and $B \subset C$. This justifies our assertion that $\subset$ is a strict semilinear ordering of Λ.

If $\omega \in \Gamma$ then right-multiplication produces a permutation $\gamma \mapsto \gamma\omega$ of Γ; this induces a permutation of Π which has the effect of mapping the equivalence relation ρ_α to $\rho_{\alpha\omega}$; consequently it induces a permutation ω^* of Λ that maps a set $\rho_\alpha(a)$ to $\rho_{\alpha\omega}(a\omega)$. This map ω^* is obviously an automorphism of the semilinearly ordered set $(\Lambda, \subset)$. Next we define a permutation t of Π as follows:

$$t : a \mapsto \begin{cases} a & \text{if } a \cap (1, \infty) \neq \emptyset, \\ a - \{1\} & \text{if } a \cap (1, \infty) = \emptyset \text{ and } 1 \in a, \\ a \cup \{1\} & \text{if } a \cap (1, \infty) = \emptyset \text{ and } 1 \notin a. \end{cases}$$

It should be clear that $t^2 = 1$ and that t preserves each of the equivalence relations ρ_α. Therefore t induces a permutation t^* of Λ which again is an automorphism of $(\Lambda, \subset)$. We define

$$g_1 := \gamma_1^*, \qquad g_2 := t^*\gamma_2^*,$$

[8]Compare S. A. Adeleke and Peter M. Neumann, "Semilinearly ordered sets, betweeness relations and their automorphism groups," Observation 2.5.

and we shall prove that g_1, g_2 generate a free subgroup G of $\mathrm{Aut}(\Lambda, \subset)$ that acts freely.

To say that G acts freely means that if $w \in G$, $\lambda \in \Lambda$ and $\lambda w = \lambda$ then $w = 1$ (that is, G acts regularly on every orbit in Λ). Now the diagram

$$
\begin{array}{ccc}
\Lambda & \xrightarrow{\;l\;} & \Gamma \\
{\scriptstyle \omega^*}\downarrow & & \downarrow{\scriptstyle \omega} \\
\Lambda & \xrightarrow{\;l\;} & \Gamma
\end{array}
$$

commutes for every $\omega \in \Gamma$ and it follows that the level map l induces a homomorphism $l_* : G \to \Gamma$ that maps g_1 to γ_1 and g_2 to γ_2. If w is a non-trivial reduced word in variables x_1, x_2 then $l_* : w(g_1, g_2) \mapsto w(\gamma_1, \gamma_2)$. Since γ_1, γ_2 are free generators of Γ we know that $w(\gamma_1, \gamma_2) \neq 1$ and therefore $w(g_1, g_2) \neq 1$. This shows that G is free. Furthermore, if $w \in G$ and $\lambda w = \lambda$ then $(\lambda l)(w l_*) = (\lambda w) l = \lambda l$, and so $w l_* = 1$ in Γ, whence, as we have just seen, $w = 1$. Thus G is free and acts freely, as we claimed.

Next we prove that if $\lambda_0 := \rho_1(\emptyset)$ and Λ_0 is the G-orbit of λ_0 then Λ_0 is semilinearly, not linearly, ordered by $\subset$. Clauses (i) and (ii) of the definition of semilinear order relations are certainly satisfied because they are true in Λ. To check clause (iii), suppose that $A, B \in \Lambda_0$, that $A \not\subset B$, $A \neq B$ and $B \not\subset A$. Then $A = \rho_\alpha(a)$, $B = \rho_\beta(b)$ for some $a, b \in \Pi$ and some $\alpha, \beta \in \Gamma$. Let $y := \max(a \cup b \cup \{\alpha, \beta\})$ and choose $n \in \mathbb{N}$ such that $\gamma \prec \gamma_1^n$. Then

$$
\lambda_0 g_1^n = \rho_1(\emptyset) g_1^n = \rho_{\gamma_1^n}(\emptyset).
$$

Now $a \equiv b \equiv \emptyset \pmod{\rho_{\gamma_1^n}}$, and, since $\alpha \prec \gamma_1^n$, $\beta \prec \gamma_1^n$ we have $\rho_\alpha < \rho_{\gamma_1^n}$, $\rho_\beta < \rho_{\gamma_1^n}$. Therefore $p_\alpha(a) \subset \rho_{\gamma_1^n}(\emptyset)$ and $\rho_\beta(b) \subset \rho_{\gamma_1^n}(\emptyset)$. Consequently, if $C := \rho_{\gamma_1^n}(\emptyset)$ then $C \in \Lambda_0$ and $A \subset C$, $B \subset C$. This shows that Λ_0 is weakly connected and is therefore semilinearly ordered. To show that Λ_0 is not linearly ordered we let $\lambda_1 := \lambda_0 g_1^{-1}$ and $\lambda_2 := \lambda_0 g_2^{-1}$. Then, since $1 \prec \gamma_1$ and $1 \prec \gamma_2$, we have

$$
\lambda_1 = \rho_1(\emptyset) g_1^{-1} = \rho_{\gamma_1^{-1}}(\emptyset),
$$

and

$$
\lambda_2 = \rho_1(\emptyset)(\gamma_2^*)^{-1}(t^*)^{-1} = \rho_{\gamma_2^{-1}}(\emptyset) t^* = \rho_{\gamma_2^{-1}}(\{1\}).
$$

Thus

$$\lambda_1 = \{\, a \in \Pi \mid a \cap (\gamma_1^{-1}, \infty) = \emptyset \,\},$$
$$\lambda_2 = \{\, a \in \Pi \mid a \cap (\gamma_2^{-1}, \infty) = \{1\} \,\},$$

and so $\lambda_1 \cap \lambda_2 = \emptyset$. Therefore Λ_0 is not linearly ordered.

As we pointed out in the remark following the statement of Theorem 2.1, we have now done enough to prove that the free group of rank 2 carries a right-invariant, strict semilinear order that is not linear, and hence to prove that (4) does not follow from (1), (2) and (3). We have set ourselves to prove a little more, however, namely, that the ordering of G is dense and limp.

We first prove limpness. Suppose that $1 < p$ and $1 < q$. If $\lambda_0 p = \rho_\alpha(a)$ and $\lambda_0 q = \rho_\beta(b)$, where $a, b \in \Pi$ and $\alpha = pl_*$, $\beta = ql_*$, then $\rho_1(\emptyset) \subset \rho_\alpha(a)$ and $\rho_1(\emptyset) \subset \rho_\beta(b)$. Therefore $\lambda_0 p = \rho_\alpha(a) = \rho_\alpha(\emptyset)$: applying q to this we find that

$$\lambda_0 pq = \rho_\alpha(\emptyset)q = \rho_{\alpha\beta}(b).$$

Now $b \equiv \emptyset \pmod{\rho_\beta}$ and so $b \cap (\beta, \infty) = \emptyset$. Furthermore, $1 \prec \alpha$ and so $\beta \prec \alpha\beta$, and therefore $b \cap (\alpha\beta, \infty) = \emptyset$. Hence $b \equiv \emptyset \pmod{\rho_{\alpha\beta}}$, from which we get that $\lambda_0 pq = \rho_{\alpha\beta}(\emptyset)$. Since $\prec$ is left-invariant on Γ we have $\alpha \prec \alpha\beta$, and so $\rho_\alpha(\emptyset) \subset \rho_{\alpha\beta}(\emptyset)$, that is, $\lambda_0 p \subset \lambda_0 pq$. Therefore $p < pq$ in G, and so $1 < pqp^{-1}$. This shows that (5) is satisfied and so $<$ is limp.

To prove that $<$ is dense we need the fourth part of Lemma 2.2. Define $d_0 := g_2$ and $d_{n+1} := d_n g_1 d_n^{-1} g_1^{-1}$ for $n > 0$, so that $d_n l_* = \delta_n$ for all n. Now

$$\lambda_0 g_2 = \rho_1(\emptyset)t^*\gamma_2^* = \rho_1(\{1\})\gamma_2^* = \rho_{\gamma_2}(\{\gamma_2\}),$$

and since $\{\gamma_2\} \cap (\gamma_2, \infty) = \emptyset$ and $1 \prec \gamma_2$, it follows that $\rho_1(\emptyset) \subset \lambda_0 g_2$, that is, that $1 < g_2$. Suppose now as inductive hypothesis that $m \geqslant 0$ and $1 < d_m$. Let $a := \emptyset d_m$ (in the natural action of G on Π). Then $\lambda_0 d_m = \rho_1(\emptyset)d_m = \rho_{\delta_m}(a)$ and since $1 < d_m$ we must have $a \cap (\delta_m, \infty) = \emptyset$. Therefore

$$\lambda_0 d_{m+1} = \rho_1(\emptyset)d_m g_1 d_m^{-1} g_1^{-1}$$
$$= \rho_{\delta_m}(a)g_1 d_m^{-1} g_1^{-1}$$
$$= \rho_{\delta_m}(\emptyset)g_1 d_m^{-1} g_1^{-1}$$
$$= \rho_{\delta_m \gamma_1}(\emptyset)\delta_m^{-1} g_1^{-1}.$$

Now $\delta_m \prec \delta_m\gamma_1$ and so $a \cap (\delta_m\gamma_1, \infty) = \emptyset$, that is, $a \equiv \emptyset$ (mod $\rho_{\delta_m\gamma_1}$). Consequently

$$\lambda_0 d_{m+1} = \rho_{\delta_m\gamma_1}(a)d_m^{-1}g_1^{-1}$$
$$= \rho_{\delta_m\gamma_1\delta_m^{-1}}(\emptyset)g_1^{-1}$$
$$= \rho_{\delta_m\gamma_1\delta_m^{-1}\gamma_1^{-1}}(\emptyset).$$

We chose δ_{m+1} to be $\delta_m\gamma_1\delta_m^{-1}\gamma_1^{-1}$ and made sure in Lemma 2.2 that this was positive in the ordering of Γ. Therefore $1 < d_{m+1}$ and so, by induction, $1 < d_n$ for all n.

Suppose that $1 < p$ in G. Let $\omega := pl_* \in \Gamma$ and choose n such that $\delta_n \prec \omega$ (as we may, by Lemma 2.2(iv)). Now $\lambda_0 d_n = \rho_{\delta_n}(\emptyset)$ since $1 < d_n$, and $\lambda_0 p = \rho_\omega(\emptyset)$ since $1 < p$. Since $\delta_n \prec \omega$ we have $\lambda_0 d_n \subset \lambda_0 p$ and so $1 < d_n < p$, that is, $1 < pd_n^{-1}$. This shows that (7) is satisfied, that is, that $<$ is dense. $\square$

III. THE ARCHIMEDEAN CONDITION

Theorem 3.1. *Let G be a group with a right-invariant, archimedean, upper semilinear ordering $<$. Then G is abelian and $<$ is linear.*

The proof depends on Frege's theorem (Theorem 1.5 above) that the ordering is limp. Since we do not expect the reader to be familiar with this result we begin with a version of Frege's proof.

Proof of Theorem 1.5. Suppose that $p, q \in P$, that is, $p, q \in G$ and $1 < p$, $1 < q$. We wish to prove that $pqp^{-1} \in P$. If $p \leqslant q$ then $1 \leqslant qp^{-1}$ and so certainly $pqp^{-1} \in P$. Therefore we may suppose that $q < p$. Suppose, if possible, that $pq < p$. If $pq^m < p$ then $pq^{m+1} < pq < p$, and so by induction we have that $pq^n < p$ for all natural numbers n. Then $q^{n+1} < pq^n < p$, and this contradicts the assumption that the order $<$ is archimedean on G. Therefore it cannot be true that $pq < p$, so $p < pq$, whence $pqp^{-1} \in P$. Thus (5) is satisfied and $<$ is limp on G. $\square$

For the proof of Theorem 3.1 we shall need the following lemma.

Lemma 3.2. *Let G be a group with a right-invariant, limp, upper semilinear ordering $<$, and let x, y be elements of G with $1 < x \leqslant y$.*

If $z := xyx^{-1}y^{-1}$ then z and z^{-1} are comparable. Moreover, if $t := \max(z, z^{-1})$ then $1 \leqslant t < y$.

Proof. From the inequality $1 < x$ by right-multiplication with $y^{-1}x^{-1}$ we get $y^{-1}x^{-1} < xy^{-1}x^{-1}$, and then left-multiplication by y gives that $x^{-1} < yxy^{-1}x^{-1} = z^{-1}$. Since $x^{-1} < 1$ and $x^{-1} < z^{-1}$, and since $<$ is an upper semilinear ordering, 1 and z^{-1} are comparable, that is, $1 \leqslant z^{-1}$ or $z^{-1} \leqslant 1$. If $1 \leqslant z^{-1}$ then $z \leqslant 1 \leqslant z^{-1}$ and if $z^{-1} \leqslant 1$ then $z^{-1} \leqslant 1 \leqslant z$. Therefore z and z^{-1} are comparable, and $1 \leqslant t$.

We have seen that $x^{-1} < z^{-1}$. Right-multiplication by z and left-multiplication by x (permissible because of limpness) gives that $z < x$. Since $x \leqslant y$ we get that $z < y$. Also, since $1 < y$ and $1 < yx$ and $<$ is limp, we have $yx < yxy$. Right-multiplication by $y^{-1}x^{-1}$ gives that $yxy^{-1}x^{-1} < y$, that is, that $z^{-1} < y$. Thus $t = \max(z, z^{-1}) < y$, as we claimed. $\square$

Proof of Theorem 3.1. Consider the assertion $A(n)$:

$$(\forall x, y, t)(1 < x \leqslant y$$
$$\&\ t = \max(xyx^{-1}y^{-1}, yxy^{-1}x^{-1}) \implies t^n < x).$$

We propose to prove this by induction on n. Certainly $A(0)$ is true, so our inductive hypothesis will be that $m \geqslant 0$ and $A(m)$ is true.

Take $x, y \in G$ with $1 < x \leqslant y$, let $z := xyx^{-1}y^{-1}$ and let $t := \max(z, z^{-1})$. Since the ordering of G is archimedean, there exists $k \in \mathbb{N}$ such that $x^k < y < x^{k+1}$. Define

$$y_1 := x, \qquad x_1 := yx^{-k}.$$

Then $1 < x_1 \leqslant y_1$. Consequently, if $z_1 := x_1y_1x_1^{-1}y_1^{-1}$ and $t_1 := \max(z_1, z_1^{-1})$ then, since we are assuming the truth of $A(m)$, we have $t_1^m < x_1$. Now

$$z_1 = (yx^{-k})x(yx^{-k})^{-1}x^{-1} = yxy^{-1}x^{-1} = z^{-1},$$

and so $t_1 = t$. Thus $t^m < yx^{-k}$, and so $t^m x^k < y$. Now $A(m)$ also tells us that that $t^m < x$ and so $t^{mk} \leqslant x^k$ (with strict inequality if $k > 0$), whence $t^{m(k+1)} \leqslant t^m x^k < y$. If $k = 0$ then $y = x$ and

$t = 1$, so that certainly $t^{m+1} < y$. If $m = 0$ then $t^{m+1} = t < y$ by Lemma 3.2. Otherwise, if $k > 0$ and $m > 0$ then $m(k+1) > m$ and so $t^{m+1} \leqslant t^{m(k+1)} < y$. Thus we know in any case that $t^{m+1} < y$. Applying this to t_1 and y_1 we see that $t_1^{m+1} < y_1$, that is, $t^{m+1} < x$. We have now shown that $A(m)$ implies $A(m+1)$ and so $A(n)$ holds for all n.

By the archimedean condition, if t is as described in $A(n)$ then $t = 1$ and so x, y commute. This then is true for all positive x, y and since these generate G, the group must be abelian. It follows directly that $<$ is a linear ordering and the proof is complete. $\square$

We are grateful to Professor P. M. Cohn for drawing to our attention Behrand's paper.[9] Our Theorem 3.1 may be inferred from Behrend's assertion (A_3) in his § 1.6, although his context and his proof are rather different from ours.

POSTSCRIPT[10]

MICHAEL DUMMETT

It was remarked in the article that, with these results, all independence problems concerning the properties (a) to (h) are in effect resolved, but not how simple a diagram can be drawn of the logical relations between them. For the present purposes, we may assume the weakest condition (a) to be satisfied, and leave it unmentioned in the implications set out below: $<$ is taken to be a right-invariant weakly connected strict upper semilinear ordering of a group G. If $<$ is archimedean, it evidently must be either dense or complete (or both): (g) $\implies$ (e) $\lor$ (f). With this exception, denseness (e) is independent of the various possible combinations of the other conditions or their negations: we may therefore set it aside, and consider only conditions (d) and (f) to (h). By Theorem 1.4, completeness (f) implies the archimedean law (g): the converse obviously fails. By Theorem 3.1, the archimedean law (g) implies commutativity (h); the failure of the converse is again obvious. Since $<$ is assumed

[9]F. A. Behrend, "A contribution to the theory of magnitudes and the foundations of analysis," *Mathematische Zeitschrift*, **63** (1956) 345–362.

[10]From *Frege and other philosophers*, Oxford: Oxford University Press, 1991, 61–64. Reprinted by kind permission of Oxford University Press and the author. ©Michael Dummett.

right-invariant, commutativity (h) trivially implies left-invariance
(d). Limpness (c) is, of course, a special case of left-invariance (d).
Linearity (b) also follows from left-invariance (d), since, if $p, q \in P$,
it holds good by left-invariance that $p < q$ if and only if $1 < p^{-1}q$, i.e.
if $p^{-1}q \in P$: condition (3) therefore implies condition (4) when $<$
is left-invariant. Conversely, linearity (b) and limpness (c) together
imply left-invariance (d): this was proved by Frege in passing from
his Theorem 637 to his Theorem 641 (Theorems 1.5 and 1.6, respectively). His argument was as follows. Assume that $<$ is linear and
limp, and suppose that $q < r$: we wish to show that $pq < pr$ for
any p. Since this is given for $p \in P$ and obvious for $p = 1$, we need
consider only the case in which $p < 1$. Since $<$ is linear,

$$pq < pr \ \lor \ pq = pr \ \lor \ pr < pq.$$

If $pq = pr$, then $q = r$, contrary to hypothesis. If $pr < pq$, then,
since $1 < p^{-1}$ and $<$ is limp, $r < q$, again contrary to hypothesis.
It follows that $pq < pr$. Thus left-invariance (d) is equivalent to the
conjunction of linearity (b) and limpness (c): (d) $\Longleftrightarrow$ (b) & (c).

From all this we obtain a diagram of the logical relations between
the conditions (b) to (d) and (f) to (h), condition (a) being assumed.

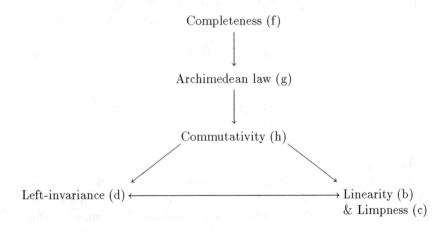

Seven possibilities are allowed by the diagram:

 (i) $<$ is complete;
 (ii) $<$ is archimedean but not complete;
 (iii) G is abelian, but $<$ non-archimedean;
 (iv) G is non-abelian, but $<$ left-invariant;
 (v) $<$ is linear but not limp;
 (vi) $<$ is limp but not linear;
 (vii) $<$ is neither linear nor limp.

In case (ii), $<$ must be dense; in all six other cases, it may or may not be.

Since it is possible to give examples of all seven cases, with dense and discrete subcases save for case (ii), it follows that no logical relations hold between the various conditions save those shown on the diagram. Cases (i) to (iv) are easily illustrated: case (iv) is just that of a non-abelian linearly ordered group, in the standard sense. An example may readily be constructed to illustrate case (v): linearity does not by itself imply left-invariance, nor, therefore, even limpness, and still does not do so when the hypothesis of denseness is added—(b) & (e) $\not\Longrightarrow$ (c). For this, G may be taken as the group of all transformations g of the set $\mathbb{R}$ of real numbers, where g has the form

$$g : z \mapsto 2^u z + p2^{-n},$$

u and p being any integers and n a non-negative integer. The set P of positive elements will consist of those transformations g for which $\sqrt{2} < \sqrt{2}g$: this has the result that $g < h$ in the group ordering if and only if $\sqrt{2}g < \sqrt{2}h$. Clearly, $<$ is dense and, in virtue of the irrationality of $\sqrt{2}$, linear. Then, where

$$g : z \mapsto 2z - 4$$
$$h : z \mapsto z - 2,$$

$g < h$, but $g^{-1} < h^{-1}$, so that $<$ is not left-invariant.

Frege's problem was whether there was an example of either case (vi) or case (vii). The construction of Theorem 2.2 provides an illustration of case (vi), the difficult one, showing that limpness and denseness do not imply linearity: (c) & (e) $\not\Longrightarrow$ (b). To get an example of case (vii), we may combine cases (v) and (vi). Where G_1, is a group with a right-invariant upper semilinear ordering $<_1$ that is limp but not linear, and G_2, a group with a right-invariant linear ordering $<_2$ that is not limp, we may take G as the direct product $G_1 \times G_2$: the ordering $<$ on G is to be defined lexicographically, so that $\langle g_1, g_2 \rangle < \langle h_1, h_2 \rangle$ if and only if either $g_1 <_1 h_1$ or $g_1 = h_1$ and $g_2 <_2 h_2$. Then $<$ will be an upper semilinear ordering, but neither limp nor linear.

Frege's theory of real numbers is of interest philosophically, as providing a far clearer insight into the salient role he allotted to the application of mathematics than does his logical construction of the natural numbers: on the basis of a general view about the relation of mathematics to its applications, he treated the applications of the real numbers as far more decisive for the way they should be defined than they are in other theories of the foundations of analysis. Mathematically, his construction of the real numbers, uncompleted because of the disaster wrought by Russell's contradiction, was a pioneering investigation of groups with orderings. His sights were set on ordered groups isomorphic to the additive group of the real numbers; but his interest in what we should call axiomatics, that is, in establishing the most economical base for each theorem he proved, led him into a painstaking investigation, only very partially anticipated by Otto Hölder, of groups not assumed to be abelian with orderings not assumed to be left-invariant. These studies threw out a problem of considerable difficulty and interest; and, in the course of them, he not only employed such subsequently familiar notions as that of a commutator, but hit on the fruitful concept of what we have labelled 'limpness,' of which, before the article in which we rounded off his enquiries, it occurred to no other mathematician to make use. It is an injustice that, in the literature on group theory, Frege is left unmentioned and denied credit for his discoveries.

16

On the Consistency of the First-order Portion of Frege's Logical System

It is well-known that Frege's logical system of his *Grundgesetze der Arithmetik* is inconsistent. However, Peter Schroeder-Heister[1] has speculated that the first-order portion of this system is consistent. On the surface, this is a somewhat surprising conjecture, because Frege's so-called "abstraction" principle is included in the first-order part of his system, and the somewhat similar abstraction principle of (first-order) naive set theory leads quickly to inconsistency. But Frege's abstraction principle is a prima facie weaker principle. Instead of assuming that any formula determines a set which satisfies that formula, it holds only that coextensive formulas must determine the same "courses of values." That is, instead of this:

(Set Theorem) $\qquad (\exists x)(y)(y \in x \equiv A)$, for any A not containing x,

we assume (roughly) this:

(Frege) $\qquad (x)(A \equiv B) \equiv \dot{x}A = \dot{x}B$, for any A, B.[2]

From *Notre Dame journal of formal logic*, **28** (1987) 161–88. Reprinted by kind permission of the editors of *Notre Dame journal of formal logic* and the author.

Work on this paper was stimulated by Peter Schroeder-Heister's "A model theoretic reconstruction of Frege's permutation argument," *Notre Dame journal of formal logic*, **28** (1987) 69–79.

[1] In the paper cited in the previous note.

It is well known that if quantification over functions is admitted into Frege's system (as Frege himself did) then it is possible to define an analogue of set membership, and the abstraction principle of naive set theory can be shown to follow from Frege's abstraction principle.[3] Russell's paradox quickly follows. But it is not obvious how to do this within the first-order portion of Frege's system. The first goal of this paper is to show that this cannot be done. Schroeder-Heister's conjecture is correct: the first-order portion of Frege's system is consistent.

The second goal of this paper is to explore the significance of the model-construction technique sketched herein for Frege's claims about the arbitrariness of the identification of truth-values with courses of values. Although Schroeder-Heister has shown that Frege's claims on this topic are false, there are some closely related claims that are true and interesting.

I. CONSISTENCY OF THE SYSTEM

To understand the following details, we need to keep in mind Frege's beliefs that: (1) truth-values are objects, not to be distinguished ontologically from other objects, and (2) terms which denote truth-values can occur syntactically in the same places that other names of objects can. Since sentences denote truth-values, this means that sentences can occur in all of the places where we would normally expect names of objects to occur, such as flanking the identity sign. This allows Frege, e.g., to use the identity sign for the material biconditional.

1.1 Syntax of the formal language L. In what follows I will use "$[X_Z^Y]$" as an abbreviation for "the result of replacing each occurrence of Y in X by an occurrence of Z." I will generally use signs of the object-language as metalinguistic names of themselves. I follow Frege's custom of using Gothic letters ($\mathfrak{a}_1, \mathfrak{a}_2, \ldots$) for variables

---

[2]In this formula $\dot{x}A$ and $\dot{x}B$ are taken to denote the "courses of values" related to formulas A and B. Frege's principle is actually broader than the one given, since he has courses of values corresponding to any open name, not just to the names that correspond to formulas in the modern sense.

[3]We can define membership as follows: $x \in y \equiv (\exists f)[f(x) \ \& \ y = \dot{z}f(z)]$. You get Russell's paradox by asking whether $\dot{z}(\sim z \in z)$ is a member of itself, using the definition and Frege's abstraction principle.

bound by the universal quantifier, and Greek letters ($\varepsilon_1, \varepsilon_2, \ldots$) for variables bound by the course-of-values abstraction symbol "'".

The syntax of the first-order part of Frege's system is this:

> The object parameters of L are: x_1, x_2, x_3, ...
>
> Every object parameter of L is a complete name of L.
>
> The function parameters of L are: f_1, f_2, f_3, ...
>
> If f_n is a function parameter of L, and A is a complete name of L, then this is a complete name of L: $f_n(A)$.
>
> If A and B are complete names of L, so are the following:

Using the horizontal:	$-A$
Using negation:	$\top A$
Using the conditional:	$(A \to B)$
Using identity:	$(A = B)$
Using universal quantification:[5]	$(\mathfrak{a}_i)[A^{x_i}_{\mathfrak{a}_i}]$
Using course-of-values abstraction:	$\dot{\varepsilon}_i[A^{x_i}_{\varepsilon_i}]$

1.2 Axiomatics. I assume that we are given a complete set of rules and axioms for the first-order predicate calculus with identity, expressed within Frege's system. I will call this the "logical" system. In addition, I will be discussing Frege's so-called "abstraction" principle governing course-of-values names. This principle states that for any complete names A and B of L the following is to be an axiom:

$$(\mathfrak{a}_i)([A^{x_n}_{\mathfrak{a}_i}] = [B^{x_n}_{\mathfrak{a}_i}]) = (\dot{\varepsilon}_j[A^{x_n}_{\varepsilon_j}] = \dot{\varepsilon}_k[B^{x_n}_{\varepsilon_k}]).^{6}$$

If we think of the formulas A and B as expressing functions, this

---

[5] As given, the rule for the universal quantifier prohibits the formation of a complete name in which some object parameter has the same subscript as a Gothic letter in whose scope it falls, and the rule for course-of-values names prohibits the formation of such a name in which an object parameter has the same subscript as that of some Greek letter in whose scope it falls. These are not important restrictions. Note that it might be truer to Frege's views about the "gappiness" of incomplete names to disallow vacuous quantification; this is not, however, relevant to the points under discussion.

[6] In order to prevent "capturing" we need to restrict this schema to instances in which x_n does not already occur within the scope of ε_j in A, or ε_k in B, or $\mathfrak{a}_i$ in either A or B.

principle tells us that A and B express coextensive functions if and only if the courses-of-values of those functions are the same. Our problem is to see whether there is a model of Frege's logical system in which this abstraction principle is true.

1.3 Interpretations. Suppose that U is a set (a "universe"), and that t and f are distinct members of U. Intuitively, U represents the class of all Fregean objects, and t and f are those objects that Frege calls "The True" and "The False." I will call σ a *basic assignment* over U if σ is any assignment of members of U to the object parameters of L, and of members of U^U to the function parameters of L. Then an *interpretation I over U of* (a portion of) L is to be a function which, given any basic assignment σ over U, produces an assignment I^σ of members of U of *all* of the complete names of (that portion of) L, and an assignment of members of U^U to all of the function parameters of (that portion of) L, where it is understood that I^σ agrees with σ on all of the parameters of L.

If I is an interpretation and σ a basic assignment, we say that I^σ makes A true if $I^\sigma[A] = t$, and makes A false if $I^\sigma[A] = f$. When we say that I alone makes A true, we mean that I^σ makes A true for any basic assignment σ. There is no presumption so far that an interpretation and a basic assignment make even the logical theorems of L true. Our task will be to show, first, how to produce an interpretation of that part of L which does not contain any course-of-values names, and which makes true all of the logical theorems of that part of L. Then we will show how to extend any such interpretation to the rest of L, including course-of-values names, so as to also make true the abstraction principle.

Suppose that I is any interpretation over U of some portion L' of L, which may contain all, some, or none of the course-of-values names of L. Assume also that L' is *syntactically grounded*, in the sense that for every name A occurring in L', if B occurs in the syntactic rule that generates A, then B is also in L'. Then corresponding to I is a unique interpretation I^* of L', determined by the following conditions (which hold for any basic assignment σ):

$$I^{*\sigma}[x_n] = I^\sigma[x_n] \quad \text{for every object parameter } x_n.$$

$$I^{*\sigma}[f_n] = I^\sigma[f_n] \quad \text{for every function parameter } f_n.$$

$$I^{*\sigma}[\dot{\varepsilon}_i A] = I^{\sigma}[\dot{\varepsilon}_i A] \quad \text{for every course-of-values name}$$
$$\text{of } L \text{ for which } I^{\sigma} \text{ is defined.}$$

$$I^{*\sigma}[f_n(A)] = I^{*\sigma}[f_n](I^{\sigma*}[A])$$

$$I^{*\sigma}[-A] = \begin{cases} t & \text{if } I^{*\sigma}[A] = t \\ f & \text{otherwise} \end{cases}$$

$$I^{*\sigma}[\top A] = \begin{cases} t & \text{if } I^{*\sigma}[A] \neq t \\ f & \text{otherwise} \end{cases}$$

$$I^{*\sigma}[(A \to B)] = \begin{cases} t & \text{if } I^{*\sigma}[A] \neq t \text{ or } I^{*\sigma}[B] = t \\ f & \text{otherwise} \end{cases}$$

$$I^{*\sigma}[(A = B)] = \begin{cases} t & \text{if } I^{*\sigma}[A] = I^{*\sigma}[B] \\ f & \text{otherwise} \end{cases}$$

$$I^{*\sigma}[(\mathfrak{a}_k)A] = \begin{cases} t & \text{if } I^{*\sigma[x_k/u]}[A_{x_k}^{\mathfrak{a}_k}] = t \text{ for every } u \in U \\ f & \text{otherwise} \end{cases}$$

(Note: $\sigma[x_k/u]$ is that assignment which is exactly like σ except that it assigns u to x_k. Recall that by the syntactic rules given above, x_k will never occur in $(\mathfrak{a}_k)A$.)

1.4 The model. Suppose that we have an interpretation I over some infinite set U for that portion of L which contains no course-of-values names:[7] I assume that it is clear that I^* makes true all of the logical theorems of the system. This part of Frege's system is so much like the ordinary (first-order) predicate calculus with identity that conventional modern methods apply. Given such an interpretation, the task is to show how to extend I to the whole of L in such a manner that the abstraction principle is satisfied.

First, define the rank of any course-of-values name $\dot{\varepsilon}_i A$ to be 1 if A contains no course-of-values names, and to be $1 + \text{rank}[B]$ if B is a course-of-values name in A whose rank is at least as high as that of any other such name in A.

---

[7]If U is not infinite then no model will be possible, since Frege's abstraction principle, together with his first-order logical principles, forces an infinite domain. (That is, it does this if we assume that $t \neq f$. Otherwise it is easy to provide a model for his system with a 1-element domain, even if the system is inconsistent.)

Now order the course-of-values names of L by rank, and, within rank, in some arbitrary manner (of order-type ω). I will use $\dot{\varepsilon}_i A_{n,m}$ to denote the mth course-of-values name of rank n.

Next, choose any countably infinite subset of U (perhaps containing all of U), and order this subset in any way you like into a countable sequence of countable sequences. I will use $u_{n,k}$ to denote the kth member of the nth sequence. It is understood that if $n \neq n'$ or $k \neq k'$ then $u_{n,k} \neq u_{n',k'}$.

We will show how to assign to each name $\dot{\varepsilon}_i A_{n,m}$ some $u_{n,k}$ as its referent (relative to each basic assignment σ). This will be done in stages. Given our initial interpretation I, we define interpretations $I_{n,m}$ in a step-by-step fashion, as follows:

Basis. $I^\sigma_{1,0} = I^{*\sigma}$ (for each σ).

Successor Step. We first extend $I_{n,m}$ to $I'_{n,m}$ by stipulating the value of $I'^\sigma_{n,m}[\dot{\varepsilon}_i A_{n,m+1}]$, for any U, as follows:

If there is some name B for which $I^\sigma_{n,m}[\dot{\varepsilon}_k B]$ is already defined, and for which $I^{\sigma[x_j/u]}_{n,m}[A^{\varepsilon_i}_{x_j}] = I^{\sigma[x_j/u]}_{n,m}[B^{\varepsilon_k}_{x_j}]$ for every $u \in U$, then $I'^\sigma_{n,m}[\dot{\varepsilon}_i A_{n,m+1}] = I^\sigma_{n,m}[\dot{\varepsilon}_k B]$. Otherwise, $I'^\sigma_{n,m}[\dot{\varepsilon}_i A_{n,m+1}]$ = the next unused $u_{n,k}$ of rank n.

(We call $u_{n,k}$ "unused" if it has not yet been assigned to any course-of-values name. Also, we assume for this account that the subscript j of x_j is chosen so as not to occur as a subscript of a Greek or Gothic letter in either A or B.)

Finally, we set $I^\sigma_{n,m+1} = I'^{*\sigma}_{n,m}$.

Limit Step. We set

$$I^\sigma_{n+1,0} = \bigcup_k I^\sigma_{n,k}.$$

(It may be verified that $I^\sigma_{n,m}$ is always a subfunction of $I^\sigma_{n,m+1}$, so the limit step does yield a single-valued function.)

To get the desired interpretation, we now set $I^\sigma_\omega = \bigcup_{n,m} I^\sigma_{n,m}$. Then I_ω, as so defined, is the desired interpretation.

Theorem. *All instances of the abstraction schema (as well as all of the other logical theorems) are true under I_ω.*

This theorem may be proved by a straightforward induction on the ordering used in the construction.[8]

II. THE ARBITRARINESS OF THE IDENTIFICATION

OF TRUTH-VALUES WITH COURSES-OF-VALUES

In the first few sections of the *Grundgesetze* Frege achieves a certain elegance by "identifying" the two truth-values t and f with the courses of values named by $\dot\varepsilon(-\varepsilon)$ and $\dot\varepsilon(\varepsilon = \top(\mathfrak{a})(\mathfrak{a} = \mathfrak{a}))$. In defense of this policy he adopts a conventionalist stance, and defends his choice by claiming that such identification is arbitrary. Schroeder-Heister has shown that this is an overgeneralization (see (3) below), and he asks about the limits of such identifications. The construction of Section I sheds some light on this issue.

Suppose that we have selected which objects (which members of U) are to be the truth-values, and that we have fixed on an interpretation and a basic assignment which jointly establish denotations for all of the names of L that do not contain course-of-values names. How does this constrain which objects must be selected to be which courses of values? A Fregean moral of the construction given above is that it hardly constrains it at all. In particular, we have the following results:

(1) It is always possible to find a model for the entire language in which neither of the truth-values is the referent of any course-of-values name relative to any σ. Just use the construction given above, leaving both t and f out of the set of objects chosen to be courses of values.

---

[8]In discussing Frege's system I have left out his definite description operator. This made the construction easier to follow. If the operator is included, the following additions should be made: First, for any course of values name $\dot\varepsilon A$, we add to the symbolism the definite description $(\iota\varepsilon)A$. Next, in giving the construction, we assume initially that L lacks definite descriptions as well as course of values names. Then, every time we stipulate the value of $I'^\sigma_{n,m}[\varepsilon_i A_{n,m+1}]$, we follow this with the stipulation that $I'^\sigma_{n,m}[\iota\varepsilon_i A_{n,m+1}]$ is to be that unique member u of U which is such that $I'^{\sigma[x_k/u]}_{n,m}[A^{\varepsilon_i}_{x_k}] = t$, if there is such a u, and otherwise it is to have the same denotation as $I'^\sigma_{n,m}[\dot\varepsilon_i An, m+1]$. (This is just a formal statement of Frege's own stipulation.)

(2) It is always possible to make Frege's choice, and to identify t as the referent of $\dot{\varepsilon}(-\varepsilon)$ and f as the referent of $\dot{\varepsilon}(\varepsilon = \top(\mathfrak{a})(\mathfrak{a} = \mathfrak{a}))$. Just choose these names as the first and second names of rank 1, and pick $u_{1,1}$ to be t and $U_{1,2}$ to be f. (The only thing that needs verifying here is that $I^{\sigma[x/u]}[-x]$ will disagree with $I^{\sigma[x/u]}[x = \top(\mathfrak{a})(\mathfrak{a} = \mathfrak{a})]$, for some $u \in U$. In fact, they will always disagree for $u = t$.) Reversing the choice is also always possible.

(3) It is not possible, in general, to select any arbitrary pair of course-of-values names and assume that the first may be assigned t as its denotation and the second assigned f. This is part of what Schroeder-Heister has shown. For example, it is never possible to arrange things so that $\dot{\varepsilon}(\varepsilon = \top(\mathfrak{a})(\mathfrak{a} = \mathfrak{a}))$ denotes t *and* $\dot{\varepsilon}_1(\varepsilon_1 = \top\dot{\varepsilon}(\varepsilon = \top(\mathfrak{a})(\mathfrak{a} = \mathfrak{a})))$ denotes f. If, for example, the construction given above is arranged so that the former course-of-values name denotes t, then the latter will be forced to denote t as well. (Note that in general the two names may receive different denotations.)

(4) However, there is an ontological analogue of this principle that is true, and this is the point that I think Frege should have made. Let me say that a course-of-values name $\dot{\varepsilon}A$ *signifies the function* g (relative to I and σ) if and only if $I^{\sigma[x/u]}[A^\varepsilon_x] = g(u)$ for every $u \in U$. Then, given *any* two functions g and h which are members of U^U, it is possible to select t as the course-of-values of g and f as the course-of-values of h. More precisely, it is possible to find an assignment σ and two course-of-value names such that the first name signifies g and denotes t (relative to I and σ), and the second signifies h and denotes f (relative to I and σ). (Just pick a σ which assigns g to f_1 and h to f_2, and choose the course-of-values names to be $\dot{\varepsilon}f_1(\varepsilon)$ and $\dot{\varepsilon}f_2(\varepsilon)$, and then carry out the construction as in (2) above.) The arbitrariness that so impressed Frege has its source in the arbitrariness of the connection between functions and their courses of values in his system. His abstraction principle requires there to be a 1-1 correlation between functions and objects, but it puts absolutely no further constraints on this correlation. This is a theme of "Über Begriff und Gegenstand,"[9] in which he holds that

---

[9]In the first-order fragment there must be a 1-1 correlation between named objects and signified functions. In the higher-order version the abstraction principle may be quantified, and it then requires a 1-1 correlation between all objects

each concept is represented by some object, but in which he never says *which* objects represent which concepts.

(5) The point just made depends on a special feature of our language L: the presence in L of function parameters. These are needed in order to make sure that the chosen functions g and h can be signified by some name in the language. For if they could not be so signified, then the construction given in section I would not apply to them. It is worth asking about a language that may have been closer to that which Frege had in mind, namely, that portion of L which contains no parameters at all. The only symbols present are the logical ones, along with bound Gothic and Greek letters. This language is limited in expressive power in a certain way. Crudely put, its formulas with one free variable each can only distinguish among the truth values and the objects named by course-of-values names; the rest are all treated alike. As a result, the complete names not containing course-of-values names have their references logically fixed, and it is totally determined which functions will be signified by course-of-value names of rank 1. In particular, if two such course-of-value names signify the same function relative to one interpretation they signify the same function relative to any interpretation, and if they signify different functions in one interpretation then they do so in all. So in such a restricted language, any two nonsynonymous course-of-values names of rank 1 may be selected as designating the two truth-values (in either order).[10] Notice that Frege's own choice is an example of this generalization. This result does not generalize to any higher ranks, for Schroeder-Heister already shows how to get a counterexample using names of ranks 1 and 2, and any course-of-values name may have its rank artificially boosted by incorporating in it superfluous course-of-values names.

APPENDIX

Let $\dot{\varepsilon}_j[A]$ and $\dot{\varepsilon}_k[B]$ be two course-of-values names. Let x_n be any object parameter which does not occur in either name. Then I call the names *synonymous* if for every I and σ:

$$I^{*\sigma[x_n/u]}[A^{\varepsilon_j}_{x_n}] = I^{*\sigma[x_n/u]}[B^{\varepsilon_k}_{x_n}] \quad \text{for every } u \in U.$$

and all functions, in violation of Cantor's theorem.

[10] A proof is given in the appendix.

The advertised result may be proved as follows. First, we have two lemmas:

1. *Suppose that A is any complete name that does not contain any course-of-values names. Then it is easy to show inductively that for any interpretations I and I', and basic assignment σ, that:*

$$\langle 1 \rangle \qquad\qquad I^{*\sigma}[A] = I'^{*\sigma}[A].$$

2. *It can also be established that if σ and σ' do not differ in what they assign to the parameters of A, then for any interpretation I:*

$$\langle 2 \rangle \qquad\qquad I^{*\sigma}[A] = I^{*\sigma'}[A].$$

Now suppose that $\dot{\varepsilon}_j A$ and $\dot{\varepsilon}_k B$ contain no parameters and no embedded course-of-values names, and suppose that they are not synonymous. This means that for some particular I, σ, and u:

$$\langle 3 \rangle \qquad\qquad I^{*\sigma[x_n/u]}[A_{x_n}^{\varepsilon_j}] \neq I^{*\sigma[x_n/u]}[B_{x_n}^{\varepsilon_k}].$$

But, since A and B contain no parameters, principle $\langle 2 \rangle$ lets us generalize this to any interpretation I. And since they contain no course-of-values names, principle $\langle 1 \rangle$ lets us generalize this to any σ. That is, we have that:

$\langle 4 \rangle$ For any I and any σ there is some $u \in U$ (actually the same u in each case, though this is not needed) such that:

$$I^{*\sigma[x_n/u]}[A_{x_n}^{\varepsilon_j}] \neq I^{*\sigma[x_n/u]}[B_{x_n}^{\varepsilon_k}].$$

But this is exactly the condition that is needed to allow us to use $\dot{\varepsilon}_j[A]$ and $\dot{\varepsilon}_k[B]$ in the construction in part 1 so as to assign them the two truth-values, as in (2) above.

17

Fregean Extensions of First-order Theories

It was shown by Terence Parsons in Chapter 16, above, that the first-order fragment of Frege's (inconsistent) logical system in the *Grundgesetze der Arithmetic* is consistent. In this note we formulate and prove a stronger version of this result for arbitrary first-order theories. We also show that a natural attempt to further strengthen our result runs afoul of Tarski's theorem on the undefinability of truth.

I. FREGEAN THEORIES

We shall call a theory T in a (countable) first-order language $\mathcal{L}$ *Fregean* if corresponding to each formula $A(x)$ with one free variable x there is a constant symbol c_A of $\mathcal{L}$ such that, for any such formulas $A(x)$, $B(x)$, the sentence

$$(*) \qquad \forall x[A(x) \leftrightarrow B(x)] \leftrightarrow c_A = c_B$$

is provable in T.

When T is the first-order fragment of Frege's system (as presented by Parsons) the scheme of sentences $(*)$ is the first-order version

From *Mathematical Logic Quarterly*, **40** (1994) 27–30. Reprinted by kind permission of the editors and the author.

of Frege's abstraction principle for predicates (concepts) with the constant c_A playing the role of the "extension" of the predicate A. We shall call this scheme *Frege's abstraction scheme*.

Our main result is the

Theorem. *If T has no finite models, then it has a conservative Fregean extension.*

Proof. We begin by introducing new constants which we shall call *special constants*: each of these will be assigned a natural number called its *level*. The special constants are defined by induction on levels as follows. Suppose that special constants of all levels $< n$ have been defined. Let $A(x)$ be a formula using symbols of $\mathcal{L}$ and just the constants of level $< n$. If $n > 0$, suppose also that $A(x)$ contains at least one special constant of level $n-1$. Then the symbol c_A is a special constant of level n called *the special constant assigned to $A(x)$*.

Let $\mathcal{L}^*$ be the language obtained by adding to $\mathcal{L}$ all special constants of all levels. Each formula $A(x)$ of $\mathcal{L}^*$ is then assigned a unique special constant: its level is the least number exceeding the levels of all special constants occurring in $A(x)$. If $A(x)$, $B(x)$ are formulas of $\mathcal{L}^*$, the sentence

$$\forall x[A(x) \leftrightarrow B(x)] \leftrightarrow c_A = c_B$$

is called the *Fregean axiom for* (c_A, c_B).

Let T^* be the theory in $\mathcal{L}^*$ obtained by adding all the Fregean axioms to T. Clearly T^* is Fregean (in $\mathcal{L}^*$).

We now claim that T^* *is a conservative extension of T*. To prove this it suffices to show that any model M of T can be expanded to a model M^* of T^* with the same domain.

To obtain M^* we provide interpretations in (the domain of) M of all special constants recursively as follows. Suppose that interpretations in M have been assigned to all special constants of level $< n$ in such a way that

    (1) all the corresponding Fregean axioms are satisfied;

    (2) the set M' of elements of M which are *not* interpretations of special constants of level $< n$ has cardinality $|M|$ (necessarily infinite).

Since M' is infinite it can be partitioned into two subsets M'' and M''' each of cardinality $|M'| = |M|$. Moreover, since M is infinite, the number of subsets of the model definable using formulas of $\mathcal{L}$ involving only special constants of level $< n$ is $\leq |M| = |M''|$. Let f be an injection of this collection of sets into M''. We define an interpretation of each special constant c_A of level n (in such a way as to preserve the truth of (1) and (2)) as follows:

(a) If for some special constant c_B of level $< n$, $A(x)$ and $B(x)$ define the same subset of M, then c_A is assigned the same interpretation as that which has (by induction hypothesis) already been assigned to c_B.

(b) Otherwise, c_A is assigned the interpretation $f(X)$, where X is the subset of M defined by $A(x)$.

(Note that in case (a) there may be more than one formula of level $< n$ defining the same subset of M as $A(x)$: but condition (1) of the induction hypothesis implies that all the corresponding special constants receive the *same* interpretation. So the stipulation in case (a) is consistent.)

This recipe furnishes interpretations in M of all the special constants. It is now easy to check that all the Fregean axioms are true in the resulting model M^*, which is therefore an expansion of M to a model of T^*.

This completes the proof of the theorem. $\square$

Corollary. *If T has an infinite model, then it has a consistent Fregean extension.*

Proof. Let T' be obtained from T by adding as axioms the sentences $d_m = d_n$ for $m \neq n$, where $\{d_m : m \in \omega\}$ is a set of new distinct constants. Since T has an infinite model, T' is consistent. Clearly T' has no finite models, so the theorem implies that T' has a conservative Fregean extension; this latter is a consistent Fregean extension of T. $\square$

We note that the assertions of the Theorem and its Corollary can be reversed: that is, any theory with a conservative (respectively consistent) Fregean extension has no finite models (respectively has an infinite model). This follows immediately from the observation that

Any model of a Fregean theory is infinite.

To prove this, take any Fregean theory T and define the sequence of formulas $A_0(x)$, $A_1(x)$, ... recursively as follows. First, set $A_0(x) \equiv (x = x)$ and for $n \geqslant 1$,

$$A_n(x) \equiv \bigwedge_{i<n} x \neq c_{A_i}.$$

For simplicity write c_n for c_{A_n}. To prove the result it suffices to show that $\vdash_T c_m \neq c_n$ for $m < n$. We do this by induction on n. Suppose that $n \geqslant 1$ and $\vdash_T c_i \neq c_m$ for all $i < m < n$. Then if $m < n$, it follows that

$$\vdash_T A_m(c_m) \wedge \neg A_n(c_m)$$

so that

$$\vdash_T \neg \forall x [A_m(x) \leftrightarrow A_n(x)].$$

The Fregean axiom for (c_m, c_n) now yields $\vdash_T c_m \neq c_n$, completing the induction step and the proof.

II. STRENGTHENING THE ABSTRACTION SCHEME

It is well-known that the inconsistency in the *Grundgesetze* arises from the fact that Frege assumes the full second-order version of the abstraction scheme. For our purposes this may be formulated as:

$$(*) \qquad \forall P \forall Q \big[\forall x [P(x) \leftrightarrow Q(x)] \leftrightarrow c(P) = c(Q) \big],$$

where P, Q are predicate variables and c is a function assigning individual (lowest level) terms to predicate variables. To see that $(*)$ is inconsistent we suppose that $\mathcal{S}$ is a second-order theory in which $(*)$ is provable, and define the formula $A(x)$ by

$$A(x) \equiv \exists P [x = c(P) \wedge \neg P(x)].$$

Then

$$\vdash_{\mathcal{S}} A(c(A)) \leftrightarrow \exists P [c(A) = c(P) \wedge \neg P(c(A))]$$
$$\leftrightarrow \exists P \big[\forall x [A(x) \leftrightarrow P(x)] \wedge \neg P(c(A)) \big]$$
$$\leftrightarrow \neg A(c(A)).$$

Thus the attempt to strengthen Frege's abstraction scheme by allowing second-order quantification *and* functional dependence of "extensions" as predicates leads to inconsistency, because it runs afoul of Russell's paradox, or, equally, Cantor's diagonalization argument. It is of interest to note that inconsistency can still arise even *without* second-order quantification, provided we insist that the functional dependence of "extensions" on predicates be *definable*. In this case, as we shall see, the inconsistency arises for what seems to be quite a different reason, namely, as a consequence of Tarski's theorem on the undefinability of truth. (A related point has been made by Peter Aczel.[1])

Let us call a theory T *Tarskian* if enough of the syntax of its language $\mathcal{L}$ can be encoded within T to ensure that T is subject to Tarski's undefinability theorem. This means that, if T is consistent, then, writing $\ulcorner A \urcorner$ for the code in $\mathcal{L}$ of a formula A, there is no formula $W(x)$ of $\mathcal{L}$ such that

$$\vdash S \leftrightarrow W(\ulcorner S \urcorner)$$

for every sentence S of $\mathcal{L}$. Note that both first-order arithmetic and set theory are Tarskian.

Now, let $\tau(x)$ be a term of $\mathcal{L}$ and consider the scheme

$$(**) \qquad \forall x[A(x) \leftrightarrow B(x)] \leftrightarrow \tau(\ulcorner A \urcorner) = \tau(\ulcorner B \urcorner),$$

obtained from Frege's abstraction scheme by replacing c_A and c_B by their "definable" counterparts $\tau(\ulcorner A \urcorner)$ and $\tau(\ulcorner B \urcorner)$. We conclude with the

Theorem. *If T is Tarskian, and there is a term τ such that $(**)$ is provable in T for all formulas $A(x)$, $B(x)$, then T is inconsistent.*

Proof. Assume that T satisfies the specified conditions. Given a sentence S, define $A(x) \equiv S \wedge (x = x)$, $B(x) \equiv (x = x)$. Then

$$\vdash_T \forall x[A(x) \leftrightarrow B(x)] \leftrightarrow S,$$

---

[1] "Frege structures and the notions of proposition, truth and set," J. Barwise et al., eds., *The Kleene symposium*, Amsterdam: North-Holland Publ. Co., 1980, pp. 31–59.

so that, by the provability of (**) in T,

$$\vdash_T S \leftrightarrow \tau(\ulcorner A \urcorner) = \tau(\ulcorner B \urcorner).$$

But the representability of the syntax of $\mathcal{L}$ within T implies that there is a formula $W(x)$ of the language of T such that $W(\ulcorner S \urcorner)$ is $\tau(\ulcorner A \urcorner) = \tau(\ulcorner B \urcorner)$ (for arbitrary S). Then $\vdash_T S \leftrightarrow W(\ulcorner S \urcorner)$ for any sentence S; since T is Tarskian, its inconsistency follows.   □

18

Saving Frege from Contradiction

In § 68 of *Die Grundlagen der Arithmetik* Frege defines the number that belongs to the concept F as the extension of the concept 'equinumerous (*gleichzahlig*) with the concept F.' In sections that follow he gives the needed definition of equinumerosity in terms of one-one correspondence, and in § 73 attempts to demonstrate that the number belonging to F is identical with that belonging to G if and only if F is equinumerous with G. In view of Hume's well-known 'standard by which we can judge of the equality and proportion of numbers,'[1] we may call the statement that the numbers belonging to F and G are equal if and only if F is equinumerous with G (or the formalization of this statement) *Hume's principle*. As we shall see, Frege's attempt to demonstrate Hume's principle, which is vital to the development of arithmetic sketched in the next ten sections of the *Grundlagen*, cannot be considered successful. We begin with a look at Frege's attempted proof before turning to our main concern, which is with two ways of repairing the damage to his work caused by the discovery of Russell's paradox.

From *Proceedings of the Aristotelian Society*, new series, **87** (1986/87) 137–51. Reprinted by courtesy of the Editor of the Aristotelian Society: ©1986/87.

[1] "When two numbers are so combin'd, as that the one has always an unite answering to every unite of the other, we pronounce them equal," *Treatise*, I, III, I.

Frege writes,

On our definition, what has to be shown is that the extension
of the concept 'equinumerous with the concept F' is the same
as the extension of the concept 'equinumerous with the concept
G,' if the concept F is equinumerous with the concept G. In
other words: it is to be proved that, for F equinumerous with
G, the following two propositions hold good universally: if the
concept H is equinumerous with the concept F, then it is also
equal to the concept G; and ... [conversely] ... .

The sophisticated definition of numbers as extensions of certain
concepts of concepts and extensive use of binary relations found in
the *Grundlagen* are evidence that Frege was there committed to the
existence of objects of all finite types: 'objects,' the items of the lowest type 0, and, for any types $t_1, \ldots, t_n$, relations of type $(t_1, \ldots, t_n)$
among items of types $t_1, \ldots, t_n$. An item of type (t), for some type
t, is called a concept. Concepts of type (0) are called 'first level
concepts'; those of type $((0))$, 'second level concepts.' The relation
borne by an object x to a concept F when x falls under F is of type
$(0, (0))$; the relation η defined below is of type $((0), 0)$. It seems clear
that Frege accepted a comprehension principle governing the existence of relations, according to which for any sequence of variables
$x_1, \ldots, x_n$ of types $t_1, \ldots, t_n$ and any predicate $A(x_1, \ldots, x_n)$ (possibly containing parameters) there is a relation of type $(t_1, \ldots, t_n)$
holding among those items of types $t_1, \ldots, t_n$ satisfying the predicate and only those. This principle can be proved from the rule of
substitution Frege used in the *Begriffschrift*. Thus, in view of the
predicate 'F is equinumerous with G' (F a first level concept parameter, G a first level concept variable), Frege concludes that there is
a second level concept under which fall all and only those first level
concepts that are equinumerous with (the value of the parameter)
F.

It also seems clear that at the time he wrote the *Grundlagen*,
Frege held that for each concept C of *whatever* type, there is a
special object $'C$, the *extension of C*. Thus extensions are objects;
and the number belonging to the first level concept F is defined by
Frege to be the extension of a certain second level concept, the one
under which fall all and only those first level concepts equinumerous

with F. We shall often abbreviate '(is) equinumerous with': eq.

The announced task of § 73 is to show that the number belonging to the concept F, NF for short, $= NG$ if F eq G. Since Frege has defined NF as $'$ eq F, what must be shown is that $'$ eq $F = {'}$ eq G under the assumption that F eq G. But almost all of § 73 is devoted to showing that if H eq F, then H eq G and observing that a similar proof shows that if H eq G, then H eq F. Frege takes it that showing these two propositions is sufficient; he writes 'in other words.' In a footnote he adds that a similar proof can be given of the converse, that F eq G if $NF = NG$. And of course we know exactly how the proof would go: 'On our definition what must be shown is that if $'$ eq $F = {'}$ eq G, in other words, if the following two propositions hold good universally: if H eq F then H eq G and if H eq G then H eq F, then F eq G. But since F eq F, by the first of these alone, F eq G.'

Why did Frege suppose that one could pass so freely between ' $'$ eq $F = {'}$ eq G' and 'for all H, H eq F iff H eq G'? It seems most implausible that any answer could be correct other than: because he thought it evident that for *any* concepts C and D of the same type (t), $'C = {'}D$ if and only if for all items X of type t, CX iff DX.

Notoriously, this assumption generates Russell's paradox (in the presence of the comprehension principle, whose validity I assume). It is noteworthy that the proof Frege gave of the inconsistency of the system of his *Grundgesetze der Arithmetik* resembles Cantor's proof that there is no one-one mapping of the power set of a set into that set rather than the version of the paradox that Russell had originally communicated to him. Of course in his second letter to Frege, well before Frege came to write the appendix to the *Grundgesetze*, where Frege's proof appears, Russell had explained to him the origins of the paradox in Cantor's work.

In the present notation, Frege's version of Russell's paradox runs: By comprehension, let R be the first level concept $[x : \exists F(x = {'}F \wedge \neg Fx)]$. Consider the object $'R$, which is the extension of R. If $\neg R \, 'R$, then since for all F, $'R = {'}F \rightarrow F \, 'R$, $R \, 'R$. So $R \, 'R$. But then for some F, $'R = {'}F$ and $\neg F \, 'R$. Thus by the principle about extensions mentioned two paragraphs back, $\forall x(Rx \leftrightarrow Fx)$. Thus $\neg R \, 'R$, contradiction.

Since Frege defines numbers as the extensions of *second* level concepts, it might be hoped that the Russell paradox does not threaten Frege's derivation of arithmetic in the *Grundlagen*, for to prove the main proposition of § 73, $'[H : H \operatorname{eq} F] = '[H : H \operatorname{eq} G]$ iff $F \operatorname{eq} G$, he needs only the principle: for any second level concepts C, D, $'C = 'D$ iff for all first level H, CH iff DH. Notice the difference between this principle—call it (VI)—and the instance of (V) in which $t = 0$ that leads to Russell's paradox: for any first level concepts F, G, $'F = 'G$ iff for all objects x, Fx iff Gx. Part of the cause of the Russell paradox is that certain extensions are in the range of the quantified variable on the right side of (V). Since this is not the case with (VI), might (VI) then be consistent?

No. Define η by: $F\eta x$ iff for some second level concept D, $x = 'D$ and DF. By comprehension one level up, let $C = [F : \exists x(\neg F\eta x \wedge Fx)]$. By comprehension at the lowest level, let $X = [x : x = 'C]$. Suppose $X\eta 'C$. By the definition of η, for some D, $'C = 'D$ and DX, whence by (VI) CX. By the definition of C, for some x, $\neg X\eta x$ and Xx. By the definition of X, $x = 'C$, and therefore $\neg X\eta 'C$. Thus $\neg X\eta 'C$, whence for every D, if $'C = 'D$ then $\neg DX$. Therefore $\neg CX$. But by the definition of C, for every x such that Xx, $X\eta x$, and since $X 'C$, $X\eta 'C$, contradiction. As with the Russell paradox, it is the assumption that $'$ is one-one that causes the trouble.

Thus not only is (V) in full generality inconsistent, so is the apparently weaker (VI). But Frege does not need the full strength of (VI) to prove that $NF = NG$ iff $F \operatorname{eq} G$. On the basis of the following proposition, 'Numbers':

$$\forall F \exists! x \forall H (H\eta x \leftrightarrow H \operatorname{eq} F))$$

he can define NF as the unique object x such that for all concepts H, $H\eta x$ iff $H \operatorname{eq} F$ and then easily prove from this definition that $NF = NG$ iff $F \operatorname{eq} G$.

Numbers expresses a proposition to whose truth Frege was committed. It is a proposition about concepts, objects couched in the language of second-order logic to which one new relation, η, has been added. ('eq' is of course definable in second-order logic in the standard way.) Thus it is involved with higher-order notions or with notions not expressible in the language of Frege's *Begriffsschrift* if at all, only in that η is a relation of concepts to objects. Notice that for any concept F the x (unique, according to Numbers) such that for all concepts H, $H\eta x$ iff H eq F will be an extension, for since F eq F, $F\eta x$, and thus for some C, $x = {}'C$ (and CF). The chief virtue of Numbers, though, is that it is formally consistent (as John Burgess,[2] Harold Hodes,[3] and the author[4] have noted).

We may see this as follows. Let the object variables in Numbers range over all natural numbers, the concept variables range over all sets of natural numbers and for all n, the n-ary relation variables range over all n-ary relations of natural numbers. (We are thus defining a 'standard' model for Numbers.) Let η be true of a set S of natural numbers and a natural number n if and only if either for some natural number m, S has m members and $n = m + 1$ or S is infinite and $n = 0$. So interpreted, Numbers is true.

For let S be an arbitrary set of natural numbers. Let $n = m + 1$ if S is finite and has m members; let $n = 0$ otherwise. Then for any set U of natural numbers, $U\eta n$ holds iff either for some m, U has m members and $n = m + 1$ or U is infinite and $n = 0$, if and only if U and S have the same number of members, if and only if U eq S holds. The uniqueness of n follows from the definition of η and the fact that S eq S holds.

Much of the interest of the proof just given lies in the fact that it can be formalized in second-order arithmetic. Let $\mathrm{Eq}(H, F)$ be the standard formula of second-order logic defining the relation 'there is a one-one correspondence between U and S.' The relation 'S is infinite and $n = 0$ or for some m, S has m members and $n = m + 1$' can defined by a formula $\mathrm{Eta}(F, x)$ of second-order arithmetic in such a way that the sentence

$$\forall F \exists! x \forall H (\mathrm{Eta}(H, x) \leftrightarrow \mathrm{Eq}(H, F))$$

is *provable* in second-order arithmetic. Thus we have a relative consistency proof: a proof of a contradiction in the result of adjoining

[2] *The philosophical review*, **93** (1984) 638–640.

[3] *The journal of philosophy*, **81** (1984) 123–49, p. 138.

[4] In "The consistency of Frege's *Foundations of arithmetic*," Chapter 8, above.

the formalization

$$\forall F \exists! x \forall H (H\eta x \leftrightarrow \text{Eq}(H, F))$$

of Numbers (with $H\eta x$ now taken as an atomic formula) to any standard axiomatic system of second-order logic could immediately ('primitive recursively') be transformed into a proof of a contradiction in second-order formal arithmetic. It is pointless to try to describe how unexpected the discovery of a contradiction in second-order arithmetic would be. Since Hume's principle is a theorem of a definitional extension of the second-order theory whose sole axiom is Numbers, it too is consistent (relative to the consistency of second-order arithmetic).

The distance between Numbers and Hume's principle is certainly not all that great: Numbers provides the justification for the introduction of the functor N, 'the number belonging to'; Numbers also follows from Hume's principle when $F\eta x$ is defined as $x = NF$. One of Numbers's minor virtues is that it encapsulates the only assumption concerning the existence of extensions that Frege actually needs. For once Frege has Hume's principle in hand, he needs nothing else.

In §§ 74–83 of the *Grundlagen*, Frege outlines the proofs of a number of propositions concerning (what we now call) the natural numbers, including the difficult theorem that every finite number has a successor. (Formalizations of) all of these can be proved in axiomatic second-order logic from Hume's principle in more or less the manner outlined in these ten sections of the *Grundlagen*. I am uncertain whether Frege was aware that Hume's principle was all he needed; his puzzling remark at the end of the *Grundlagen* about attaching no decisive importance to the introduction of extensions of concepts may be taken as some evidence that he knew this.

It's a pity that Russell's paradox has obscured Frege's accomplishment in the *Grundlagen*. It's utterly remarkable that the whole of arithmetic can be deduced in second-order logic from this one simple principle, which might appear to be nothing more than a definition. Of course, Hume's principle isn't a definition, since 'NF' and 'NG' are intended to denote objects in the range of the first-order variables. (Cf. Wright's book *Frege's conception of numbers as objects*.[5]) And as Frege's work shows, Hume's principle is much

---

[5] Aberdeen: Aberdeen University Press, 1983.

more powerful than we might have supposed it to be, implying, with the aid of second-order logic, the whole of second-order arithmetic (while failing to imply $\perp$).

In fact, that Hume's principle is consistent can easily come to seem like a matter of purest luck. Suppose we do for isomorphism of (binary) relations what we have just done for the notion of equinumerosity of concepts: adjoin to second-order logic an axiom

(OrdType) $\bar{R} = \bar{S} \leftrightarrow R \text{ iso } S,$

with $\bar{}$ a function sign that takes a binary relation variable and makes a term of the type of object variables, and $R \text{ iso } S$ some formula expressing the order-isomorphism of the relations that are values of the variables R and S. In other words, suppose we introduce in the obvious way what Cantor called 'order types' and Russell 'relation numbers.' It would, I imagine, be the obvious guess that if Hume's principle is consistent, then so is OrdType, which states that the order types of two relations are the same iff the relations are order-isomorphic.

In §§ 85 and 86 of the *Grundlagen* Frege takes Cantor to task for having appealed to 'inner intuition' instead of providing definitions of *Number* and *following in a series*. Frege adds that he thinks he can imagine how these two concepts could be made precise. One would have liked to see Frege's account of Cantor's notions; one cannot but suspect that in order to reproduce Cantor's theory of ordinal numbers, Frege would have derived OrdType from a (possible tacit) appeal to (V).

Doing so would have landed him in trouble deeper than any he was in in the *Grundlagen*, however, and not just because of the appeal to (V). For the guess that OrdType is consistent if Hume's principle is consistent is wrong. As Hodes has also observed, OrdType leads to a contradiction via the reasoning of the Burali-Forti paradox. Thus although Numbers is consistent, a principle no less definitional in appearance and rather similar in content turns out to be inconsistent. In view of the inconsistency of (VI) and OrdType, the consistency of Hume's principle is sheer luck.

To show that arithmetic follows from Hume's principle, or its near relation Numbers, is to give a profound analysis of arithmetic, but it is not to base arithmetic on a principle strikingly like Frege's Rule (V). We know from Russell's and Cantor's paradoxes that there can be no function from (first level) concepts to objects that assigns different objects to concepts under which different objects fall. Identifying concepts under which the same objects fall, we may say that there is no one-one function from concepts into objects. But the function denoted by N is a particularly non-one-one function. With the exception of $[x : x \neq x]$, every concept shares its number with infinitely many other concepts. One might wonder whether one could base arithmetic on a function assigning objects to concepts which, though necessarily not one-one, fails to be one-one at only one of its values. We'll see how to do this below.

In the appendix to the second volume of his *Grundgesetze*, Frege asks

> Is it always permissible to speak of the extension of a concept, of a class? And if not, how do we recognize the exceptional cases? Can we always infer from the extension of one concept's coinciding with that of a second that every object which falls under the first concept also falls under the second? These are the questions raised by Mr. Russell's communication.

Before showing how Russell's paradox could be deduced in the system of the *Grundgesetze*, he declares

> Thus there is no alternative but to recognize the extensions of concepts, or classes, as objects in the full and proper sense of the word, while conceding that our interpretation hitherto of the words 'extension of a concept' is in need of correction.

After showing that his rule (Vb) leads to Russell's paradox, Frege proves that every function from concepts to objects assigns the same value to some pair of concepts under which different objects fall. He observes that the proof is 'carried out without the use of propositions or notations whose justification is in any way doubtful' and adds that

> this simply does away with extensions in the generally received sense of the term. We may not say that in general the expression 'the extension of one concept coincides with that of another' means the same as the expression 'every object falling under the first also falls under the second and conversely.'

Frege then proposes a repair. In place of the defective rule (V), assume (V′), which we may put: the extensions of F and G are identical iff the same objects *other than those extensions* fall under F and G. He remarks that 'Obviously this cannot be taken as defining the extension of a concept but merely as stating the distinctive property of this second level function.'

It is well known that Frege's proposed repair fails. A particularly useful discussion of the failure is found in Resnik's book *Frege and the philosophy of mathematics*.[6] I want to consider an alternative repair to the *Grundgesetze* suggested by the second question asked in its appendix: How do we recognize the exceptional cases?

Frege does not in fact offer an answer to his question. Although he does discuss certain exceptional cases in the appendix, they are not the ones referred to in his question, which are, presumably, the concepts that lack an extension in the customary sense of the term. The exceptions Frege discusses are not concepts but certain objects, namely, extensions of concepts.

But there is a simple answer that Frege might have given, one that uses only such notions as were available in 1908. Identification of the exceptional concepts will suggest a replacement for rule (V) which Frege might well have found perfectly acceptable, and which seems no more ad hoc than Frege's own rule (V′). The defect Russell revealed could have been repaired rather early, and by a patch that is really quite simple and closely related to ideas found in Cantor's work, some of which, at least, was familiar to Frege.

I shall not discuss the question whether the repair vindicates logicism. I doubt that anything can do that. I merely wish to claim that the repair I shall give should have been no less acceptable to Frege than the one he actually offered.

We'll begin the description of the repair with a bit of stipulation. Let's detach the existence of extensions from the term 'coextensive' and say that a concept F is *coextensive with* a concept G if and only if all objects that fall under F fall under G and vice versa. Five more definitions follow, of 'subconcept,' 'goes into,' 'V,' 'small' and 'similar.'

Let us call a concept F a *subconcept of* a concept G if every

---

[6]Ithaca: Cornell University Press, 1980.

object that falls under F falls under G. Let us say that a concept F *goes into* G if F is equinumerous with a subconcept of G. If F is a subconcept of G, then F goes into G; if F goes into G and G goes into H, then F goes into H. It can be shown that if F and G go into each other, then they are equinumerous.

Let V be the concept, $[x : x = x]$, *identical with itself*. And let us say that a concept F is *small* if V does not go into F. V is not small. If F goes into G and G is small, then F is small; thus any subconcept of a small concept is small and any concept equinumerous or coextensive with a small concept is small. Let us say that F is *similar to* G iff (F is small $\vee$ G is small $\rightarrow$ F is coextensive with G).

We want now to see that *is similar to* is an equivalence relation. Reflexivity and symmetry are obvious. As for transitivity, suppose that F is similar to G and G to H. If F is small, then F is coextensive with G (for F is similar to G), thus G is small, and then G is coextensive with H; thus F is coextensive with H. And in like manner, but going the other way, if H is small, F is coextensive with H. Thus F is similar to H.

We now suppose that associated with each concept F, there is an object $*F$, which I will call the *subtension* of F, and that as extensions were supposed to be in one-one correspondence with equivalence classes of the equivalence relation *coextensive with*, so subtensions are in one-one correspondence with equivalence classes of the equivalence relation *similar to*; thus the principle (New V) holds:

$$*F = *G \text{ iff } F \text{ is similar to } G.$$

In view of the 'Julius Caesar problem' it may be uncertain whether (New V) can be taken as *defining* subtensions, but like (V) and unlike (V′) it does not merely state the distinctive property of a certain second level function. (V) and our replacement (New V) explain in a non-circular way, as (V′) did not, when objects given as extensions or subtensions of concepts are identical; the statements of the identity conditions do not contain expressions explicitly referring to those very extensions or subtensions.

Moreover, (New V) enables us to define the 'exceptional' concepts quite easily, as those that are not small. For it follows from (New

V) that for every concept F, if F is small, then for every concept G, $*F = *G$ iff F is coextensive with G. Furthermore if F is not small then there is a concept G not coextensive with F but such that $*F = *G$; of course any such G will itself fail to be small. (Since F is not small, F is equinumerous with V; but as we shall see, V is equinumerous with $V - 0$ (defined below). Thus F is equinumerous with one of its proper subconcepts G; since G is not small, $*F = *G$.)

We must now make it plausible that arithmetic can be developed in second-order logic from (New V) alone. There are many ways to do this; perhaps the easiest is to develop 'finite set theory' from (New V) taking the development of arithmetic from finite set theory for granted.

Following Frege, let us say that $x \in$ (is a member of) y, if for some F, $y = *F$ and Fx. And let us call an object y a *set* if $y = *G$ for some *small* concept G. Thus if y is a set and for some concept H, $y = *H$, then $x \in y$ iff Hx.

Again, *à la* Frege, let $0 = *[x : x \neq x]$. Since 0 is an object, $[x : x \neq x]$ is small and 0 is therefore a set. For all x, not: $x \in 0$. $*V$, however, is not a set. Therefore there are at least two objects. Thus for any object y, the concept $[x : x = y]$ is small; let $\{y\} = *[x : x = y]$. For any object y, $\{y\}$ is a set. (So $\{*V\}$ is a set even though $*V$ is not.)

For any concept F and any object y, let $F + y$ be the concept $[x : Fx \lor x = y]$ and $F - y$ the concept $[x : Fx \land x \neq y]$. We now want to see that if F is small, so is $F + y$, for any object y.

We first observe since $0 \neq \{z\}$, V goes into $V - 0$ via the map which sends each object x into $\{x\}$. Suppose that $F + y$ is not small. Then V goes into $F + y$ via the map φ which, switching one or two values of φ if necessary, we may assume sends 0 to y. Then $V - 0$ goes into F via (a restriction of) φ. Since V goes into $V - 0$, V goes into F and F is not small. It follows that if F is small, so is $F + y$.

For any objects z, w, let $z + w = *[x : x \in z \lor x = w]$. Then if z is a set, so is $z + w$; $x \in z + w$ iff $x \in z \lor x = w$.

As in *Grundlagen* § 83, we may define $HF = [x : \forall F(F0 \land \forall z \forall w (Fz \land Fw \to Fz + w) \to Fx)]$. An induction principle for HF follows directly: to show that all HF objects fall under a certain concept F, it suffices to show that 0 does, and that $z + w$ does whenever z and w do. Thus all HF objects together with all of their members are HF sets.

The axioms of (second-order) General Set Theory are:

Extensionality: $\forall x \forall y (\forall z (z \in x \leftrightarrow z \in y) \rightarrow x = y)$,

Adjunction: $\forall w \forall z \exists y \forall x (x \in y \leftrightarrow x \in z \vee x = w)$, and all

Separation: $\forall F \forall z \exists y \forall x (x \in y \leftrightarrow x \in z \wedge Fx)$.

These axioms all hold when relativized to HF. For extensionality, note that two HF sets coincide if the same HF sets belong to both; separation is easily proved by induction on z. Second-order arithmetic can now be deduced in the usual way from General Set Theory. It is of some interest to note that the relativizations of the remaining axioms of Zermelo-Fraenkel set theory plus Choice minus Infinity can also be deduced from (New V).

Note also that the derivation of General Set Theory from (New V) is quite elementary, not much more difficult than it would have been from (V). We have had to check that certain subtensions were sets, but these checks were easily made. And although equinumerosity figures in the definition of the key notion of smallness, the Schröder-Bernstein theorem, or the technique of its proof, is nowhere used.

The hereditarily finite sets are the members of the smallest set A containing all finite subsets of A. Of course the null set $\emptyset$ is hereditarily finite, as are $\{\emptyset\}$, $\{\{\emptyset\}\}$, $\{\emptyset, \{\emptyset\}\}$, etc. An alternative characterization of the hereditarily finite sets is that they are the members of the smallest set containing $\emptyset$ and containing $z \cup \{w\}$ whenever it contains z and w. Our construction shows that the hereditarily finite sets can be seen as 'constructed from' the relation *is similar to* as the finite cardinals arise from the relation *is equinumerous with*, and as extensions were supposed to arise from *is coextensive with*. Truth-values arise in a similar manner from *is materially equivalent to*, via the axiom: $Vp = Vq \leftrightarrow (p \leftrightarrow q)$.

When the natural numbers or the hereditarily finite sets are thus 'constructed' from equinumerosity or similarity, other objects are constructed too. We have already met the non-set $*V$. On the construction of the *Grundlagen*, along with the usual natural numbers some funny numbers arise, among them the number NV of things there are, the number $N[x : \exists Fx = NF]$ of numbers there are, and the number of finite numbers there are. Frege acknowledged the last of these, dubbing it ∞_1, but he must admit all of them if he wants to define 0 as the number belonging to the concept *not identical with itself*. (It is consistent with Numbers that all three are distinct; it is also consistent that they are all identical.)

It is often said that Zermelo-Fraenkel set theory is motivated by a doctrine of 'limitation of size': a collection is a set if it is 'small' or 'not too big,' a collection being 'too big' if it is equinumerous with the collection of *all* sets. The notion of smallness is sometimes taken as to motivate the axioms of set theory: it is thought that if certain sets are small, then certain other sets formed from them by various operations will also be small. (Michael Hallett has effectively criticized the thought that the power set operation produces small sets from small sets.[7]) In most treatments of set theory, the idea of smallness is left at the motivational level. Our construction explicitly incorporates it into our axiom (New V) governing subtensions. Another respect in which our construction differs from that of ZF or its class-theoretic relatives is the combination of a 'universal' object with the absence of a complement operation: for every x, $x \in *V$; but if for every $x \neq 0$, $x \in y$, then $0 \in y$ also.

It follows from (New V) that if F is small, then $*F = *G$ if and only if F and G are coextensive; if neither F nor G is small, then $*F = *G$ (for F and G then satisfy the definition of 'similar'). Our construction, as we have noted, concentrates the non-one-oneness of the function $*$ in a single value, the object $*V$. A theorem of set theory throws some light on the question how non-one-one any function like $*$ from concepts to objects must be. It follows from the Zermelo-König inequality (which can be proved in ZF plus the axiom of choice) that for any infinite set x and function f from the power set of x into x, there is a member a of x such that there are at least as many subsets y of x such that $fy = a$ as there are subsets of x altogether. Thus (higher-order set theory implies that) any attempt to assign concepts (classes) to objects must assign to some one object as many concepts as there are concepts altogether. There is then a clear sense in which the failure of $*$ to be a one-one function is no worse than necessary and the replacement of extensions $'F$

---

[7]In his *Cantorian set theory and limitation of size*, Oxford: Oxford University Press, 1984.

by subtensions $*F$ is a minimal departure from the project of the *Grundgesetze*. The theorem also shows that project not to have been a near miss.

Although I have given an informal sketch of the derivation of General Set Theory from (New V), it is to be emphasized that this derivation can be carried out formally in axiomatic second-order logic in which the sole axiom (other than the standard axioms of second-order logic) is (New V). (Of course the rules of formation will guarantee that for each concept variable F, $*F$ is a term of the type of object variables.)

There remains a matter not yet attended to: the consistency of (New V). It should now be no surprise that (New V) is consistent (if second-order arithmetic is). Indeed, it is quite simple to provide a standard second-order model for (New V).

As in the proof of the consistency of Numbers, let the object variables range over all natural numbers. Since the model is standard and its domain is countably infinite, a subset X of the domain satisfies 'is small' if and only if X is finite. We must now supply a suitable function τ from sets of natural numbers to natural numbers with which to interpret $*$.

Let D be some one-one map of all finite sets of natural numbers into the natural numbers. (The best known such D is given by: $D(X)$ = the number whose binary numeral, for every number x, contains a 1 at the 2^x's place iff $x \in X$.) Then if X and Y are finite sets of natural numbers, $1 + D(X) = 1 + D(Y)$ iff $X = Y$.

For any set X of natural numbers, let $\tau(X) = 0$ if X is infinite and $= 1 + D(X)$ if X is finite. Then $\tau(X) = \tau(Y)$ iff either X and Y are both infinite or $X = Y$, iff (X or Y is finite $\rightarrow X = Y$). Thus (New V) does indeed have a standard model: it is true over the natural numbers when $*$ is interpreted by τ. Utilizing the particular function D defined above, we can convert the foregoing argument into a proof of the consistency of (New V) (relative to that of second-order arithmetic) in the usual way.

How then does (New V) prevent Russell's paradox? Let's try to re-derive it: By comprehension, let R be the first level concept $[x : \exists F(x = *F \wedge \neg Fx)]$. If $\neg R * R$, then since for all F, $*R = *F \rightarrow F * R$, $R * R$. So $R * R$. So for some F, $*R = *F$ and $\neg F * R$. But we cannot show that $\forall x(Rx \leftrightarrow Fx)$ unless we can show that

F or R is small, and this there is no way of doing if second-order arithmetic is consistent. The unsurprising conclusion is that R is not small. It is more interesting to note that since every number fails to fall under at least one concept of which it is the number, the Russellian number $N[x : \exists F(x = NF \wedge \neg Fx)]$ is (provably) identical with $N[x : \exists Fx = NF]$, the number of numbers.

A piece of mathematics carried out in an inconsistent theory need not be vitiated by the inconsistency of the theory: it may be possible to develop the mathematics in a suitable proper subtheory. The development of arithmetic outlined in the *Grundlagen* can be carried out in the consistent theory obtained by adding Numbers to the system of *Begriffsschrift* as well as in the inconsistent system of the *Grundgesetze*. Consistent systems similar to, but stronger than, (New V) plus second-order logic can readily be given, e.g., by replacing 'small' by 'countable.' It would be of some interest to find out how much of the mathematics done in the *Grundgesetze* can be reproduced in such systems.[8]

[8]Research for this paper was carried out under a grant from the National Science Foundation.

Index of Frege's Writings

General Index

a priori/a posteriori, 52–67
Abel, Niels, 110, 152
absolute infinite, 92
abstract objects, 127–129;
  reference to, 16–17. *See also*
  reference
abstraction: definition by,
  17–20, 127, 134; in Aristotle, 334; logical, 7, 403–404;
  psychological, 78–80, 133
abstraction operator, 198, 398
abstraction principle, 422–
  428, 436. *See also* Axiom V;
  Basic Law V; Rule (V)
Aczel, Peter, 436
Adeleke, Samson, 20n, 400n,
  406n, 413n
analytic/synthetic, 31–32, 42–
  45, 52–67
Anscombe, Elizabeth, 191n,
  202n
ancestral of a relation, 72,
  80–84, 165, 273, 298, 305,
  306, 399

application operator, 262
Archimedes, 345, 350–351,
  353, 379
archimedean law, 392, 394,
  401-402, 410-11, 416-18,
  420. *See also* axiom of
  Archimedes
Aristotle, 334–337, 338n, 347,
  354
Aspray, W., 82n, 143n, 355n
associative law for addition,
  51–52
Ausderungsschema, 173
Austin, John, 5n, 44n, 69n,
  116n, 182n
axiom of Archimedes, 345,
  356
axiom of choice, 9n, 94–95
axiom of infinity, 3, 9, 231
axiom of ordered pairs, 242
Axiom V of *Grundgesetze*,
  143, 186, 194, 259–261,
  263n, 286–287, 296, 322.
  See also Basic Law V; New